Signals and Communication Technology

More information about this series at http://www.springer.com/series/4748

Dong Yu · Li Deng

Automatic Speech Recognition

A Deep Learning Approach

 Springer

Dong Yu
Microsoft Research
Bothell
USA

Li Deng
Microsoft Research
Redmond, WA
USA

ISSN 1860-4862
ISBN 978-1-4471-5778-6
DOI 10.1007/978-1-4471-5779-3

ISSN 1860-4870   (electronic)
ISBN 978-1-4471-5779-3   (eBook)

Library of Congress Control Number: 2014951663

Springer London Heidelberg New York Dordrecht

Printed on acid-free paper

Springer is part of Springer Science+Business Media (www.springer.com)

To my wife and parents

Dong Yu

To Lih-Yuan, Lloyd, Craig, Lyle, Arie, and Axel

Li Deng

Foreword

This is the first book on automatic speech recognition (ASR) that is focused on the deep learning approach, and in particular, deep neural network (DNN) technology. The landmark book represents a big milestone in the journey of the DNN technology, which has achieved overwhelming successes in ASR over the past few years. Following the authors' recent book on "Deep Learning: Methods and Applications", this new book digs deeply and exclusively into ASR technology and applications, which were only relatively lightly covered in the previous book in parallel with numerous other applications of deep learning. Importantly, the background material of ASR and technical detail of DNNs including rigorous mathematical descriptions and software implementation are provided in this book, invaluable for ASR experts as well as advanced students.

One unique aspect of this book is to broaden the view of deep learning from DNNs, as commonly adopted in ASR by now, to encompass also deep generative models that have advantages of naturally embedding domain knowledge and problem constraints. The background material did justice to the incredible richness of deep and dynamic generative models of speech developed by ASR researchers since early 90's, yet without losing sight of the unifying principles with respect to the recent rapid development of deep discriminative models of DNNs. Comprehensive comparisons of the relative strengths of these two very different types of deep models using the example of recurrent neural nets versus hidden dynamic models are particularly insightful, opening an exciting and promising direction for new development of deep learning in ASR as well as in other signal and information processing applications. From the historical perspective, four generations of ASR technology have been recently analyzed. The 4th Generation technology is really embodied in deep learning elaborated in this book, especially when DNNs are seamlessly integrated with deep generative models that would enable extended knowledge processing in a most natural fashion.

All in all, this beautifully produced book is likely to become a definitive reference for ASR practitioners in the deep learning era of 4th generation ASR. The book masterfully covers the basic concepts required to understand the ASR field as a whole, and it also details in depth the powerful deep learning methods that have shattered the field in 2 recent years. The readers of this book will become articulate in the new state-of-the-art of ASR established by the DNN technology, and be poised to build new ASR systems to match or exceed human performance.

By Sadaoki Furui, President of Toyota Technological Institute at Chicago, and Professor at the Tokyo Institute of Technology.

Preface

Automatic Speech Recognition (ASR), which is aimed to enable natural human–machine interaction, has been an intensive research area for decades. Many core technologies, such as Gaussian mixture models (GMMs), hidden Markov models (HMMs), mel-frequency cepstral coefficients (MFCCs) and their derivatives, n-gram language models (LMs), discriminative training, and various adaptation techniques have been developed along the way, mostly prior to the new millenium. These techniques greatly advanced the state of the art in ASR and in its related fields. Compared to these earlier achievements, the advancement in the research and application of ASR in the decade before 2010 was relatively slow and less exciting, although important techniques such as GMM–HMM sequence discriminative training were made to work well in practical systems during this period.

In the past several years, however, we have observed a new surge of interest in ASR. In our opinion, this change was led by the increased demands on ASR in mobile devices and the success of new speech applications in the mobile world such as voice search (VS), short message dictation (SMD), and virtual speech assistants (e.g., Apple's Siri, Google Now, and Microsoft's Cortana). Equally important is the development of the deep learning techniques in large vocabulary continuous speech recognition (LVCSR) powered by big data and significantly increased computing ability. A combination of a set of deep learning techniques has led to more than 1/3 error rate reduction over the conventional state-of-the-art GMM–HMM framework on many real-world LVCSR tasks and helped to pass the adoption threshold for many real-world users. For example, the word accuracy in English or the character accuracy in Chinese in most SMD systems now exceeds 90 % and even 95 % on some systems.

Given the recent surge of interest in ASR in both industry and academia we, as researchers who have actively participated in and closely witnessed many of the recent exciting deep learning technology development, believe the time is ripe to write a book to summarize the advancements in the ASR field, especially those during the past several years.

Along with the development of the field over the past two decades or so, we have seen a number of useful books on ASR and on machine learning related to ASR, some of which are listed here:

- Deep Learning: Methods and Applications, by Li Deng and Dong Yu (June 2014)
- Automatic Speech and Speaker Recognition: Large Margin and Kernel Methods, by Joseph Keshet, Samy Bengio (January 2009)
- Speech Recognition Over Digital Channels: Robustness and Standards, by Antonio Peinado and Jose Segura (September 2006)
- Pattern Recognition in Speech and Language Processing, by Wu Chou and Biing-Hwang Juang (February 2003)
- Speech Processing—A Dynamic and Optimization-Oriented Approach, by Li Deng and Doug O'Shaughnessy (June 2003)
- Spoken Language Processing: A Guide to Theory, Algorithm and System Development, by Xuedong Huang, Alex Acero, and Hsiao-Wuen Hon (April 2001)
- Digital Speech Processing: Synthesis, and Recognition, Second Edition, by Sadaoki Furui (June 2001)
- Speech Communications: Human and Machine, Second Edition, by Douglas O'Shaughnessy (June 2000)
- Speech and Language Processing—An Introduction to Natural Language Processing, Computational Linguistics, and Speech Recognition, by Daniel Jurafsky and James Martin (April 2000)
- Speech and Audio Signal Processing, by Ben Gold and Nelson Morgan (April 2000)
- Statistical Methods for Speech Recognition, by Fred Jelinek (June 1997)
- Fundamentals of Speech Recognition, by Lawrence Rabiner and Biing-Hwang Juang (April 1993)
- Acoustical and Environmental Robustness in Automatic Speech Recognition, by Alex Acero (November 1992).

All these books, however, were either published before the rise of deep learning for ASR in 2009 or, as our 2014 overview book, were focused on less technical aspects of deep learning for ASR than is desired. These earlier books did not include the new deep learning techniques developed after 2010 with sufficient technical and mathematical detail as would be demanded by ASR or deep learning specialists. Different from the above books and in addition to some necessary background material, our current book is mainly a collation of research in most recent advances in deep learning or discriminative and hierarchical models, as applied specific to the field of ASR. Our new book presents insights and theoretical foundation of a series of deep learning models such as deep neural network (DNN), restricted Boltzmann machine (RBM), denoising autoencoder, deep belief network, recurrent neural network (RNN) and long short-term memory (LSTM) RNN, and their application in ASR through a variety of techniques including the DNN-HMM hybrid system, the tandem and bottleneck systems, multi-task and transfer learning, sequence-discriminative training, and DNN adaptation. The book further discusses practical considerations, tricks, setups, and speedups on applying the deep learning models and related techniques in building real-world real-time ASR systems. To set the background, our book also includes two chapters that introduce GMMs and HMMs with their variants. However, we omit details of the GMM–HMM techniques that do not directly relate to the theme of the book—the hierarchical modeling or deep learning approach. Our book is thus complementary to, instead of replacement of, the published books listed above on many of similar topics. We believe this book will be of interest to advanced graduate students, researchers, practitioners, engineers, and scientists in speech processing and machine learning fields. We hope our book not only provides reference to many of the techniques used in the filed but also ignites new ideas to further advance the field.

During the preparation of the book, we have received encouragement and help from Alex Acero, Geoffrey Zweig, Qiang Huo, Frank Seide, Jasha Droppo, Mike Seltzer, and Chin-Hui Lee. We also thank Springer editors, Agata Oelschlaeger and Kiruthika Poomalai, for their kind and timely help in polishing up the book and for handling its publication.

Seattle, USA, July 2014   Dong Yu
                                                                            Li Deng

Contents

Part II   Deep Neural Networks

Acronyms

ADMM	Alternating directions method of multipliers
AE-BN	Autoencoder bottleneck
ALM	Augmented Lagrangian multiplier
AM	Acoustic model
ANN	Artificial neural network
ANN-HMM	Artificial neural network-hidden Markov model
ASGD	Asynchronous stochastic gradient descent
ASR	Automatic speech recognition
BMMI	Boosted maximum mutual information
BP	Backpropagation
BPTT	Backpropagation through time
CD	Contrastive divergence
CD-DNN-HMM	Context-dependent-deep neural network-hidden Markov model
CE	Cross entropy
CHiME	Computational hearing in multisource environments
CN	Computational network
CNN	Convolutional neural network
CNTK	Computational network toolkit
CT	Conservative training
DAG	Directed acyclic graph
DaT	Device-aware training
DBN	Deep belief network
DNN	Deep neural network
DNN-GMM-HMM	Deep neural network-Gaussian mixture model–hidden Markov model
DNN-HMM	Deep neural network-hidden Markov model
DP	Dynamic programming
DPT	Discriminative pretraining

EBW	Extended Baum–Welch algorithm
EM	Expectation-maximization
fDLR	Feature-space discriminative linear regression
fMLLR	Feature-space maximum likelihood linear regression
FSA	Feature-space speaker adaptation
F-smoothing	Frame-smoothing
GMM	Gaussian mixture model
GPGPU	General-purpose graphical processing units
HDM	Hidden dynamic model
HMM	Hidden Markov model
HTM	Hidden trajectory model
IID	Independent and identically distributed
KLD	Kullback–Leibler divergence
KL-HMM	Kullback–Leibler divergence-based HMM
LBP	Layer-wise backpropagation
LHN	Linear hidden network
LIN	Linear input network
LM	Language model
LON	Linear output network
LSTM	Long short-term-memory (recurrent neural network)
LVCSR	Large vocabulary continuous speech recognition
LVSR	Large vocabulary speech recognition
MAP	Maximum a posteriori
MBR	Minimum Bayesian risk
MFCC	Mel-frequency cepstral coefficient
MLP	Multi-layer perceptron
MMI	Maximum mutual information
MPE	Minimum phone error
MSE	Mean square error
MTL	Multitask learning
NAT	Noise adaptive training
NaT	Noise-aware training
NCE	Noise contrastive estimation
NLL	Negative log-likelihood
oDLR	Output-feature discriminative linear regression
PCA	Principal component analysis
PLP	Perceptual linear prediction
RBM	Restricted Boltzmann machine
ReLU	Rectified linear unit
RKL	Reverse Kullback–Leibler divergence
RNN	Recurrent neural network
ROVER	Recognizer output voting error reduction
RTF	Real-time factor

SaT	Speaker-aware training
SCARF	Segmental conditional random field
SGD	Stochastic gradient descent
SHL-MDNN	Shared-hidden-layer multilingual DNN
SIMD	Single instruction multiple data
SKL	Symmetric Kullback–Leibler divergence
sMBR	State minimum Bayesian risk
SMD	Short message dictation
SVD	Singular value decomposition
SWB	Switchboard
UBM	Universal background model
VS	Voice search
VTLN	Vocal tract length normalization
VTS	Vector Taylor series
WTN	Word transition network

Symbols

General Mathematical Operators

$\mathbf{x}$	A vector
x_i	The i-th element of $\mathbf{x}$
$|x|$	Absolute value of x
$\|\mathbf{x}\|$	Norm of vector $\mathbf{x}$
$\mathbf{x}^T$	Transpose of vector $\mathbf{x}$
$\mathbf{a}^T\mathbf{b}$	Inner product of vectors $\mathbf{a}$ and $\mathbf{b}$
$\mathbf{ab}^T$	Outer product of vectors $\mathbf{a}$ and $\mathbf{b}$
$\mathbf{a} \bullet \mathbf{b}$	Element-wise product of vectors $\mathbf{a}$ and $\mathbf{b}$
$\mathbf{a} \otimes \mathbf{b}$	Cross product of vectors $\mathbf{a}$ and $\mathbf{b}$
$\mathbf{A}$	A matrix
A_{ij}	The element value at the i-th row and j-th column of matrix $\mathbf{A}$
$tr(\mathbf{A})$	Trace of matrix $\mathbf{A}$
$\mathbf{A} \otimes \mathbf{B}$	Khatri-Rao product of matrices $\mathbf{A}$ and $\mathbf{B}$
$\mathbf{A} \oslash \mathbf{B}$	Element-wise division of $\mathbf{A}$ and $\mathbf{B}$
$\mathbf{A} \circ \mathbf{B}$	Inner product of vectors applied on matrices $\mathbf{A}$ and $\mathbf{B}$ column-wise
$\mathbf{A} \odot \mathbf{B}$	Inner product of vectors applied on matrices $\mathbf{A}$ and $\mathbf{B}$ row-wise
$\mathbf{A}^{-1}$	Inverse of matrix $\mathbf{A}$
$\mathbf{A}^\dagger$	Pseudoinverse of matrix $\mathbf{A}$
$\mathbf{A}^\alpha$	Element-wise power of matrix $\mathbf{A}$
$vec(\mathbf{A})$	The vector formed by concatenating columns of $\mathbf{A}$
$\mathbf{I}_n$	$n \times n$ identity matrix
$\mathbf{1}_{m,n}$	$m \times n$ matrix with all 1's
$\mathbb{E}$	Statistical expectation operator
$\mathbb{V}$	Statistical covariance operator
$\langle \mathbf{x} \rangle$	Average of vector $\mathbf{x}$
$\odot$	Convolution operator
$\mathbf{H}$	Hessian matrix
$\mathbf{J}$	Jacobian matrix

$p(\mathbf{x})$ Probability density function of random vector $\mathbf{x}$
$P(x)$ Probability of x
∇ Gradient operator

More Specific Mathematical Symbols

$\mathbf{w}^*$ Optimal $\mathbf{w}$
$\hat{\mathbf{w}}$ Estimated value of $\mathbf{w}$
$\mathbf{R}$ Correlation matrix
Z Partition function
$\mathbf{v}$ Visible units in a network
$\mathbf{h}$ Hidden units in a network
$\mathbf{o}$ Observation (feature) vector
$\mathbf{y}$ Output prediction vector
ε Learning rate
θ Threshold
λ Regularization parameter
$\mathcal{N}(\mathbf{x}; \mu, \Sigma)$ Random variable $\mathbf{x}$ follows a Gaussian distribution with mean vector μ and covariance matrix Σ
μ_i i-th component of the mean vector μ
σ_i^2 The i-th variance component
c_m Gaussian mixture component weight for the m-th Gaussian
$a_{i,j}$ HMM transition probability from state i to state j
$b_i(\mathbf{o})$ HMM emission probability for observation $\mathbf{o}$ at state i
Λ The entire model parameter set
$\mathbf{q}$ HMM state sequence
π HMM initial state probability

Chapter 1
Introduction

Abstract Automatic speech recognition (ASR) is an important technology to enable and improve the human–human and human–computer interactions. In this chapter, we introduce the main application areas of ASR systems, describe their basic architecture, and then introduce the organization of the book.

1.1 Automatic Speech Recognition: A Bridge for Better Communication

Automatic speech recognition (ASR) has been an active research area for over five decades. It has always been considered as an important bridge in fostering better human–human and human–machine communication. In the past, however, speech never actually became an important modality in the human–machine communication. This is partly because the technology at that time was not good enough to pass the usable bar for most real world users under most real usage conditions, and partly because in many situations alternative communication modalities such as keyboard and mouse significantly outperform speech in the communication efficiency, restriction, and accuracy.

In the recent years, speech technology started to change the way we live and work and became one of the primary means for humans to interact with some devices. This trend started due to the progress made in several key areas. First, Moor's law continues to function. The computational power available today, through multi-core processors, general purpose graphical processing units (GPGPUs), and CPU/GPU clusters, is several orders of magnitude more than that available just a decade ago. This makes training of more powerful yet complex models possible. These more computation-demanding models, which are the topic of this book, significantly reduced the error rates of the ASR systems. Second, we can now access to much more data than before, thanks to the continued advance of the Internet and the cloud computing. By building models on big data collected from the real usage scenarios, we can eliminate many model assumptions made before and make systems more robust. Third, mobile devices, wearable devices, intelligent living room devices, and in-vehicle infotainment systems became popular. On these devices and systems, alternative interaction

© Springer-Verlag London 2015
D. Yu and L. Deng, *Automatic Speech Recognition*,
Signals and Communication Technology, DOI 10.1007/978-1-4471-5779-3_1

Fig. 1.1 Components in a typical speech to speech translation system

modalities such as keyboard and mouse are less convenient than that in the personal computers. Speech, which is the natural way of human–human communication and a skill that majority of people already have, thus becomes a more favorable interaction modality on these devices and systems.

In the recent years, there are many applications in which speech technology plays an important role. These applications can be classified as applications that help improve human–human communication (HHC) and that help improve human–machine communication (HMC).

1.1.1 Human–Human Communication

Speech technology can remove barriers between human–human interactions. In the past, people who speak different languages need a human interpreter to be able to talk to each other. This sets a significant limitation on who people can communicate with and when the communication can happen. For example, people who cannot speak Chinese often find it difficult to travel in China alone. This barrier, however, can be alleviated by speech-to-speech (s2s) translation systems, of which a recent demo from Microsoft Research can be found at [2]. Besides being used by travelers, s2s translation systems can also be integrated into communication tools such as Skype to allow people who speak different languages to freely communicate with each other remotely. Figure 1.1 illustrates the key components in a typical s2s translation system in which speech recognition is at the first stage of the pipeline.

Speech technology can also help HHC in other ways. For example, in the unified messaging systems, the speech transcription sub-system can be used to convert voice messages left by a caller into text. The transcribed text can then be easily sent to the recipient through emails, instant messaging, or short message, and conveniently consumed by the recipient. In another example, the ASR technology can be used to dictate short messages to reduce the effort needed for the users to send short messages to their friends. The speech recognition technology can also be used to recognize and index speeches and lectures so that users can easily find the information that are interesting to them.

1.1.2 Human–Machine Communication

Speech technologies can also greatly improve HMC. The most popular applications in this category include voice search, personal digital assistant, gaming, living room interaction systems, and in-vehicle infotainment systems:

- Voice search (VS) [24–26] applications allow users to search for information such as restaurants, driving directions, and product reviews directly through speech. They significantly reduce effort needed for users to input the search queries. Nowadays, voice search applications have been very popular in mobile devices such as iPhone, Windows Phone, and Android Phone.
- Personal digital assistance (PDA) has been prototyped for a decade. However, it became popular only recently after Apple released the Siri system in iPhone. Since then, many other companies released similar products. PDA knows the information stored in your mobile devices, some world knowledge, and users' interaction history with the system, and thus can serve users better. The example tasks that a PDA can do include dialing a phone number, scheduling a meeting, asking for an answer, and searching for a music, all by directly issuing a voice command.
- The gaming experience can be greatly improved if games are integrated with speech technology. For example, in some of Microsoft's Xbox games, players can talk to the cartoon characters to ask for information or issue commands.
- The living room interaction systems and in-vehicle infotainment systems [23] are very similar in functionality. These systems allow users to interact with them through speech so that users can play music, ask for information, or control the systems. However, since these systems are used under different conditions, they encounter different design challenges.

All the applications and systems discussed in this sub-section are examples of spoken language systems [11]. As shown in Fig. 1.2, spoken language systems often include one or more of four major components: a speech recognition component that converts speech into text; a spoken language understanding component that finds semantic information in the spoken words; a text-to-speech component that conveys spoken information; and a dialog manager that communicates with applications and other three components. All these components are important to build a successful spoken language system. In this book, we only focus on the ASR component.

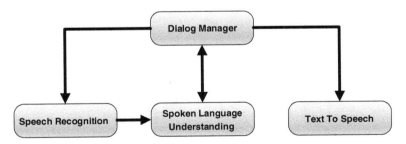

Fig. 1.2 Components in a typical spoken language system

1.2 Basic Architecture of ASR Systems

The typical architecture of an ASR system is illustrated in Fig. 1.3. As indicated in the figure, the ASR system has four main components: Signal processing and feature extraction, acoustic model (AM), language model (LM), and hypothesis search.

The signal processing and feature extraction component takes as input the audio signal, enhances the speech by removing noises and channel distortions, converts the signal from time-domain to frequency-domain, and extracts salient feature vectors that are suitable for the following acoustic models. The acoustic model integrates knowledge about acoustics and phonetics, takes as input the features generated from the feature extraction component, and generates an AM score for the variable-length feature sequence. The language model estimates the probability of a hypothesized word sequence, or LM score, by learning the correlation between words from a (typically text) training corpora. The LM score often can be estimated more accurately if the prior knowledge about the domain or task is known. The hypothesis search component combines AM and LM scores given the feature vector sequence and the hypothesized word sequence, and outputs the word sequence with the highest score as the recognition result. In this book, we focus on the AM component.

The two main issues to deal with by the AM component are the variable-length feature vectors and variability in the audio signals. The variable length feature problem is often addressed by techniques such as dynamic time warping (DTW) and hidden Markov model (HMM) [18], which we will describe in Chap. 3. The variability in the audio signals is caused by complicated interaction of speaker characteristics (e.g., gender, illness, or stress), speech style and rate, environment noise, side talks, channel distortion (e.g., microphone difference), dialect differences, and nonnative accents. A successful speech recognition system must contend with all of these acoustic variability.

As we move from the constrained tasks to the real world applications described in Sect. 1.1, additional challenges incur. As illustrated in Fig. 1.4, a practical ASR system, nowadays, needs to deal with huge (millions) vocabulary, free-style conversation, noisy far field spontaneous speech and mixed languages.

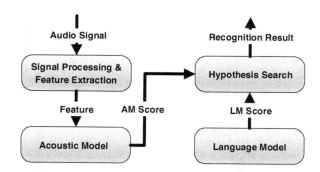

Fig. 1.3 Architecture of ASR systems

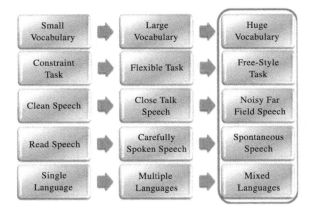

Fig. 1.4 The ASR problems we work on today (*right* column) are much more difficult than what we have worked on in the past due to the demand from the real world applications

In the past, the dominant state-of-the-art ASR system typically uses the mel-frequency cepstral coefficient (MFCC) [4] or relative spectral transform-perceptual linear prediction (RASTA-PLP) [7] as the feature vectors and the Gaussian mixture model (GMM)-HMM as the acoustic model. In the 1990s, these GMM-HMM AMs are trained using the maximum likelihood (ML) training criterion. In 2000s, the sequence discriminative training algorithms such as minimum classification error (MCE) [13] and minimum phone error (MPE) [16] were proposed and further improved the ASR accuracy.

In recent years, discriminative hierarchical models such as deep neural networks (DNNs) [3, 9] became feasible and significantly reduced error rates, thanks to the continued improvements in the computation power, availability of large training set, and better understanding of these models. For example, the context-dependent (CD)-DNN-HMM achieved one third of error rate reduction on the Switchboard conversational transcription task over the conventional GMM-HMM systems trained with sequence discriminative criteria [22].

In this book, we describe the new development of these discriminative hierarchical models, including DNN, convolutional neural network (CNN), and recurrent neural network (RNN). We discuss the theoretical foundations of these models as well as the practical tricks that make the system work. Since we are most familiar with our own work, the book will mainly focus on our own work, but will cover related work conducted by other researchers when appropriate.

1.3 Book Organization

We organize the book into five parts. We devote Part I to the conventional GMM-HMM systems as well as the related mathematical models and variants, extracting the content from the established books of [5, 10, 19] and the tutorial materials of [12, 17]. Part II introduces DNN, its initialization and training algorithms. Part III discusses the DNN-HMM hybrid system for speech recognition, the tricks to improve the training and evaluation speed of DNNs, and the sequence-discriminative training algorithms. Part IV describes DNNs from the joint representation learning and model optimization point of view. Based on this view, we introduce the tandem system for speech recognition and the adaptation techniques for DNNs. Part V covers the advanced models such as multilingual DNN, recurrent neural networks, and computational networks. We summarize the book in Chap. 15 with an outline of the key milestones in the history of developing deep learning-based ASR systems and provide some thoughts we have in the future directions of ASR research.

Compared with our recent book of [6], the current book goes into much greater technical depth and is confined exclusively within speech recognition without spreading into other deep learning applications.

1.3.1 Part I: Conventional Acoustic Models

Chapters 2 and 3 provide you with a basic theoretic foundation on the conventional GMM-HMM acoustic model. These chapters prepare you to understand the discriminative hierarchical models we will introduce in the rest of the book.

Chapter 2 discusses Gaussian mixture models, the principle of maximum likelihood, and the expectation-maximization [15] training algorithm. Chapter 3 introduces hidden Markov model, the most prominent technique used in modern speech recognition systems. It explains how HMMs handle the variable-length signal problem, and describe the forward-backward training algorithm and the Viterbi decoding algorithm. GMM-HMM forms the foundation of the modern speech recognition systems before the prevailing of the CD-DNN-HMM systems we focus on in this book.

1.3.2 Part II: Deep Neural Networks

Chapters 4 and 5 provide you with an in-depth look at deep neural networks. We highlight techniques that are proven to work well in building real systems and explain in details how and why these techniques work from both theoretic and practical perspectives.

Chapter 4 introduces deep neural networks, the famous back-propagation algorithm [14, 20], and various practical tricks to train a DNN effectively and efficiently. Chapter 5 discusses advanced DNN initialization techniques, including the generative pretraining and discriminative pretraining [21]. We focus on the restricted Boltzmann machine (RBM) [8], noisy auto-encoder [1], and the relationship between two.

1.3.3 Part III: DNN-HMM Hybrid Systems for ASR

In Chaps. 6–8, we follow the same spirit to discuss proven techniques in integrating DNNs with HMMs for speech recognition.

Chapter 6 describes the DNN-HMM hybrid model [3], in which the HMM is used to model the sequential property of the speech signal and DNN is used to model the emission probabilities in the HMM. Chapter 7 discusses the practical implementation tricks that improve the speed of the training and decoding of the DNN-HMM systems. Chapter 8 discusses the sequence-discriminative training algorithms that further improve the recognition accuracy of DNN-HMM hybrid systems.

1.3.4 Part IV: Representation Learning in Deep Neural Networks

In Chaps. 9–11 we look at DNNs from a different angle and discuss how this new angle suggests other approaches that DNN can be used for speech recognition.

Chapter 9 describes the joint feature learning and model optimization point of view of DNNs. We argue that the DNN can be separated at arbitrary hidden layer and consider all layers below it as the feature transformation and all layers above it as the classification model. Chapter 10 introduces the tandem architecture and bottleneck systems, in which DNNs are served as a separate feature extractors that fed into the conventional GMM-HMM. Chapter 11 shows different techniques with which DNNs can be effectively adapted.

1.3.5 Part V: Advanced Deep Models

From Chaps. 12 to 14 we introduce several advanced deep models. In Chap. 12 we describe DNN-based multitask and transfer learning with which the feature representations are shared and transferred across related tasks. We will use multilingual and crosslingual speech recognition as the main example, which uses a shared-hidden-layer DNN architecture, to demonstrate these techniques. In Chap. 13 we illustrate recurrent neural network, including long short-term memory neural networks, for speech recognition. In Chap. 14 we introduce computational network, a unified framework for describing arbitrary learning machines, such as deep neural networks (DNNs), computational neural networks (CNNs), recurrent neural networks (RNNs) including the version with long short-term memory (LSTM), logistic regression, and maximum entropy model, which can be illustrated as a series of computational steps.

References

1. Bengio, Y., Lamblin, P., Popovici, D., Larochelle, H.: Greedy layer-wise training of deep networks. In: Proceedings of the Neural Information Processing Systems (NIPS), pp. 153–160 (2006)
2. Clayton, S.: Microsoft research shows a promising new breakthrough in speech translation technology. http://blogs.technet.com/b/next/archive/2012/11/08/microsoft-research-shows-a-promising-new-breakthrough-in-speech-translation-technology.aspx (2012)
3. Dahl, G.E., Yu, D., Deng, L., Acero, A.: Context-dependent pre-trained deep neural networks for large-vocabulary speech recognition. IEEE Trans. Audio, Speech Lang. Process. **20**(1), 30–42 (2012)
4. Davis, S., Mermelstein, P.: Comparison of parametric representations for monosyllabic word recognition in continuously spoken sentences. IEEE Trans. Acoust. Speech Signal Process. **28**(4), 357–366 (1980)
5. Deng, L., O'Shaughnessy, D.: Speech Processing—A Dynamic and Optimization-Oriented Approach. Marcel Dekker Inc, New York (2003)
6. Deng, L., Yu, D.: Deep Learning: Methods and Applications. NOW Publishers, Delft (2014)
7. Hermansky, H.: Perceptual linear predictive (PLP) analysis of speech. J. Acoust. Soc. Am. **87**, 1738 (1990)
8. Hinton, G.: A practical guide to training restricted Boltzmann machines. Technical Report UTML TR 2010-003, University of Toronto (2010)
9. Hinton, G., Deng, L., Yu, D., Dahl, G.E., Mohamed, A.R., Jaitly, N., Senior, A., Vanhoucke, V., Nguyen, P., Sainath, T.N., et al.: Deep neural networks for acoustic modeling in speech recognition: the shared views of four research groups. IEEE Signal Process. Mag. **29**(6), 82–97 (2012)
10. Huang, X., Acero, A., Hon, H.W.: Spoken Language Processing: A Guide to Theory, Algorithm, and System Development. Prentice Hall, Englewood Cliffs (2001)
11. Huang, X., Acero, A., Hon, H.W., et al.: Spoken Language Processing, vol. 18. Prentice Hall, Englewood Cliffs (2001)
12. Huang, X., Deng, L.: An overview of modern speech recognition. In: Indurkhya, N., Damerau, F.J. (eds.) Handbook of Natural Language Processing, 2nd edn. CRC Press, Taylor and Francis Group, Boca Raton (2010). ISBN 978-1420085921
13. Juang, B.H., Hou, W., Lee, C.H.: Minimum classification error rate methods for speech recognition. IEEE Trans. Speech Audio Process. **5**(3), 257–265 (1997)
14. LeCun, Y., Bottou, L., Orr, G.B., Müller, K.R.: Efficient backprop. In: Neural Networks: Tricks of the Trade, pp. 9–50. Springer (1998)
15. Moon, T.K.: The expectation-maximization algorithm. IEEE Signal Process. Mag. **13**(6), 47–60 (1996)
16. Povey, D., Woodland, P.C.: Minimum phone error and I-smoothing for improved discriminative training. In: Proceedings of International Conference on Acoustics, Speech and Signal Processing (ICASSP), vol. 1, pp. I–105 (2002)
17. Rabiner, L.: A tutorial on hidden markov models and selected applications in speech recognition. Proc. IEEE **77**(2), 257–286 (1989)
18. Rabiner, L., Juang, B.H.: An introduction to hidden markov models. IEEE ASSP Mag. **3**(1), 4–16 (1986)
19. Rabiner, L., Juang, B.H.: Fundamentals of Speech Recognition. Prentice-Hall, Upper Saddle River (1993)
20. Rumelhart, D.E., Hintont, G.E., Williams, R.J.: Learning representations by back-propagating errors. Nature **323**(6088), 533–536 (1986)
21. Seide, F., Li, G., Chen, X., Yu, D.: Feature engineering in context-dependent deep neural networks for conversational speech transcription. In: Proceedings of IEEE Workshop on Automatic Speech Recognition and Understanding (ASRU), pp. 24–29 (2011)

22. Seide, F., Li, G., Yu, D.: Conversational speech transcription using context-dependent deep neural networks. In: Proceedings of Annual Conference of International Speech Communication Association (INTERSPEECH), pp. 437–440 (2011)
23. Seltzer, M.L., Ju, Y.C., Tashev, I., Wang, Y.Y., Yu, D.: In-car media search. IEEE Signal Process. Mag. **28**(4), 50–60 (2011)
24. Wang, Y.Y., Yu, D., Ju, Y.C., Acero, A.: An introduction to voice search. IEEE Signal Process. Mag. **25**(3), 28–38 (2008)
25. Yu, D., Ju, Y.C., Wang, Y.Y., Zweig, G., Acero, A.: Automated directory assistance system-from theory to practice. In: Proceedings of Annual Conference of International Speech Communication Association (INTERSPEECH), pp. 2709–2712 (2007)
26. Zweig, G., Chang, S.: Personalizing model [M] for voice-search. In: Proceedings of Annual Conference of International Speech Communication Association (INTERSPEECH), pp. 609–612 (2011)

Part I
Conventional Acoustic Models

Chapter 2
Gaussian Mixture Models

Abstract In this chapter we first introduce the basic concepts of random variables and the associated distributions. These concepts are then applied to Gaussian random variables and mixture-of-Gaussian random variables. Both scalar and vector-valued cases are discussed and the probability density functions for these random variables are given with their parameters specified. This introduction leads to the Gaussian mixture model (GMM) when the distribution of mixture-of-Gaussian random variables is used to fit the real-world data such as speech features. The GMM as a statistical model for Fourier-spectrum-based speech features plays an important role in acoustic modeling of conventional speech recognition systems. We discuss some key advantages of GMMs in acoustic modeling, among which is the easy way of using them to fit the data of a wide range of speech features using the EM algorithm. We describe the principle of maximum likelihood and the related EM algorithm for parameter estimation of the GMM in some detail as it is still a widely used method in speech recognition. We finally discuss a serious weakness of using GMMs in acoustic modeling for speech recognition, motivating new models and methods that form the bulk part of this book.

2.1 Random Variables

The most basic concept in probability theory and in statistics is the random variable. A scalar random variable is a real-valued function or variable, which takes its value based on the outcome of a random experiment. A vector-valued random variable is a set of scalar random variables, which may either be related to or be independent of each other. Since the experiment is random, the value assumed by the random variable is random as well. A random variable can be understood as a mapping from a random experiment to a variable. Depending on the nature of the experiment and of the design of the mapping, a random variable can take either discrete values, continuous values, or a mix of discrete and continuous values. We, hence, see the names of discrete random variable, continuous random variable, or hybrid random variable. All possible values, which may be assumed by a random variable, are

© Springer-Verlag London 2015
D. Yu and L. Deng, *Automatic Speech Recognition*,
Signals and Communication Technology, DOI 10.1007/978-1-4471-5779-3_2

sometimes called its domain. In this as well as a few other later chapters, we use the same notations to describe random variables and other concepts as those adopted in [16].

The fundamental characterization of a continuous-valued random variable, x, is its distribution or the probability density function (PDF), denoted generally by $p(x)$. The PDF for a continuous random variable at $x = a$ is defined by

$$p(a) \doteq \lim_{\Delta a \to 0} \frac{P(a - \Delta a < x \le a)}{\Delta a} \ge 0 \qquad (2.1)$$

where $P(\cdot)$ denotes the probability of the event.

The cumulative distribution function of a continuous random variable x evaluated at $x = a$ is defined by

$$P(a) \doteq P(x \le a) = \int_{-\infty}^{a} p(x)dx. \qquad (2.2)$$

A PDF has to satisfy the property of normalization:

$$P(x \le \infty) = \int_{-\infty}^{\infty} p(x)dx = 1. \qquad (2.3)$$

If the normalization property is not held, we sometimes call the PDF an improper density or unnormalized distribution.

For a continuous random vector $\mathbf{x} = (x_1, x_2, \ldots, x_D)^{\mathrm{T}} \in \mathscr{R}^D$, we can similarly define their joint PDF of $p(x_1, x_2, \ldots, x_D)$. Further, a marginal PDF for each of the random variable x_i in the random vector $\mathbf{x}$ is defined by

$$p(x_i) \doteq \int \int_{all\ x_j:\ x_j \neq x_i} \cdots \int p(x_1, \ldots, x_D)\, dx_1 \ldots dx_{i-1} dx_{i+1} \ldots dx_D. \qquad (2.4)$$

It has the same properties as the PDF for a scalar random variable.

2.2 Gaussian and Gaussian-Mixture Random Variables

A scalar continuous random variable x is normally or Gaussian distributed if its PDF is

$$p(x) = \frac{1}{(2\pi)^{1/2}\sigma} \exp\left[-\frac{1}{2}\left(\frac{x-\mu}{\sigma}\right)^2\right] \doteq \mathcal{N}(x; \mu, \sigma^2),$$
$$(-\infty < x < \infty; \sigma > 0) \tag{2.5}$$

An equivalent notation for the above is

$$x \sim \mathcal{N}(\mu, \sigma^2),$$

denoting that random variable x obeys a normal distribution with mean μ and variance σ^2. With the use of the precision parameter r, a Gaussian PDF can also be written as

$$p(x) = \sqrt{\frac{r}{2\pi}} \exp\left[-\frac{r}{2}(x-\mu)^2\right]. \tag{2.6}$$

It is a simple exercise to show that for a Gaussian random variable x, $E(x) = \mu$, $var(x) = \sigma^2 = r^{-1}$.

The normal random vector $\mathbf{x} = (x_1, x_2, \ldots, x_D)^T$, also called multivariate or vector-valued Gaussian random variable, is defined by the following joint PDF:

$$p(\mathbf{x}) = \frac{1}{(2\pi)^{D/2}|\boldsymbol{\Sigma}|^{1/2}} \exp\left[-\frac{1}{2}(\mathbf{x}-\boldsymbol{\mu})^T\boldsymbol{\Sigma}^{-1}(\mathbf{x}-\boldsymbol{\mu})\right] \doteq \mathcal{N}(\mathbf{x}; \boldsymbol{\mu}, \boldsymbol{\Sigma}) \tag{2.7}$$

An equivalent notation is $\mathbf{x} \sim \mathcal{N}(\boldsymbol{\mu} \in \mathcal{R}^D, \boldsymbol{\Sigma} \in \mathcal{R}^{D \times D})$. It is also straight forward to show that for a multivariate Gaussian random variable, the expectation and covariance matrix are given by $E(\mathbf{x}) = \boldsymbol{\mu}$; $E[(\mathbf{x}-\bar{\mathbf{x}})(\mathbf{x}-\bar{\mathbf{x}})^T] = \boldsymbol{\Sigma}$.

The Gaussian distribution is commonly used in many engineering and science disciplines including speech recognition. The popularity arises not only from its highly desirable computational properties, but also from its ability to approximate many naturally occurring real-world data, thanks to the law of large numbers.

Let us now move to discuss the Gaussian-mixture random variable with the distribution called mixture of Gaussians. A scalar continuous random variable x has a Gaussian-mixture distribution if its PDF is specified by

$$p(x) = \sum_{m=1}^{M} \frac{c_m}{(2\pi)^{1/2}\sigma_m} \exp\left[-\frac{1}{2}\left(\frac{x-\mu_m}{\sigma_m}\right)^2\right] \tag{2.8}$$

$$= \sum_{m=1}^{M} c_m \mathcal{N}(x; \mu_m, \sigma_m^2) \quad (-\infty < x < \infty; \sigma_m > 0; c_m > 0)$$

where the positive mixture weights sum to unity: $\sum_{m=1}^{M} c_m = 1$.

The most obvious property of Gaussian mixture distribution is its multimodal one ($M > 1$ in Eq. 2.8), in contrast to the unimodal property of the Gaussian distribution where $M = 1$. This makes it possible for a mixture Gaussian distribution to adequately describe many types of physical data (including speech data) exhibiting multimodality poorly suited for a single Gaussian distribution. The multimodality in data may come from multiple underlying causes each being responsible for one particular mixture component in the distribution. If such causes are identified, then the mixture distribution can be decomposed into a set of cause-dependent or context-dependent component distributions.

It is easy to show that the expectation of a random variable x with the mixture Gaussian PDF of Eq. 2.8 is $E(x) = \sum_{m=1}^{M} c_m \mu_m$. But unlike a (uni-modal) Gaussian distribution, this simple summary statistic is not very informative unless all the component means, $\mu_m, m = 1, \ldots, M$, in the Gaussian-mixture distribution are close to each other.

The multivariate generalization of the mixture Gaussian distribution has the joint PDF of

$$p(\mathbf{x}) = \sum_{m=1}^{M} \frac{c_m}{(2\pi)^{D/2} |\boldsymbol{\Sigma}_m|^{1/2}} \exp\left[-\frac{1}{2}(\mathbf{x} - \boldsymbol{\mu}_m)^{\mathrm{T}} \boldsymbol{\Sigma}_m^{-1}(\mathbf{x} - \boldsymbol{\mu}_m)\right]$$

$$= \sum_{m=1}^{M} c_m \mathcal{N}(\mathbf{x}; \boldsymbol{\mu}_m, \boldsymbol{\Sigma}_m), \qquad (c_m > 0). \tag{2.9}$$

The use of this multivariate mixture Gaussian distribution has been one key factor contributing to improved performance of many speech recognition systems (prior to the rise of deep learning); e.g., [14, 23, 24, 27]. In most applications, the number of mixture components, M, is chosen a priori according to the nature of the problem, although attempts have been made to sidestep such an often difficult problem of finding the "right" number; e.g., [31].

In using the multivariate mixture Gaussian distribution of Eq. 2.8, if the variable x's dimensionality, D, is large (say, 40, for speech recognition problems), then the use of full (nondiagonal) covariance matrices ($\boldsymbol{\Sigma}_m$) would involve a large number of parameters (on the order of $M \times D^2$). To reduce the number of parameters, one can opt to use diagonal covariance matrices for $\boldsymbol{\Sigma}_m$. Alternatively, when M is large, one can also constrain all covariance matrices to be the same; i.e., "tying" $\boldsymbol{\Sigma}_m$ for all mixture components, m. An additional advantage of using diagonal covariance matrices is significant simplification of computations needed for the applications of the Gaussian-mixture distributions. Reducing full covariance matrices to diagonal ones may have seemed to impose uncorrelatedness among data vector components. This has been misleading, however, since a mixture of Gaussians each with a diagonal covariance matrix can at least effectively describe the correlations modeled by one Gaussian with a full covariance matrix.

2.3 Parameter Estimation

The Gaussian-mixture distributions we just discussed contain a set of parameters. In the multivariate case of Eq. 2.8, the parameter set consists of $\boldsymbol{\Theta} = \{c_m, \boldsymbol{\mu}_m, \boldsymbol{\Sigma}_m\}$. The parameter estimation problem, also called learning, is to determine the values of these parameters from a set of data typically assumed to be drawn from the Gaussian-mixture distribution.

It is common to think of Gaussian mixture modeling and the related parameter estimation as a missing data problem. To understand this, let us assume that the data points under consideration have "membership," or the component of the mixture, in one of the individual Gaussian distributions we are using to model the data. At the start, this membership is unknown, or missing. The task of parameter estimation is to *learn* appropriate parameters for the distribution, with the connection to the data points being represented as their membership in the individual Gaussian distributions.

Here, we focus on maximum likelihood methods for parameter estimation of the Gaussian-mixture distribution, and the expectation maximization (EM) algorithm in particular. The EM algorithm is the most popular technique used to estimate the parameters of a mixture given a fixed number of mixture components, and it can be used to compute the parameters of any parametric mixture distribution. It is an iterative algorithm with two steps: an expectation or E-step and a maximization or M-step. We will cover the general statistical formulation of the EM algorithm, based on [5], in more detail in Chap. 3 on hidden Markov models, and here we only discuss its practical use for the parameter estimation problem related to the Gaussian mixture distribution.

The EM algorithm is of particular appeal for the Gaussian mixture distribution as the main topic of this chapter, where closed-form expressions in the M-step are available as expressed in the following iterative fashion[1]:

$$c_m^{(j+1)} = \frac{1}{N} \sum_{t=1}^{N} h_m^{(j)}(t), \tag{2.10}$$

$$\boldsymbol{\mu}_m^{(j+1)} = \frac{\sum_{t=1}^{N} h_m^{(j)}(t) x^{(t)}}{\sum_{t=1}^{N} h_m^{(j)}(t)}, \tag{2.11}$$

$$\boldsymbol{\Sigma}_m^{(j+1)} = \frac{\sum_{t=1}^{N} h_m^{(j)}(t)[\mathbf{x}^{(t)} - \boldsymbol{\mu}_m^{(j)}][x^{(t)} - \boldsymbol{\mu}_m^{(j)}]^T}{\sum_{t=1}^{N} h_m^{(j)}(t)}, \tag{2.12}$$

where the posterior probabilities (also called the membership responsibilities) computed from the E-step are given by

---

[1] Detailed derivation of these formulae can be found in [1], which we omit here. Related derivations for similar but more general models can be found in [2, 3, 6, 15, 18].

$$h_m^{(j)}(t) = \frac{c_m^{(j)} \mathcal{N}(\mathbf{x}^{(t)}; \boldsymbol{\mu}_m^{(j)}, \boldsymbol{\Sigma}_m^{(j)})}{\sum_{i=1}^{n} c_i^{(j)} \mathcal{N}(\mathbf{x}^{(t)}; \boldsymbol{\mu}_i^{(j)}, \boldsymbol{\Sigma}_i^{(j)})}. \tag{2.13}$$

That is, on the basis of the current (denoted by superscript j above) estimate for the parameters, the conditional probability for a given observation $\mathbf{x}^{(t)}$ being generated from mixture component m is determined for each data sample point at $t = 1, \ldots, N$, where N is the sample size. The parameters are then updated such that the new component weights correspond to the average conditional probability and each component mean and covariance is the component specific weighted average of the mean and covariance of the entire sample set.

It has been well established that each successive EM iteration will not decrease the likelihood, a property not shared by most other gradient based maximization techniques. Further, the EM algorithm naturally embeds within it constraints on the probability vector, and for sufficiently large sample sizes positive definiteness of the covariance iterates. This is a key advantage since explicitly constrained methods incur extra computational costs to check and maintain appropriate values. Theoretically, the EM algorithm is a first-order one and as such converges slowly to a fixed-point solution. However, convergence in likelihood is rapid even if convergence in the parameter values themselves is not. Another disadvantage of the EM algorithm is its propensity to spuriously identify local maxima and its sensitivity to initial values. These problems can be addressed by evaluating EM at several initial points in the parameter space although this may become computationally costly. Another popular approach to address these issues is to start with one Gaussian component and split the Gaussian components after each epoch.

In addition to the EM algorithm discussed above for parameter estimation that is rested on maximum likelihood or data fitting, other types of estimation aimed to perform discriminative estimation or learning have been developed for Gaussian or Gaussian mixtures, as special cases of the related but more general statistical models such as the Gaussian HMM and its Gaussian-mixture counterpart; e.g., [22, 25, 26, 33].

2.4 Mixture of Gaussians as a Model for the Distribution of Speech Features

When speech waveforms are processed into compressed (e.g., by taking logarithm of) short-time Fourier transform magnitudes or related cepstra, the Gaussian-mixture distribution discussed above is shown to be quite appropriate to fit such speech features when the information about the temporal order is discarded. That is, one can use the Gaussian-mixture distribution as a *model* to represent frame-based speech features. We use the Gaussian mixture model (GMM) to refer to the use of the Gaussian-mixture distribution for representing the data distribution. In this case and in the remainder of this book, we generally use *model* or computational model to refer to a form of mathematical abstraction of aspects of some realistic physical process (such as the human speech process), following the guiding principles detailed in [9]. Such models are established often with necessary simplification and approximation aimed at mathematical or computational tractability. The tractability is crucial in making the mathematical abstraction amenable to computer or algorithmic implementation for practical engineering applications (such as speech analysis and recognition).

Both inside and outside the speech recognition domain, the GMM is commonly used for modeling the data and for statistical classification. GMMs are well known for their ability to represent arbitrarily complex distributions with multiple modes. GMM-based classifiers are highly effective with widespread use in speech research, primarily for speaker recognition, denoising speech features, and speech recognition. For speaker recognition, the GMM is directly used as a universal background model (UBM) for the speech feature distribution pooled from all speakers [4, 28, 32, 34]. In speech feature denoising or noise tracking applications, the GMM is used in a similar way and as a prior distribution [10–13, 19, 21]. In speech recognition applications, the GMM is integrated into the doubly stochastic model of HMM as its output distribution conditioned on a state, a topic which will be discussed in a great detail in Chap. 3.

When speech sequence information is taken into account, the GMM is no longer a good model as it contains no sequence information. A class of more general models, called the hidden Markov models (HMM) to be discussed in Chap. 3, captures the sequence information. Given a fixed state of the HMM, the GMM remains a reasonably good model for the PDF of speech feature vectors allocated to the state.

GMMs have several distinct advantages that make them suitable for modeling the PDFs over speech feature vectors associated with each state of an HMM. With enough components, they can model PDFs to any required level of accuracy, and they are easy to fit to data using the EM algorithm described in Sect. 2.3. A huge amount of research has gone into finding ways of constraining GMMs to increase their evaluation speed and to optimize the tradeoff between their flexibility and the amount of training data required to avoid overfitting. This includes the development of parameter-tied or semi-tied GMMs and subspace GMMs.

Beyond the use of the EM algorithm for parameter estimation of the GMM parameters, the speech recognition accuracy obtained by a GMM-based system (which is interfaced with the HMM) has been drastically improved if the GMM parameters are discriminatively learned after they have been generatively trained by EM to maximize its probability of generating the observed speech features in the training data. This is especially true if the discriminative objective function used for training is closely related to the error rate on phones, words, or sentences. The accuracy can also be improved by augmenting (or concatenating) the input speech features with tandem or bottleneck features generated using neural networks, which we will discuss in a later chapter. GMMs had been very successful in modeling speech features and in acoustic modeling for speech recognition for many years (until around year 2010–2011 when deep neural networks were shown to outperform the GMMs).

Despite all their advantages, GMMs have a serious shortcoming. That is, GMMs are statistically inefficient for modeling data that lie on or near a nonlinear manifold in the data space. For example, modeling the set of points that lie very close to the surface of a sphere only requires a few parameters using an appropriate model class, but it requires a very large number of diagonal Gaussians or a fairly large number of full-covariance Gaussians. It is well-known that speech is produced by modulating a relatively small number of parameters of a dynamical system [7, 8, 17, 20, 29, 30]. This suggests that the true underlying structure of speech is of a much lower dimension than is immediately apparent in a window that contains hundreds of coefficients. Therefore, other types of model, which can capture better properties of speech features, are expected to work better than GMMs for acoustic modeling of speech. In particular, the new models should more effectively exploit information embedded in a large window of frames of speech features than GMMs. We will return to this important problem of characterizing speech features after discussing a model, the HMM, for characterizing temporal properties of speech in the next chapter.

References

1. Bilmes, J.: A gentle tutorial of the EM algorithm and its application to parameter estimation for Gaussian mixture and hidden Markov models. Technical Report, TR-97-021, ICSI (1997)
2. Bilmes, J.: What HMMs can do. IEICE Trans. Inf. Syst. **E89-D**(3), 869–891 (2006)
3. Bishop, C.: Pattern Recognition and Machine Learning. Springer, Heidelberg (2006)
4. Dehak, N., Kenny, P., Dehak, R., Dumouchel, P., Ouellet, P.: Front-end factor analysis for speaker verification. IEEE Trans. Audio, Speech Lang. Process. **19**(4), 788–798 (2011)
5. Dempster, A.P., Laird, N.M., Rubin, D.B.: Maximum-likelihood from incomplete data via the EM algorithm. J. Royal Stat. Soc. Ser. B. **39**, 1–38 (1977)
6. Deng, L.: A generalized hidden markov model with state-conditioned trend functions of time for the speech signal. Signal Process. **27**(1), 65–78 (1992)
7. Deng, L.: Computational models for speech production. In: Computational Models of Speech Pattern Processing, pp. 199–213. Springer, New York (1999)
8. Deng, L.: Switching dynamic system models for speech articulation and acoustics. In: Mathematical Foundations of Speech and Language Processing, pp. 115–134. Springer, New York (2003)
9. Deng, L.: Dynamic Speech Models—Theory, Algorithm, and Applications. Morgan and Claypool, New York (2006)
10. Deng, L., Acero, A., Plumpe, M., Huang, X.: Large vocabulary speech recognition under adverse acoustic environment. In: Proceedings of International Conference on Spoken Language Processing (ICSLP), pp. 806–809 (2000)
11. Deng, L., Droppo, J.: A. Acero: recursive estimation of nonstationary noise using iterative stochastic approximation for robust speech recognition. IEEE Trans. Speech Audio Process. **11**, 568–580 (2003)
12. Deng, L., Droppo, J., Acero, A.: A Bayesian approach to speech feature enhancement using the dynamic cepstral prior. In: Proceedings of International Conference on Acoustics, Speech and Signal Processing (ICASSP), vol. 1, pp. I-829–I-832 (2002)
13. Deng, L., Droppo, J., Acero, A.: Enhancement of log mel power spectra of speech using a phase-sensitive model of the acoustic environment and sequential estimation of the corrupting noise. IEEE Trans. Speech Audio Process. **12**(2), 133–143 (2004)

14. Deng, L., Kenny, P., Lennig, M., Gupta, V., Seitz, F., Mermelsten, P.: Phonemic hidden markov models with continuous mixture output densities for large vocabulary word recognition. IEEE Trans. Acoust, Speech Signal Process. **39**(7), 1677–1681 (1991)
15. Deng, L., Mark, J.: Parameter estimation for markov modulated poisson processes via the em algorithm with time discretization. In: Telecommunication Systems (1993)
16. Deng, L., O'Shaughnessy, D.: Speech Processing—A Dynamic and Optimization-Oriented Approach. Marcel Dekker Inc, New York (2003)
17. Deng, L., Ramsay, G., Sun, D.: Production models as a structural basis for automatic speech recognition. Speech Commun. **33**(2–3), 93–111 (1997)
18. Deng, L., Rathinavelu, C.: A Markov model containing state-conditioned second-order non-stationarity: application to speech recognition. Comput. Speech Lang. **9**(1), 63–86 (1995)
19. Deng, L., Wang, K., Acero, A., Hon, H., Droppo, J., Boulis, C., Wang, Y., Jacoby, D., Mahajan, M., Chelba, C., Huang, X.: Distributed speech processing in mipad's multimodal user interface. IEEE Trans. Audio Speech Lang. Process. **20**(9), 2409–2419 (2012)
20. Divenyi, P., Greenberg, S., Meyer, G.: Dynamics of Speech Production and Perception. IOS Press, Washington (2006)
21. Frey, B., Deng, L., Acero, A., Kristjansson, T.: Algonquin: iterating laplaces method to remove multiple types of acoustic distortion for robust speech recognition. In: Proceedings of European Conference on Speech Communication and Technology (EUROSPEECH) (2000)
22. He, X., Deng, L.: Discriminative Learning for Speech Recognition: Theory and Practice. Morgan and Claypool, New York (2008)
23. Huang, X., Acero, A., Hon, H.W., et al.: Spoken Language Processing. Prentice Hall, Englewood Cliffs (2001)
24. Huang, X., Deng, L.: An overview of modern speech recognition. In: Indurkhya, N., Damerau, F.J. (eds.) Handbook of Natural Language Processing, 2nd edn. CRC Press, Taylor and Francis Group, Boca Raton, FL (2010). ISBN 978-1420085921
25. Jiang, H., Li, X.: Discriminative learning in sequential pattern recognition—a unifying review for optimization-oriented speech recognition. IEEE Signal Process. Mag. **27**(3), 115–127 (2010)
26. Jiang, H., Li, X., Liu, C.: Large margin hidden markov models for speech recognition. IEEE Trans. Audio, Speech Lang. Process. **14**(5), 1584–1595 (2006)
27. Juang, B.H., Levinson, S.E., Sondhi, M.M.: Maximum likelihood estimation for mixture multivariate stochastic observations of markov chains. In: IEEE International Symposium on Information Theory vol. 32(2), pp. 307–309 (1986)
28. Kenny, P.: Joint factor analysis of speaker and session variability: theory and algorithms. CRIM, Montreal, (Report) CRIM-06/08-13 (2005)
29. King, S., Frankel, J., Livescu, K., McDermott, E., Richmond, K., Wester, M.: Speech production knowledge in automatic speech recognition. J. Acoust. Soc. Am. **121**, 723–742 (2007)
30. Lee, L.J., Fieguth, P., Deng, L.: A functional articulatory dynamic model for speech production. In: Proceedings of International Conference on Acoustics, Speech and Signal Processing (ICASSP), vol. 2, pp. 797–800. Salt Lake City (2001)
31. Rasmussen, C.E.: The infinite gaussian mixture model. In: Proceedings of Neural Information Processing Systems (NIPS) (1999)
32. Reynolds, D., Rose, R.: Robust text-independent speaker identification using gaussian mixture speaker models. IEEE Trans. Speech Audio Process. **3**(1), 72–83 (1995)
33. Xiao, L., Deng, L.: A geometric perspective of large-margin training of Gaussian models. IEEE Signal Process. Mag. **27**, 118–123 (2010)
34. Yin, S.C., Rose, R., Kenny, P.: A joint factor analysis approach to progressive model adaptation in text-independent speaker verification. IEEE Trans. Audio Speech Lang. Process. **15**(7), 1999–2010 (2007)

Chapter 3
Hidden Markov Models and the Variants

Abstract This chapter builds upon the reviews in the previous chapter on aspects of probability theory and statistics, including random variables and Gaussian mixture models, and extends the reviews to the Markov chain and the hidden Markov sequence or model (HMM). Central to the HMM is the concept of state, which is itself a random variable typically taking discrete values. Extending from a Markov chain to an HMM involves adding uncertainty or a statistical distribution on each of the states in the Markov chain. Hence, an HMM is a doubly-stochastic process, or probabilistic function of a Markov chain. When the state of the Markov sequence or HMM is confined to be discrete and the distributions associated with the HMM states do not overlap, we reduce it to a Markov chain. This chapter covers several key aspects of the HMM, including its parametric characterization, its simulation by random number generators, its likelihood evaluation, its parameter estimation via the EM algorithm, and its state decoding via the Viterbi algorithm or a dynamic programming procedure. We then provide discussions on the use of the HMM as a generative model for speech feature sequences and its use as the basis for speech recognition. Finally, we discuss the limitations of the HMM, leading to its various extended versions, where each state is made associated with a dynamic system or a hidden time-varying trajectory instead of with a temporally independent stationary distribution such as a Gaussian mixture. These variants of the HMM with state-conditioned dynamic systems expressed in the state-space formulation are introduced as a generative counterpart of the recurrent neural networks to be described in detail in Chap. 13.

3.1 Introduction

In the previous chapter, we reviewed aspects of probability theory and basic statistics, where we introduced the concept of random variables and the associated concept of probability distributions. We then discussed Gaussian and mixture-of-Gaussian random variables and their vector-valued or multivariate versions. All these concepts and examples are *static*, meaning that they do not have the temporal dimension making the length or the dimension of the random vectors variable according to how long the temporal sequence is. For a *static* section of the speech signal, its

© Springer-Verlag London 2015
D. Yu and L. Deng, *Automatic Speech Recognition*,
Signals and Communication Technology, DOI 10.1007/978-1-4471-5779-3_3

features based on spectral magnitudes (e.g., cepstra) can be well characterized by the multivariate distribution of mixture of Gaussians. This gives rise to the Gaussian mixture model (GMM) of speech features for a short-term or static speech sound pattern.

In this chapter, we will extend the concept of the random variable to the (discrete-time) random sequence, which is a collection of random variables indexed by uniformly spaced discrete times with a variable length. For the general statistical characterization of random sequences, see Chap. 3 of [43], but in this chapter we will extract only the part on Markov sequences as the most commonly used class of general random sequences. The concept of state is essential to a Markov sequence. When the state of the Markov sequence is confined to be discrete, we have a Markov chain, where all possible values taken by the discrete state variable constitutes the (discrete) state space and which we will cover in Sect. 3.2.

When each discrete state value is generalized to be a new random variable (discrete or continuous), the Markov chain is then generalized to the (discrete or continuous) hidden Markov sequence, or the hidden Markov model (HMM) when it is used to characterize or approximate statistical properties of real-world data sequences. In Sect. 3.3, we first parameterize the HMM in terms of transition probabilities of the underlying Markov chain and of the distributional parameters in the static PDFs given a fixed state. We then show how an HMM can be simulated via probabilistic sampling. Efficient computation of the likelihood of an observation sequence given the HMM is covered in detail, as this is an important but nonobvious element in applying the HMM to speech recognition and other practical problems.

Then, in Sect. 3.4, we first provide background information on the EM algorithm for maximum likelihood estimation of the parameters in general statistical models that contain hidden or latent random variables. We then apply the EM algorithm to solve the learning or parameter-estimation problem for the HMM (as well as the GMM, which can be viewed as a special case of the HMM). The resulting algorithm for the HMM learning is the celebrated Baum-Welch algorithm, widely used in speech recognition and other applications involving the HMM. Step-by-step derivations of the E-step in the Baum-Welch algorithm are given, which provides the conditional probabilities of an HMM state given the input training data. The M-step for the estimation of the transition probabilities of the Markov chain and of the mean vectors and covariance matrices in the Gaussian HMM is also given with step-by-step derivation.

We use Sect. 3.5 to present the celebrated Viterbi algorithm for optimally decoding the HMM state sequence given the input sequence data. The technique of dynamic programming, which is the underlying principle of the Viterbi algorithm, is described.

Finally, in Sect. 3.6, we connect the HMM as a statistical model to practical speech problems. The discussion starts with the HMM's capability as an elegant generative model for speech feature sequences; e.g., [3–5, 72]. The reasonably good match between the HMM and speech data enables this generative model to be used for the classification task of speech recognition via the use of Bayes rule [41, 57]. An analysis of the weaknesses of the HMM as a generative model for speech motivates its extensions to several variants, where the temporal independence and stationarity in the distribution of the observed speech data conditioned on each HMM state is replaced by more realistic, nonstationary, and temporally correlated dynamic systems with latent or hidden structure [15, 27, 44, 77, 80, 98]. The mathematical formalism of such dynamic systems expressed as the state-space model naturally bridges these HMM variants to the recurrent neural networks to be presented later in Chap. 13 of this book.

3.2 Markov Chains

A Markov chain is a discrete-state Markov sequence, a special case of a general Markov sequence. The state space of a Markov chain is of a discrete nature and is finite: $q_t \in \{s^{(j)}, j = 1, 2, \ldots, N\}$. Each of these discrete values is associated with a state in the Markov chain. Because of the one-to-one correspondence between state $s^{(j)}$ and its index j, we often use the two interchangeably.

A Markov chain, $\mathbf{q}_1^T = q_1, q_2, \ldots, q_T$, is completely characterized by the transition probabilities, defined by

$$P(q_t = s^{(j)} | q_{t-1} = s^{(i)}) \doteq a_{ij}(t), \quad i, j = 1, 2, \ldots, N \tag{3.1}$$

and by the initial state-distribution probabilities. If these transition probabilities are independent of time t, then we have a homogeneous Markov chain.

The transition probabilities of a (homogeneous) Markov chain are often conveniently put into a matrix form:

$$A = [a_{ij}], \quad \text{where} \quad a_{ij} \geq 0 \ \forall i, j; \quad \text{and} \quad \sum_{j=1}^{N} a_{ij} = 1 \ \forall i, \tag{3.2}$$

which is called the transition matrix of the Markov chain. Given the transition probabilities of a Markov chain, the state-occupation probability

$$p_j(t) \doteq P[q_t = s^{(j)}]$$

can be easily computed. The computation is recursive according to

$$p_i(t + 1) = \sum_{j=1}^{N} a_{ji} p_j(t), \quad \forall i. \tag{3.3}$$

If the state-occupation distribution of a Markov chain asymptotically converges: $p_i(t) \to \bar{\pi}(s^{(i)})$ as $t \to \infty$, we then call $\bar{\pi}(s^{(i)})$ a stationary distribution of the Markov chain. For a Markov chain to have a stationary distribution, its transition probabilities, a_{ij}, have to satisfy

$$\bar{\pi}(s^{(i)}) = \sum_{j=1}^{N} a_{ji}\bar{\pi}(s^{(j)}), \quad \forall i. \tag{3.4}$$

The stationary distribution of a Markov chain plays an important role in a class of powerful statistical methods collectively named Markov Chain Monte Carlo (MCMC) methods. These methods are used to simulate (i.e., to sample or to draw) arbitrarily complex distributions, enabling one to carry out many difficult statistical inference and learning tasks, which would otherwise be mathematically intractable. The theoretical foundation of the MCMC methods is the asymptotic convergence of a Markov chain to its stationary distribution, $\bar{\pi}(s^{(i)})$. That is, regardless of the initial distribution, the Markov chain is an asymptotically unbiased draw from $\bar{\pi}(s^{(i)})$. Therefore, in order to sample from an arbitrarily complex distribution, $p(s)$, one can construct a Markov chain, by designing appropriate transition probabilities, a_{ij}, so that its stationary distribution is $\bar{\pi}(s) = p(s)$.

Three other interesting and useful properties of a Markov chain can be easily derived. First, the state duration in a Markov chain is an exponential or geometric distribution: $p_i(d) = C (a_{ii})^{d-1}$, where the normalizing constant is $C = 1 - a_{ii}$. Second, the mean state duration is

$$\bar{d}_i = \sum_{d=1}^{\infty} d\, p_i(d) = \sum_{d=1}^{\infty}(1 - a_{ii})(a_{ii})^{d-1} = \frac{1}{1 - a_{ii}}. \tag{3.5}$$

Finally, the probability for an arbitrary observation sequence of a Markov chain, which is a finite-length state sequence $\mathbf{q}_1^T$, can be easily evaluated. This is simply the product of the transition probabilities traversing the Markov chain: $P(\mathbf{q}_1^T) = \bar{\pi}_{q_1} \prod_{t=1}^{T-1} a_{q_t q_{t+1}}$, where $\bar{\pi}_{s_1}$ is the initial state-occupation probability at $t = 1$.

3.3 Hidden Markov Sequences and Models

Let us view the Markov chain discussed above as an information source capable of generating observational output sequences. Then, we can call the Markov chain an observable Markov sequence because its output has one-to-one correspondence to a state. That is, each state corresponds to a deterministically observable variable or event. There is no randomness in the output in any given state. This lack of randomness makes the Markov chain too restrictive to describe many real-world informational sources, such as speech feature sequences, in an adequate manner.

Extension of the Markov chain to embed randomness, which overlaps among the states in the Markov chain gives rise to a hidden Markov sequence. This extension is accomplished by associating an observation probability distribution with each state in the Markov chain. The Markov sequence thus defined is a doubly embedded random sequence whose underlying Markov chain is not directly observable, hence a hidden sequence. The underlying Markov chain in the hidden Markov sequence can be observed only through a separate random function characterized by the observation probability distributions.

Note that if the observation probability distributions do not overlap across the states, then the underlying Markov chain would not be hidden. This is because, despite the randomness embedded in the states, any observational value over a fixed range specific to a state would be able to map uniquely to this state. In this case, the hidden Markov sequence essentially reduces to a Markov chain. Some excellent and more detailed exposition on the relationship between a Markov chain and its probabilistic function, or a hidden Markov sequence, can be found in [103, 104].

When a hidden Markov sequence is used to describe a physical, real-world informational source, i.e., to approximate the statistical characteristics of such a source, we often call it a hidden Markov model (HMM). One very successful practical use of the HMM has been in speech processing applications, including speech recognition, speech synthesis, and speech enhancement; e.g., [1, 12, 17, 46–48, 66, 71, 81, 83, 103, 111, 120, 124, 126, 128]. In these applications, the HMM is used as a powerful model to characterize the temporally nonstationary, spatially variable, but regular, learnable patterns of the speech signal. One key aspect of the HMM as the acoustic model of speech is its sequentially arranged Markov states, which permit the use of piecewise stationarity for approximating the globally nonstationary properties of speech feature sequences. Very efficient algorithms have been developed to optimize the boundaries of the local quasi-stationary temporal regimes, which we will discuss in Sect. 3.6.

3.3.1 Characterization of a Hidden Markov Model

We now give a formal characterization of a hidden Markov sequence model or HMM in terms of its basic elements and parameters.

1. Transition probabilities, $\mathbf{A} = [a_{ij}]$, $i, j = 1, 2, \ldots, N$, of a homogeneous Markov chain with a total of N states

$$a_{ij} = P(q_t = j | q_{t-1} = i), \quad i, j = 1, 2, \ldots, N. \tag{3.6}$$

2. Initial Markov chain state-occupation probabilities: $\pi = [\pi_i]$, $i = 1, 2, \ldots, N$, where $\pi_i = P(q_1 = i)$.
3. Observation probability distribution, $P(\mathbf{o}_t | s^{(i)})$, $i = 1, 2, \ldots, N$. If $\mathbf{o}_t$ is discrete, the distribution associated with each state gives the probabilities of symbolic observations $\{\mathbf{v}_1, \mathbf{v}_2, \ldots, \mathbf{v}_K\}$:

$$b_i(k) = P[\mathbf{o}_t = \mathbf{v}_k | q_t = i], \quad i = 1, 2, \ldots, N. \tag{3.7}$$

If the observation probability distribution is continuous, then the parameters, Λ_i, in the PDF characterize state i in the HMM.

The most common and successful PDF used in speech processing for characterizing the continuous observation probability distribution in the HMM is a multivariate mixture Gaussian distribution for vector-valued observation ($\mathbf{o}_t \in \mathscr{R}^D$):

$$b_i(\mathbf{o}_t) = \sum_{m=1}^{M} \frac{c_{i,m}}{(2\pi)^{D/2}|\boldsymbol{\Sigma}_{i,m}|^{1/2}} \exp\left[-\frac{1}{2}(\mathbf{o}_t - \boldsymbol{\mu}_{i,m})^{\mathsf{T}} \boldsymbol{\Sigma}_{i,m}^{-1}(\mathbf{o}_t - \boldsymbol{\mu}_{i,m})\right] \quad (3.8)$$

In this Gaussian-mixture HMM, the parameter set Λ_i comprises scalar mixture weights, $c_{i,m}$, Gaussian mean vectors, $\boldsymbol{\mu}_{i,m} \in \mathscr{R}^{\mathscr{D}}$, and Gaussian covariance matrices, $\boldsymbol{\Sigma}_{i,m} \in \mathscr{R}^{D \times D}$.

When the number of mixture components is reduced to one: $M = 1$, the state-dependent output PDF reverts to a (uni-modal) Gaussian:

$$b_i(\mathbf{o}_t) = \frac{1}{(2\pi)^{D/2}|\boldsymbol{\Sigma}_i|^{1/2}} \exp\left[-\frac{1}{2}(\mathbf{o}_t - \boldsymbol{\mu}_i)^{\mathsf{T}} \boldsymbol{\Sigma}_i^{-1}(\mathbf{o}_t - \boldsymbol{\mu}_i)\right] \quad (3.9)$$

and the corresponding HMM is commonly called a (continuous-density) Gaussian HMM.

Given the model parameters, one convenient way of characterizing a Gaussian HMM is to view it as a generative device producing a sequence of observational data, $\mathbf{o}_t, t = 1, 2, \ldots, T$. In this view, the data at each time t is generated from the model according to

$$\mathbf{o}_t = \boldsymbol{\mu}_i + \mathbf{r}_t(\boldsymbol{\Sigma}_i), \quad (3.10)$$

where state i at a given time t is determined by the evolution of the Markov chain characterized by a_{ij}, and

$$\mathbf{r}_t(\boldsymbol{\Sigma}_i) = \mathcal{N}(0, \boldsymbol{\Sigma}_i) \quad (3.11)$$

is a zero-mean, Gaussian, IID (independent and identically distributed) residual sequence, which is generally state dependent as indexed by i. Because the residual sequence $\mathbf{r}_t(\boldsymbol{\Sigma}_i)$ is IID, and because $\boldsymbol{\mu}_i$ is constant (i.e., not time-varying) given state i, their sum, which gives the observation $\mathbf{o}_t$ is thus also IID given the state. Therefore, the HMM discussed above would produce locally or piecewise stationary sequences. Since the temporal locality in question is confined within state occupation of the HMM, we sometimes use the term stationary-state HMM to explicitly denote such a property.

One simple way to extend a stationary-state HMM so that the observation sequence is no longer state-conditioned IID is as follows. We can modify the constant term $\boldsymbol{\mu}_i$ in Eq. 3.10 to explicitly make it time-varying:

$$\mathbf{o}_t = \mathbf{g}_t(\Lambda_i) + \mathbf{r}_t(\boldsymbol{\Sigma}_i), \quad (3.12)$$

where parameters Λ_i in the deterministic time-trend function $\mathbf{g}_t(\Lambda_i)$ is dependent on state i in the Markov chain. This gives rise to the trended (Gaussian) HMM [18, 23, 24, 32, 45, 50, 62, 69, 85, 120, 126], a special version of a nonstationary-state HMM where the first-order statistics (mean) are time-varying and thus violating a basic condition of wide-sense stationarity.

3.3.2 Simulation of a Hidden Markov Model

When we view a hidden Markov sequence or an HMM as a model for the information source, which has been explicitly depicted in Eq. 3.10, sometimes it is desirable to use this model to generate its samples. This is the problem of simulating the HMM given appropriate values for all model parameters: $\{A, \pi, B\}$ for a discrete HMM or $\{A, \pi, \Lambda\}$ for a continuous-density HMM. The result of the simulation is to produce an observation sequence, $\mathbf{o}_1^T = \mathbf{o}_1, \mathbf{o}_2, \ldots, \mathbf{o}_T$, which obeys the statistical law embedded in the HMM. A simulation process is described in Algorithm 3.1.

Algorithm 3.1 Draw Samples from an HMM.

1: **procedure** DRAWFROMHMM($A, \pi, P(\mathbf{o}_t|s^{(i)})$)

$\qquad\qquad\qquad\qquad\qquad\qquad\qquad\qquad\qquad$ ▷ A is the transition probability

$\qquad\qquad\qquad\qquad\qquad\qquad\qquad$ ▷ π is the initial state occupation probability

$\qquad\qquad\qquad$ ▷ $P(\mathbf{o}_t|s^{(i)})$ is the observation probability given a state (either Eq. 3.7 if discrete or Eq. 3.8 if continuous

2: $\qquad$ Select an initial state $q_1 = s^{(i)}$ by drawing from the discrete distribution π

3: $\qquad$ **for** $t \leftarrow 1; t \leq T; t \leftarrow t + 1$ **do**

4: $\qquad\qquad$ Draw an observation $\mathbf{o}_t$ based on $P(\mathbf{o}_t|s^{(i)})$

5: $\qquad\qquad$ Make a Markov-chain transition from the current state $q_t = s^{(i)}$ to a new state $q_{t+1} = s^{(j)}$ according to the transition probability a_{ij}, and assign $i \leftarrow j$.

6: $\qquad$ **end for**

7: **end procedure**

3.3.3 Likelihood Evaluation of a Hidden Markov Model

Likelihood evaluation is a basic task needed for speech processing applications involving an HMM that uses a hidden Markov sequence to approximate vectorized speech features.

Let $\mathbf{q}_1^T = (q_1, \ldots, q_T)$ be a finite-length sequence of states in a Gaussian-mixture HMM or GMM-HMM, and let $P(\mathbf{o}_1^T, \mathbf{q}_1^T)$ be the joint likelihood of the observation sequence $\mathbf{o}_1^T = (\mathbf{o}_1, \ldots, \mathbf{o}_T)$ and the state sequence $\mathbf{q}_1^T$. Let $P(\mathbf{o}_1^T|\mathbf{q}_1^T)$ denote the likelihood that the observation sequence $\mathbf{o}_1^T$ is generated by the model conditioned on the state sequence $\mathbf{q}_1^T$.

In the Gaussian-mixture HMM, the conditional likelihood $P(\mathbf{o}_1^T|\mathbf{q}_1^T)$ is in the form of

$$P(\mathbf{o}_1^T|\mathbf{q}_1^T) = \prod_{t=1}^{T} b_i(\mathbf{o}_t) = \prod_{t=1}^{T} \sum_{m=1}^{M} \frac{c_{i,m}}{(2\pi)^{D/2}|\boldsymbol{\Sigma}_{i,m}|^{1/2}} \exp\left[-\frac{1}{2}(\mathbf{o}_t - \boldsymbol{\mu}_{i,m})^{\mathrm{T}} \boldsymbol{\Sigma}_{i,m}^{-1}(\mathbf{o}_t - \boldsymbol{\mu}_{i,m})\right]$$

(3.13)

On the other hand, the probability of state sequence $\mathbf{q}_1^T$ is just the product of transition probabilities, i.e.,

$$P(\mathbf{q}_1^T) = \pi_{q_1} \prod_{t=1}^{T-1} a_{q_t q_{t+1}}.$$

(3.14)

In the remaining of the chapter, for notational simplicity, we consider the case where the initial state distribution has the probability of one in the starting state: $\pi_{q_1} = 1$.

Note that the joint likelihood $P(\mathbf{o}_1^T, \mathbf{q}_1^T)$ can be obtained by the product of likelihoods in Eqs. 3.13 and 3.14:

$$P(\mathbf{o}_1^T, \mathbf{q}_1^T) = P(\mathbf{o}_1^T|\mathbf{q}_1^T)P(\mathbf{q}_1^T).$$

(3.15)

In principle, the total likelihood for the observation sequence can be computed by summing the joint likelihoods in Eq. 3.15 over all possible state sequences $\mathbf{q}_1^T$:

$$P(\mathbf{o}_1^T) = \sum_{\mathbf{q}_1^T} P(\mathbf{o}_1^T, \mathbf{q}_1^T).$$

(3.16)

However, the amount of this computation is exponential in the length of the observation sequence, T, and hence the naive computation of $P(\mathbf{o}_1^T)$ is not tractable. In the next section, we will describe the forward-backward algorithm [6], which computes $P(\mathbf{o}_1^T)$ for the HMM with complexity linear in T.

3.3.4 An Algorithm for Efficient Likelihood Evaluation

To describe the algorithm, we first define the forward probabilities by

$$\alpha_t(i) = P(q_t = i, \mathbf{o}_1^t), \quad t = 1, \ldots, T,$$

(3.17)

and the backward probabilities by

$$\beta_t(i) = P(\mathbf{o}_{t+1}^T|q_t = i), \quad t = 1, \ldots, T - 1,$$

(3.18)

both for each state i in the Markov chain. The forward and backward probabilities can be calculated recursively from

$$\alpha_t(j) = \sum_{i=1}^{N} \alpha_{t-1}(i)a_{ij}b_j(\mathbf{o}_t), \quad t = 2, 3, \ldots, T; \quad j = 1, 2, \ldots, N \quad (3.19)$$

$$\beta_t(i) = \sum_{j=1}^{N} \beta_{t+1}(j)a_{ij}b_j(\mathbf{o}_{t+1}), \quad t = T-1, T-2, \ldots, 1; \quad i = 1, 2, \ldots, N$$
$$(3.20)$$

Proofs of these recursions are given immediately after this subsection. The starting value for the α recursion is, according to the definition in Eq. 3.17,

$$\alpha_1(i) = P(q_1 = i, \mathbf{o}_1) = P(q_1 = i)P(\mathbf{o}_1|q_1) = \pi_i b_i(\mathbf{o}_1), \quad i = 1, 2, \ldots, N$$
$$(3.21)$$

and that for the β recursion is chosen as

$$\beta_T(i) = 1, \quad i = 1, 2, \ldots, N, \quad (3.22)$$

so as to provide the correct values for β_{T-1} according to the definition in Eq. 3.18.
To compute the total likelihood $P(\mathbf{o}_1^T)$ in Eq. 3.16, we first compute

$$\begin{aligned}
P(q_t = i, \mathbf{o}_1^T) &= P(q_t = i, \mathbf{o}_1^t, \mathbf{o}_{t+1}^T) \\
&= P(q_t = i, \mathbf{o}_1^t)P(\mathbf{o}_{t+1}^T|\mathbf{o}_1^t, q_t = i) \\
&= P(q_t = i, \mathbf{o}_1^t)P(\mathbf{o}_{t+1}^T|q_t = i) \\
&= \alpha_t(i)\beta_t(i),
\end{aligned} \quad (3.23)$$

for each state i and $t = 1, 2, \ldots, T$ using definitions in Eqs. 3.17 and 3.18. Note that $P(\mathbf{o}_{t+1}^T|\mathbf{o}_1^t, q_t = i) = P(\mathbf{o}_{t+1}^T|q_t = i)$ because the observations are IID given the state in the HMM. Given this, $P(\mathbf{o}_1^T)$ can be computed as

$$P(\mathbf{o}_1^T) = \sum_{i=1}^{N} P(q_t = i, \mathbf{o}_1^T) = \sum_{i=1}^{N} \alpha_t(i)\beta_t(i). \quad (3.24)$$

Taking $t = T$ in Eq. 3.24 and using Eq. 3.22 lead to

$$P(\mathbf{o}_1^T) = \sum_{i=1}^{N} \alpha_T(i). \quad (3.25)$$

Thus, strictly speaking, the β recursion is not necessary for the forward scoring computation, and hence the algorithm is often called the forward algorithm.

However, the β computation is a necessary step for solving the model parameter estimation problem, which will be covered in the next section.

3.3.5 Proofs of the Forward and Backward Recursions

Proofs of the recursion formulas, Eqs. 3.19 and 3.20, are provided here, using the total probability theorem, Bayes rule, and using the Markov property and conditional independence property of the HMM.

For the forward probability recursion, we have

$$
\begin{aligned}
\alpha_t(j) &= P(q_t = j, \mathbf{o}_1^t) \\
&= \sum_{i=1}^{N} P(q_{t-1} = i, \ q_t = j, \mathbf{o}_1^{t-1}, \mathbf{o}_t) \\
&= \sum_{i=1}^{N} P(q_t = j, \mathbf{o}_t | q_{t-1} = i, \mathbf{o}_1^{t-1}) P(q_{t-1} = i, \mathbf{o}_1^{t-1}) \\
&= \sum_{i=1}^{N} P(q_t = j, \mathbf{o}_t | q_{t-1} = i) \alpha_{t-1}(i) \\
&= \sum_{i=1}^{N} P(\mathbf{o}_t | q_t = j, q_{t-1} = i) P(q_t = j | q_{t-1} = i) \alpha_{t-1}(i) \\
&= \sum_{i=1}^{N} b_j(\mathbf{o}_t) a_{ij} \alpha_{t-1}(i).
\end{aligned}
\tag{3.26}
$$

For the backward probability recursion, we have

$$
\begin{aligned}
\beta_t(i) &= P(\mathbf{o}_{t+1}^T | q_t = i) \\
&= \frac{P(\mathbf{o}_{t+1}^T, q_t = i)}{P(q_t = i)} \\
&= \frac{\sum_{j=1}^{N} P(\mathbf{o}_{t+1}^T, q_t = i, \ q_{t+1} = j)}{P(q_t = i)} \\
&= \frac{\sum_{j=1}^{N} P(\mathbf{o}_{t+1}^T | q_t = i, q_{t+1} = j) P(q_t = i, q_{t+1} = j)}{P(q_t = i)} \\
&= \sum_{j=1}^{N} P(\mathbf{o}_{t+1}^T | q_{t+1} = j) \frac{P(q_t = i, \ q_{t+1} = j)}{P(q_t = i)}
\end{aligned}
$$

$$= \sum_{j=1}^{N} P(\mathbf{o}_{t+2}^{T}, \mathbf{o}_{t+1} | q_{t+1} = j) a_{ij}$$

$$= \sum_{j=1}^{N} P(\mathbf{o}_{t+2}^{T} | q_{t+1} = j) P(\mathbf{o}_{t+1} | q_{t+1} = j) a_{ij}$$

$$= \sum_{j=1}^{N} \beta_{t+1}(j) b_j(\mathbf{o}_{t+1}) a_{ij}. \tag{3.27}$$

3.4 EM Algorithm and Its Application to Learning HMM Parameters

3.4.1 Introduction to EM Algorithm

Despite many unrealistic aspects of the HMM as a model for speech feature sequences, one of the most important reasons for its wide-spread use in speech recognition is the Baum-Welch algorithm developed in 1960s [6], which is a prominent instance of the highly popular EM (Expectation-Maximization) algorithm [22], for efficient training of the HMM parameters from data. In this section, we describe first the general principle of the EM algorithm. Then, we move to its application to the HMM parameter estimation problem, where the special method of EM becomes known as the Baum-Welch algorithm. For tutorial material on the EM and its basic applications, see [10, 12, 42, 66, 94].

When there are hidden or latent random variables in a statistical model, maximum likelihood estimation is often difficult and the EM algorithm often becomes effective. Let us denote the complete data by $\mathbf{y} = \{\mathbf{o}, \mathbf{h}\}$ where $\mathbf{o}$ is observed data (e.g., speech feature sequence data) and $\mathbf{h}$ is hidden random variables (e.g., unobserved HMM state sequence). Here, we consider the problem of finding an estimate for the unknown parameter θ, which requires maximization of the log-likelihood function, $\log p(\mathbf{o}; \theta)$. However, we may find that this is either too difficult or there are difficulties in finding an expression for the PDF itself. In such circumstances, an iterative solution is possible if the *complete* data, $\mathbf{y}$, can be found such that the PDF in terms of $\mathbf{y}$ is much easier to express in closed form and to maximize. In general we can find a mapping from the complete to the incomplete or partial data: $\mathbf{o} = \mathbf{g}(\mathbf{y})$. However, this is usually not evident until one is able to define what the complete data set is. Unfortunately defining what constitutes the complete data is usually an arbitrary procedure that is highly problem specific and often requires some ingenuity on the part of the algorithm designer.

As a way to motivate the EM algorithm, we wish to overcome the computational difficulty of direct optimization of the PDF on the partially observed data **o**. To accomplish this, we supplement the available data **o** with *imaginary* missing, unobserved, or hidden data **h**, to form the complete data **y**. The hope is that with a clever choice of the hidden data **h**, we can work on the complete data **y** rather than on the original partial data **o** to make the optimization easier for the log likelihood of **o**.

Once we have identified the complete data **y**, even though an expression for log $p(\mathbf{y}; \theta)$ can now be derived easily, we cannot directly maximize log $p(\mathbf{y}; \theta)$ with respect to θ since **y** is unavailable. However, we observed **o** and if we further assume that we have a good *guessed* estimate for θ, then we can consider the expected value of log $p(\mathbf{y}; \theta)$ conditioned on what we have observed, or the following conditional expectation:

$$Q(\theta|\theta_0) = E_{h|o}[\log p(\mathbf{y}; \theta)|\mathbf{o}; \theta_0] = E[\log p(\mathbf{o}, \mathbf{h}; \theta)|\mathbf{o}; \theta_0] \qquad (3.28)$$

and we attempt to maximize this expectation to yield, not the maximum likelihood estimate, but the next *best* estimate for θ given the previously available estimate of θ_0.

Using Eq. 3.28 for computing the conditional expectation when hidden vector **h** is continuous, we have

$$Q(\theta|\theta_0) = \int p(\mathbf{h}|\mathbf{o}; \theta_0) \log p(\mathbf{y}; \theta)d\mathbf{h}. \qquad (3.29)$$

When the hidden vector **h** is discrete (i.e., taking only discrete values), Eq. 3.28 is used to evaluate the conditional expectation:

$$Q(\theta|\theta_0) = \sum_{\mathbf{h}} P(\mathbf{h}|\mathbf{o}; \theta_0) \log p(\mathbf{y}; \theta) \qquad (3.30)$$

where $P(\mathbf{h}|\mathbf{o}; \theta_0)$ is a conditional distribution given the initial parameter estimate θ_0, and the summation is over all possible discrete-valued vectors that **h** may take.

Given the initial parameters θ_0, the EM algorithm iterates alternating between the E-step, which finds an appropriate expression for the conditional expectation and sufficient statistics for its computation, and the M-step, which maximizes the conditional expectation, until either the algorithm converges or other stopping criteria are met.

Convergence of the EM algorithm is guaranteed (under mild conditions) in the sense that the average log-likelihood of the complete data does not decrease at each iteration, that is

$$Q(\theta|\theta_{k+1}) \geq Q(\theta|\theta_k)$$

with equality when θ_k is already an maximum-likelihood estimate.

The main properties of the EM algorithm are:

- It gives only a local, rather than the global, optimum in the likelihood of partially observed data.
- An initial value for the unknown parameter is needed, and as with most iterative procedures a good initial estimate is required for desirable convergence and a good maximum-likelihood estimate.
- The selection of the complete data set is arbitrary.
- Even if $\log p(\mathbf{y}; \theta)$ can usually be easily expressed in closed form, finding the closed-form expression for the expectation is usually hard.

3.4.2 Applying EM to Learning the HMM—Baum-Welch Algorithm

We now discuss how maximum-likelihood parameter estimation and, in particular, the EM algorithm is applied to solve the learning problem for the HMM. As introduced in the preceding section, the EM algorithm is a general iterative technique for maximum likelihood estimation, with local optimality in general, when hidden variables exist. When such hidden variables take the form of a Markov chain, the EM algorithm becomes the Baum-Welch algorithm. Below we use a Gaussian HMM as the example to describe steps involved in deriving E-step and M-step computations, where the complete data in the general case of EM above consists of the observation sequence and the hidden Markov-chain state sequence; i.e., $\mathbf{y} = [\mathbf{o}_1^T, \mathbf{q}_1^T]$.

Each iteration in the EM algorithm consists of two steps for any incomplete data problem including the current HMM parameter estimation problem. In the E (expectation) step of the Baum-Welch algorithm, the following conditional expectation, or the auxiliary function $Q(\theta|\theta_0)$, need to be computed:

$$Q(\theta|\theta_0) = E[\log P(\mathbf{o}_1^T, \mathbf{q}_1^T|\theta)|\mathbf{o}_1^T, \theta_0], \qquad (3.31)$$

where the expectation is taken over the "hidden" state sequence $\mathbf{q}_1^T$. For the EM algorithm to be of utility, $Q(\theta|\theta_0)$ has to be sufficiently simplified so that the M (maximization) step can be carried out easily. Estimates of the model parameters are obtained in the M step via maximization of $Q(\theta|\theta_0)$, which is in general much simpler than direct procedures for maximizing $P(\mathbf{o}_1^T|\theta)$.

An iteration of the above two steps will lead to maximum likelihood estimates of model parameters with respect to the objective function $P(\mathbf{o}_1^T|\theta)$. This is a direct consequence of Baum's inequality [6], which asserts that

$$\log\left(\frac{P(\mathbf{o}_1^T|\theta)}{P(\mathbf{o}_1^T|\theta_0)}\right) \geq Q(\theta|\theta_0) - Q(\theta_0|\theta_0) = 0.$$

We now carry out the E- and M-steps for the Gaussian HMM below, including detailed derivations.

3.4.2.1 E-Step

The goal of the E-step is to simplify the conditional expectation $Q(\theta|\theta_0)$ into a form suitable for direct maximization in the M-step. To proceed, we first explicitly write out the $Q(\theta|\theta_0)$ function in terms of expectation over state sequences $\mathbf{q}_1^T$ in the form of a weighted sum

$$
\begin{aligned}
Q(\theta|\theta_0) &= E[\log P(\mathbf{o}_1^T, \mathbf{q}_1^T|\theta)|\mathbf{o}_1^T, \theta_0] \\
&= \sum_{\mathbf{q}_1^T} P(\mathbf{q}_1^T|\mathbf{o}_1^T, \theta_0) \log P(\mathbf{o}_1^T, \mathbf{q}_1^T|\theta),
\end{aligned}
\tag{3.32}
$$

where θ and θ_0 denote the HMM parameters in the current and the immediately previous EM iterations, respectively. To simplify the writing, denote by $N_t(i)$ the quantity

$$
-\frac{D}{2}\log(2\pi) - \frac{1}{2}\log|\boldsymbol{\Sigma}_i| - \frac{1}{2}(\mathbf{o}_t - \boldsymbol{\mu}_i)^T \boldsymbol{\Sigma}_i^{-1}(\mathbf{o}_t - \boldsymbol{\mu}_i).
$$

which is logarithm of the Gaussian PDF associated with state i.

We now use $P(\mathbf{q}_1^T) = \prod_{t=1}^{T-1} a_{q_t q_{t+1}}$ and $P(\mathbf{o}_1^T, \mathbf{q}_1^T) = P(\mathbf{o}_1^T|\mathbf{q}_1^T)P(\mathbf{q}_1^T)$. These lead to

$$
\log P(\mathbf{o}_1^T, \mathbf{q}_1^T|\theta) = \sum_{t=1}^{T} N_t(q_t) + \sum_{t=1}^{T-1} \log a_{q_t q_{t+1}}
$$

and the conditional expectation in Eq. 3.32 can be rewritten as

$$
Q(\theta|\theta_0) = \sum_{\mathbf{q}_1^T} P(\mathbf{q}_1^T|\mathbf{o}_1^T, \theta_0) \sum_{t=1}^{T} N_t(q_t) + \sum_{\mathbf{q}_1^T} P(\mathbf{q}_1^T|\mathbf{o}_1^T, \theta_0) \sum_{t=1}^{T-1} \log a_{q_t q_{t+1}}.
\tag{3.33}
$$

To simplify $Q(\theta|\theta_0)$ here, we write the first term in Eq. 3.33 as

$$
Q_1(\theta|\theta_0) = \sum_{i=1}^{N} \left\{ \sum_{\mathbf{q}_1^T} P(\mathbf{q}_1^T|\mathbf{o}_1^T, \theta_0) \sum_{t=1}^{T} N_t(q_t) \right\} \delta_{q_t, i},
\tag{3.34}
$$

and the second term as

$$
Q_2(\theta|\theta_0) = \sum_{i=1}^{N} \sum_{j=1}^{N} \{ \sum_{\mathbf{q}_1^T} P(\mathbf{q}_1^T|\mathbf{o}_1^T, \theta_0) \sum_{t=1}^{T-1} \log a_{q_t q_{t+1}} \} \delta_{q_t, i} \delta_{q_{t+1}, j},
\tag{3.35}
$$

where δ indicates the Kronecker delta function. Let us examine Eq. 3.34 first. By exchanging summations and using the obvious fact that

$$\sum_{\mathbf{q}_1^T} P(\mathbf{q}_1^T|\mathbf{o}_1^T, \theta_0)\delta_{q_t,i} = P(q_t = i|\mathbf{o}_1^T, \theta_0),$$

we can simplify Q_1 into

$$Q_1(\theta|\theta_0) = \sum_{i=1}^{N}\sum_{t=1}^{T} P(q_t = i|\mathbf{o}_1^T, \theta_0)N_t(i). \tag{3.36}$$

After carrying out similar steps for $Q_2(\theta|\theta_0)$ in Eq. 3.35 we obtain a similar simplification

$$Q_2(\theta|\theta_0) = \sum_{i=1}^{N}\sum_{j=1}^{N}\sum_{t=1}^{T-1} P(q_t = i, \ q_{t+1} = j|\mathbf{o}_1^T, \theta_0) \log a_{ij}. \tag{3.37}$$

We note that in maximizing $Q(\theta|\theta_0) = Q_1(\theta|\theta_0) + Q_2(\theta|\theta_0)$, the two terms can be maximized independently. That is, $Q_1(\theta|\theta_0)$ contains only the parameters in Gaussians, while $Q_2(\theta|\theta_0)$ involves just the parameters in the Markov chain. Also, in maximizing $Q(\theta|\theta_0)$, the weights in Eqs. 3.36 and 3.37, or $\gamma_t(i) = P(q_t = i|\mathbf{o}_1^T, \theta_0)$ and $\xi_t(i, j) = P(q_t = i, q_{t+1} = j|\mathbf{o}_1^T, \theta_0)$, respectively, are treated as known constants due to their conditioning on θ_0. They can be computed efficiently via the use of the forward and backward probabilities discussed earlier.

The posterior state transition probabilities in the Gaussian HMM are

$$\xi_t(i, j) = \frac{\alpha_t(i)\beta_{t+1}(j)a_{ij}\exp(N_{t+1}(j))}{P(\mathbf{o}_1^T|\theta_0)}, \tag{3.38}$$

for $t = 1, 2, \ldots, T - 1$. (Note that $\xi_T(i, j)$ has no definition.) The posterior state occupancy probabilities can be obtained by summing $\xi_t(i, j)$ over all the destination states j according to

$$\gamma_t(i) = \sum_{j=1}^{N} \xi_t(i, j), \tag{3.39}$$

for $t = 1, 2, \ldots, T - 1$. $\gamma_T(i)$ can be obtained by its very definition:

$$\gamma_T(i) = P(q_T = i|\mathbf{o}_1^T, \theta_0) = \frac{P(q_T = i, \mathbf{o}_1^T|\theta_0)}{P(\mathbf{o}_1^T|\theta_0)} = \frac{\alpha_T(i)}{P(\mathbf{o}_1^T|\theta_0)}. \tag{3.40}$$

Note for the left-to-right HMM, $\gamma_T(i)$ has a value of one for $i = N$ and of zero otherwise.

Further, we note that the summations in Eqs. 3.36 and 3.37 are taken over states i or over state pairs i, j, which is significantly simpler than the summations over state sequences $\mathbf{q}_1^T$ as in the unsimplified forms of $Q_1(\theta|\theta_0)$ and $Q_2(\theta|\theta_0)$ in Eq. 3.33. Equations 3.36 and 3.37 are the simplistic form of the auxiliary objective function, which can be maximized in the M-step discussed next.

3.4.2.2 M-Step

The reestimation formulas for the transition probabilities of the Markov chain in the Gaussian HMM can be easily established by setting $\frac{\partial Q_2}{\partial a_{ij}} = 0$, for Q_2 in Eq. 3.37 and for $i, j = 1, 2, \ldots, N$, subject to the constraint $\sum_{j=1}^{N} a_{ij} = 1$. The standard Lagrange multiplier procedure leads to the reestimation formula of

$$\hat{a}_{ij} = \frac{\sum_{t=1}^{T-1} \xi_t(i, j)}{\sum_{t=1}^{T-1} \gamma_t(i)}, \tag{3.41}$$

where $\xi_t(i, j)$ and $\gamma_t(i)$ are computed according to Eqs. 3.38 and 3.39.

To derive the reestimation formulas for the parameters in the state-dependent Gaussian distributions, we first remove optimization-independent terms and factors in Q_1 in Eq. 3.36. Then we have an equivalent objective function of

$$Q_1(\boldsymbol{\mu}_i, \boldsymbol{\Sigma}_i) = \sum_{i=1}^{N} \sum_{t=1}^{Tr} \gamma_t(i) \left(\mathbf{o}_t - \boldsymbol{\mu}_i\right)^{\mathsf{T}} \boldsymbol{\Sigma}_i^{-1} \left(\mathbf{o}_t - \boldsymbol{\mu}_i\right) - \frac{1}{2} \log |\boldsymbol{\Sigma}_i|. \tag{3.42}$$

The reestimation formula for the covariance matrices are obtained by solving

$$\frac{\partial Q_1}{\partial \boldsymbol{\Sigma}_i} = 0, \tag{3.43}$$

for $i = 1, 2, \ldots, N$.

For solving it, we employ the trick of variable transformation: $\mathbf{K} = \boldsymbol{\Sigma}^{-1}$ (we omit the state index i for simplicity), and we treat Q_1 as a function of $\mathbf{K}$. Then, the derivative of $\log |\mathbf{K}|$ (a term in Eq. 3.36) with respect to $\mathbf{K}$'s (l, m)-th entry, k_{lm}, is the (l, m)-th entry of $\boldsymbol{\Sigma}$, or σ_{lm}. We now can reduce $\frac{\partial Q_1}{\partial k_{lm}} = 0$ to

$$\sum_{t=1}^{T} \gamma_t(i) \left\{ \frac{1}{2}\sigma_{lm} - \frac{1}{2}(\mathbf{o}_t - \boldsymbol{\mu}_i)_l (\mathbf{o}_t - \boldsymbol{\mu}_i)_m \right\} = 0 \tag{3.44}$$

for each entry: $l, m = 1, 2, \ldots, D$. Writing this result in a matrix form, we obtain the compact reestimation formula for the covariance matrix in state i as follows:

$$\hat{\mathbf{\Sigma}}_i = \frac{\sum_{t=1}^{T} \gamma_t(i)(\mathbf{o}_t - \hat{\boldsymbol{\mu}}_i)(\mathbf{o}_t - \hat{\boldsymbol{\mu}}_i)^\mathsf{T}}{\sum_{t=1}^{T} \gamma_t(i)} \tag{3.45}$$

for each state: $i = 1, 2, \ldots, N$, where $\hat{\boldsymbol{\mu}}_i$ is the re-estimate of the mean vectors in the Gaussian HMM in state i, whose reestimation formula is straightforward to derive and has the following easily interpretable form:

$$\hat{\boldsymbol{\mu}}_i = \frac{\sum_{t=1}^{T} \gamma_t(i)\mathbf{o}_t}{\sum_{t=1}^{T} \gamma_t(i)} \tag{3.46}$$

The above derivation is for the single-Gaussian HMM. The EM algorithm for the GMM-HMM can be similarly derived by considering the Gaussian component of each frame at each state as another hidden variable. In Chap. 6 we will describe the deep neural network (DNN)-HMM hybrid system in which the observation probability is estimated using a DNN.

3.5 Viterbi Algorithm for Decoding HMM State Sequences

3.5.1 Dynamic Programming and Viterbi Algorithm

Dynamic Programming (DP) is a divide-and-conquer method for solving complex problems by breaking them down into simpler sub-problems [7, 108]. It was originally developed by R. Bellman in 1950s [7]. The foundation of DP was laid by the Bellman optimality principle. The principle stipulates that: "In an optimization problem concerning multiple stages of interrelated decisions, whatever the initial state (or condition) and the initial decisions are, the remaining decisions must constitute an optimal rule of choosing decisions with regards to the state that results from the first decision".

As an example, we discuss the optimality principle in Markov decision processes. A Markov decision process is characterized by two sets of parameters. The first set is the transition probabilities of

$$P_{ij}^k(n) = P(\text{state}_j, \text{stage}_{n+1} | \text{state}_i, \text{stage}_n, \text{decision}_k),$$

where the current state of the system is dependent only on the state of the system at the previous stage and the decision taken at that stage (Markov property). The second set of parameters provide rewards defined by:

$R_i^k(n) = $ reward at stage n and at state i when decision k is chosen.

Let us define $F(n, i)$ to be the average of the total reward at state n and state i when the optimal decision is taken. This can be computed by DP using the following recursion given by the optimality principle:

$$F(n, i) = \max_k \{R_i^k(n) + \sum_j P_{ij}^k(n) F(n + 1, j)\}. \tag{3.47}$$

In particular, when $n = N$ (at the final stage), the total reward at state i is

$$F(N, i) = \max_k R_i^k(N). \tag{3.48}$$

The optimal decision sequence can be traced back after the end of this recursive computation.

We see in the above that in applying DP, various stages (e.g., stage $1, 2, \ldots, n,$ $\ldots, N$ in the above example) in the optimization process must be identified. We are required at each stage to make optimal decision(s). There are several states (indexed by i in the above example) of the system associated with each stage. The decision (indexed by k) taken at a given stage changes the problem from the current stage n to the next stage $n + 1$ according to the transition probability $P_{ij}^k(n)$.

If we apply the DP technique to finding the optimal path, then the Bellman optimality principle can be alternatively stated as follows: "The optimal path from nodes A to C through node B must consist of the optimal path from A to B concatenated with the optimal path from B to C". The implication of this optimality principle is tremendous. That is, in order to find the best path from node A via a "predecessor" node B, there will be no need to reconsider all the partial paths leading from A to B. This significantly reduces the path search effort compared with the brute-force search or exhaustive search. While it may be unknown whether the "predecessor" node B is on the best path or not, many candidates can be evaluated and the correct one be ultimately determined via a backtracking procedure in DP. The Bellman optimality principle is the essence of a very popular optimization technique in speech processing applications involving the HMM, which we describe below.

3.5.2 Dynamic Programming for Decoding HMM States

One fundamental computational problem associated with the HMM discussed so far in this chapter is to find, in an efficient manner, the best sequence of the HMM states given an arbitrary sequence of observations $\mathbf{o}_1^T = \mathbf{o}_1, \mathbf{o}_2, \ldots, \mathbf{o}_T$. This is a complex T-stage path-finding optimization problem, and is directly suited for the DP solution. The DP technique used for such purposes is also called the Viterbi algorithm, developed originally for optimal convolution-code channel decoding in digital communication.

To illustrate the Viterbi algorithm as an optimal path-finding technique, we can use the two-dimensional grid, also called the trellis diagram, for a left-to-right HMM. A node in the trellis diagram is associated with both a time frame t on the horizontal axis and an HMM state i on the vertical axis.

For a given HMM characterized by the state transitional probabilities a_{ij} and by the state-conditioned output probability distributions $b_i(\mathbf{o}_t)$, let $\delta_i(t)$ represent the maximal value of the joint likelihood of the partial observation sequence $\mathbf{o}_1^t$ up to time t, and the associated HMM state sequence while in state i at time t. That is,

$$\delta_i(t) = \max_{q_1, q_2, \dots, q_{t-1}} P(\mathbf{o}_1^t, q_1^{t-1}, q_t = i). \tag{3.49}$$

Note that each $\delta_i(t)$ defined here is associated with a node in the trellis diagram. Each increment of time corresponds to reaching a new stage in DP. At the final stage $t = T$, we have the objective function of $\delta_i(T)$, which is accomplished via all the previous stages of computation for $t \leq T - 1$. Based on the DP optimality principle, the optimal partial likelihood of Eq. 3.50 at the processing stage of $t + 1$ can be computed using the following functional equation as a recursion:

$$\delta_j(t+1) = \max_i \delta_i(t) a_{ij} b_j(\mathbf{o}_{t+1}), \tag{3.50}$$

for each state j. Each state at this processing stage is a hypothesized "precursor" node in the global optimal path. All such nodes except one will be eventually eliminated after the backtracking operation. The essence of DP used here is that we only need to compute the quantities of $\delta_j(t+1)$ as individual nodes in the trellis, removing the need to keep track of a very large number of partial paths from the initial stage to the current $(t+1)$th stage, which would be required for the exhaustive search. The optimality is guaranteed, due to the DP optimality principle, with the computation only linearly, rather than geometrically, increasing with the length T of the observation data sequence.

Besides the key recursion of Eq. 3.50, the complete Viterbi algorithm requires additional steps of recursion initialization, recursion termination, and path backtracking. The complete algorithm is described in Algorithm 3.2 with initial state probabilities π_i. The result of the Viterbi algorithm is P^*, the maximum joint likelihood of the observation and state sequence, together with $q^*(t)$, the corresponding state transition path.

The optimal state transition path found by the above Viterbi algorithm for a left-to-right HMM is equivalent to the information required to determine the optimal segmentation of the HMM states. The concept of state segmentation is most relevant to a left-to-right HMM commonly used in speech modeling and recognition, as each state in such an HMM is typically associated with a reasonably large number of consecutive time frames in the observation sequence. This is so because the observations cannot be easily assigned back to earlier states due to the left-to-right constraint and because the last frame must be accounted for by the right-most state in the left-to-right HMM.

Note that this same Viterbi algorithm can be applied to single-Gaussian HMM, GMM-HMM and even the DNN-HMM we will describe in Chap. 6.

Algorithm 3.2 Viterbi algorithm for decoding HMM state sequence.

1: **procedure** VITERBIDECODE($A = [a_{ij}], \pi, b_j(\mathbf{o}_t)$)
$\qquad\qquad\qquad\qquad\qquad\qquad\qquad\qquad\qquad\qquad\qquad$ ▷ A is the transition probability
$\qquad\qquad\qquad\qquad\qquad\qquad\qquad\qquad\qquad$ ▷ π is the initial state occupation probability
$\qquad\qquad\qquad\qquad$ ▷ $b_j(\mathbf{o}_t)$ is observation probability given HMM state j for observation vector $\mathbf{o}_t$
2: $\quad \delta_i(1) \leftarrow \pi_i b_i(\mathbf{o}_1)$ $\qquad\qquad\qquad\qquad\qquad\qquad\qquad\qquad\qquad$ ▷ Initialize at $t = 1$
3: $\quad \psi_i(1) \leftarrow 0$ $\qquad\qquad\qquad\qquad\qquad\qquad\qquad\qquad\qquad\qquad$ ▷ Initialize at $t = 1$
4: $\quad$ **for** $t \leftarrow 2; t \leq T; t \leftarrow t + 1$ **do** $\qquad\qquad\qquad\qquad\qquad\qquad$ ▷ Forward recursion
5: $\qquad \delta_j(t) \leftarrow \max_i \delta_i(t-1) a_{ij} b_j(\mathbf{o}_t)$
6: $\qquad \psi_j(t) \leftarrow \arg \max_{1 \leq i \leq N} \delta_i(t-1) a_{ij}$
7: $\quad$ **end for**
8: $\quad P^* \leftarrow \max_{1 \leq i \leq N}[\delta_i(T)]$
9: $\quad q(T) \leftarrow \max_{1 \leq i \leq N}[\delta_i(T)]$ $\qquad\qquad\qquad\qquad\qquad$ ▷ Initialize backtracking
10: $\quad$ **for** $t \leftarrow T - 1; t \geq 1; t \leftarrow t - 1$ **do** $\qquad\qquad\qquad$ ▷ Backward state tracking
11: $\qquad q^*(t) \leftarrow \psi_{q^*(t+1)}(t+1)$
12: $\quad$ **end for**
$\qquad\qquad$ Return optimal HMM state path $q^*(t), \quad 1 \leq t \leq T$
13: **end procedure**

3.6 The HMM and Variants for Generative Speech Modeling and Recognition

The popularity of the HMM in speech recognition stems from its ability as a generative sequence model of acoustic features of speech. See excellent reviews of the HMM for selected speech modeling and recognition applications in [3, 4, 72, 103–105]. One most interesting and unique problem in speech modeling and in the related speech recognition application lies in the nature of variable length in acoustic-feature sequences. This unique characteristic of speech rests primarily in its temporal dimension. That is, the actual values of the speech feature are correlated lawfully with the elasticity in the temporal dimension. As a consequence, even if two word sequences are identical, the acoustic data of speech features typically have distinct lengths. For example, different acoustic samples from the same sentence usually contain different data dimensionality, depending on how the speech sounds are produced and in particular how fast the speaking rate is. Further, the discriminative cues among separate speech classes are often distributed over a reasonably long temporal span, which often crosses neighboring speech units. Other special aspects of speech include class-dependent acoustic cues. These cues are often expressed over diverse time spans that would benefit from different lengths of analysis windows in speech analysis and feature extraction.

Conventional wisdom posits that speech is a one-dimensional temporal signal in contrast to image and video as higher dimensional signals. This view is simplistic and does not capture the essence and difficulties of the speech recognition problem. Speech is best viewed as a two-dimensional signal, where the spatial (or frequency or tonotopic) and temporal dimensions have vastly different characteristics, in contrast to images where the two spatial dimensions tend to have similar properties. The "spatial" dimension in speech is associated with the frequency distribution and related transformations, capturing a number of variability types including primarily those arising from environments, speakers, accent, speaking style, and speaking rate. The latter induces correlations between spatial and temporal dimensions, and the environment factors include microphone characteristics, speech transmission channel, ambient noise, and room reverberation.

The temporal dimension in speech, and in particular its correlation with the spatial or frequency-domain properties of speech, constitutes one of the unique challenges for speech recognition. The HMM addresses this challenge to a limited extent. In this section, a selected set of advanced generative models, as various extensions of the HMM, will be described that are aimed to address the same challenge, where Bayesian approaches are used to provide temporal constraints as prior knowledge about aspects of the physical process of human speech production.

3.6.1 GMM-HMMs for Speech Modeling and Recognition

In speech recognition, one most common generative learning approach is based on the Gaussian-mixture-model based hidden Markov models, or GMM-HMM; e.g., [12, 39, 74, 103, 105]. As discussed earlier, a GMM-HMM is a statistical model that describes two dependent random processes, an observable process, and a hidden Markov process. The observation sequence is assumed to be *generated* by each hidden state according to a Gaussian mixture distribution. A GMM-HMM is parameterized by a vector of state prior probabilities, the state transition probability matrix, and by a set of state-dependent parameters in Gaussian mixture models. In terms of modeling speech, a state in the GMM-HMM is typically associated with a subsegment of a phone in speech. One important innovation in the use of HMMs for speech recognition is the introduction of context-dependent states (e.g., [40, 70]), motivated by the desire to reduce output variability of speech feature vectors associated with each state, a common strategy for "detailed" generative modeling. A consequence of using context dependency is a vast expansion of the HMM state space, which, fortunately, can be controlled by regularization methods such as state tying. It turns out that such context dependency also plays a critical role in the recent advance of speech recognition in the area of discrimination-based deep learning [16, 19, 20, 110, 119], to be discussed in later chapters of this book.

The introduction of the HMM and the related statistical methods to speech recognition in mid 1970s [3, 72] can be regarded as the most significant paradigm shift in the field, as discussed and analyzed in [4, 5]. One major reason for this early success is the highly efficient EM algorithm [6], which we described earlier in this chapter. This maximum likelihood method, often called Baum-Welch algorithm, had been a principal way of training the HMM-based speech recognition systems until 2002, and is still one major step (among many) in training these systems nowadays. It is interesting to note that Baum-Welch algorithm serves as one major motivating example for the later development of the more general EM algorithm [22]. The goal of maximum likelihood or EM method in training GMM-HMM speech recognizers is to minimize the empirical risk with respect to the joint likelihood loss involving a sequence of linguistic labels and a sequence of acoustic data of speech, often extracted at the frame level. In large-vocabulary speech recognition systems, it is normally the case that word-level labels are provided, while state-level labels are latent. Moreover, in training GMM-HMM-based speech recognition systems, parameter tying is often used as a type of regularization. For example, similar acoustic states of the triphones can share the same Gaussian mixture model.

The use of the generative model of HMMs for representing the (piecewise stationary) dynamic speech pattern and the use of EM algorithm for training the tied HMM parameters constitute one most prominent and successful example of generative learning in speech recognition. This success has been firmly established by the speech community, and has been widely spread to machine learning and related communities. In fact, the HMM has become a standard tool not only in speech recognition, but also in machine learning as well as their related fields such as bioinformatics and natural language processing. For many machine learning as well as speech recognition researchers, the success of HMMs in speech recognition is a bit surprising due to the well-known weaknesses of the HMM in modeling speech dynamics. The remaining part of this section is aimed to address ways of using more advanced dynamic generative models and related techniques for speech modeling and recognition.

3.6.2 Trajectory and Hidden Dynamic Models for Speech Modeling and Recognition

Despite great success of GMM-HMMs in speech modeling and recognition, their weaknesses, such as the conditional independence and piecewise stationary assumptions, have been well known for speech modeling and recognition applications since early days [15, 23, 24, 32, 46, 51, 95, 96]. Since early 1990s, speech recognition researchers have begun the development of statistical models that capture more realistic dynamic properties of speech in the temporal dimension than HMMs do. This class of extended HMM models have been variably called stochastic segment model [95, 96], trended or nonstationary-state HMM [18, 23, 32], trajectory segmental model [69, 95], trajectory HMM [124, 126], stochastic trajectory model [62], hidden dynamic model [15, 25, 29, 44, 87–89, 98, 107], buried Markov model [11, 13, 14], structured speech model, and hidden trajectory model [29, 50–52, 117, 118, 128], depending on different "prior knowledge" applied to the temporal structure of speech and on various simplifying assumptions to facilitate the model implementation. Common to all these beyond-HMM model variants is some temporal dynamic structure built into the models. Based on the nature of such structure, we can classify these models into two main categories. In the first category are the models focusing on temporal correlation structure at the "surface" acoustic level. The second category consists of deep hidden or latent dynamics, where the underlying speech production mechanisms are exploited as a prior to represent the temporal structure that accounts for the visible speech pattern. When the mapping from the hidden dynamic layer to the visible layer is limited to be linear and deterministic, then the generative hidden dynamic models in the second category reduce to the first category.

The temporal span in many of the generative dynamic/trajectory models above is often controlled by a sequence of linguistic labels, which segment the full sentence into multiple regions from left to right; hence segment models.

In general, the trajectory or segmental models with hidden or latent dynamics make use of the switching state space formulation, well studied in the literature; e.g., [28, 53, 54, 61, 79, 92, 106]. These models exploit temporal recursion to define the hidden dynamics, $\mathbf{z}(k)$, which may correspond to articulatory movements during human speech production. Each discrete region or segment, s, of such dynamics is characterized by the s-dependent parameter set Λ_s, with the "state noise" denoted by $\mathbf{w}_s(k)$. The memory-less nonlinear mapping function is exploited to link the hidden dynamic vector $\mathbf{z}(k)$ to the observed acoustic feature vector $\mathbf{o}(k)$, with the "observation noise" denoted by $\mathbf{v}_s(k)$, and parameterized also by segment-dependent parameters. This pair of "state equation" and "observation equation" below form a general state-space switching nonlinear dynamic system model:

$$\mathbf{z}(k) = \mathbf{q}_k[\mathbf{z}(k-1), \Lambda_s] + \mathbf{w}_s(k-1) \tag{3.51}$$

$$\mathbf{o}(k') = \mathbf{r}_{k'}[\mathbf{z}(k'), \Omega_{s'}] + \mathbf{v}_{s'}(k'), \tag{3.52}$$

where subscripts k and k' denote that functions $\mathbf{q}[.]$ and $\mathbf{r}[.]$ are time varying and may be asynchronous with each other. In the mean time, s or s' denotes the dynamic region that is correlated with discrete linguistic categories either in terms of allophone states as in the standard GMM-HMM system (e.g., [40, 70, 103]) or in terms of atomic units constructed from articulation-motivated phonological features (e.g., [26, 47, 59, 77, 86, 111]).

The speech recognition literature has reported a number of studies on switching nonlinear state space models, both theoretical and experimental. The specific forms of the functions of $\mathbf{q}_k[\mathbf{z}(k-1), \Lambda_s]$ and $\mathbf{r}_{k'}[\mathbf{z}(k'), \Omega_{s'}]$ and their parameterization are determined by prior knowledge based on the understanding temporal properties of speech. In particular, state equation 3.51 takes into account the temporal elasticity in spontaneous speech and its correlation with the "spatial" properties in hidden speech dynamics such as articulatory positions or vocal tract resonance frequencies. For example, these latent variables do not oscillate within each phone-bound temporal region. Observation Eq. 3.52 incorporates knowledge about forward, nonlinear mapping from articulation to acoustics, an intensely studied subject in speech production and speech analysis research [33–35, 76].

When nonlinear functions of $\mathbf{q}_k[\mathbf{z}(k-1), \Lambda_s]$ and $\mathbf{r}_{k'}[\mathbf{z}(k'), \Omega_{s'}]$ are reduced to linear functions (and when synchrony between the two equations are eliminated), the switching nonlinear dynamic system model is reduced to its linear counterpart, or the switching linear dynamic system. This simplified system can be viewed as a hybrid of the standard HMM and linear dynamical systems, one associated with each HMM state. The general mathematical description of the switching linear dynamic system can be written as

$$\mathbf{z}(k) = \mathbf{A}_s \mathbf{z}(k-1) + \mathbf{B}_s \mathbf{w}_s(k) \tag{3.53}$$

$$\mathbf{o}(k) = \mathbf{C}_s \mathbf{z}(k) + \mathbf{v}_s(k). \tag{3.54}$$

where subscript s denotes the left-to-right HMM state or the region of the switching state in the linear dynamics. There has been an interesting set of work on the switching linear dynamic system applied to speech recognition. The early set of studies have been reported in [95, 96] for generative speech modeling and for speech recognition applications. More recent studies [53, 92] applied linear switching dynamic systems to noise-robust speech recognition and explored several approximate inference techniques. The study reported in [106] applied another approximate inference technique, a special type of Gibbs sampling, to a speech recognition problem.

3.6.3 The Speech Recognition Problem Using Generative Models of HMM and Its Variants

Toward the end of this chapter, let us focus on a discussion on issues related to using generative models such as the standard HMM and its extended versions just described for discriminative classification problems such as speech recognition. More detailed discussions on this important topic can be found in [41, 57, 127]. In particular, we have omitted in this chapter the topic of discriminative learning of the generative model of HMM, very important in the development of ASR based on the GMM-HMM and related architectures. We leave the readers to a plethora of literature on this topic in [2, 9, 17, 18, 48, 56, 63–67, 73, 90, 91, 99–102, 109, 112, 114, 115, 121–123, 125]. Another important topic, which we also omitted in this chapter, is the use of the generative, GMM-HMM-based models for integrating the modeling of the effects of noise in ASR. The ability to naturally carry out such integrated modeling is one of the strengths of statistical generative models such as the GMM and HMM, which we also leave the readers to the literature including many review articles [1, 30, 31, 36–38, 48, 49, 55, 58, 60, 75, 81–84, 92, 113].

A generative statistical model characterizes joint probabilities of input data and their corresponding labels, which can be decomposed into the prior probability of the labels (e.g., speech class tags) and the probability of class-conditioned data (e.g., acoustic features of speech). Via Bayes rule, the posterior probability of class labels given the data can be easily determined and used as the basis for the decision rule for classification. One key factor for the success of such generative modeling approach in classification tasks is how good the model is to the true data distribution. The HMM has been shown to be a reasonably good model for the statistical distribution of sequence data of speech acoustics, especially in its temporal characteristics. As a result, the HMM has become a popular model for speech recognition since mid 1980s.

However, several weaknesses of the standard HMM as a generative model for speech have been well understood, including, among others, the temporal independence of speech data conditioned on each HMM state and lack of lawful correlation between the acoustic features and ways in which speech sounds are produced (e.g., speaking rate and style). These weaknesses have motivated extensions of the HMM in several ways, some discussed in this section. The main thread of these extensions include the replacement of the Gaussian or Gaussian-mixture-like, independent and identical distributions associated with each HMM state by more realistic, temporally correlated dynamic systems or nonstationary trajectory models, both of which contain latent, continuous-valued dynamic structure.

During the development of these hidden trajectory and hidden dynamic models for speech recognition, a number of machine learning techniques, notably approximate variational inference and learning techniques [61, 79, 97, 116], have been usefully applied with modifications and improvement to suit the speech-specific properties and speech recognition applications. However, the success has mostly been limited to relatively small tasks. We can identify four main sources of difficulties (as well as new opportunities) in successfully applying these types of generative models to large-scale speech recognition. First, scientific knowledge on the precise nature of the underlying articulatory speech dynamics and its deeper articulatory control mechanisms is far from complete. Coupled with the need for efficient computation in training and decoding for speech recognition applications, such knowledge was forced to be again simplified, reducing the modeling power and precision further. Second, most of the work in this area has been placed within the generative learning setting, having a goal of providing parsimonious accounts (with small parameter sets) for speech variations due to contextual factors and coarticulation. In contrast, the recent joint development of deep learning methods, which we will cover in several later chapters of this book, combines generative and discriminative learning paradigms and makes use of massive instead of parsimonious parameters. There appears to be a huge potential for synergy of research here, especially in the light of the recent progress on variational inference expected to improve the quality of deep, generative modeling and learning [8, 21, 68, 78, 93]. Third, most of the hidden trajectory or hidden dynamic models have focused on only isolated aspects of speech dynamics rooted in deep human production mechanisms, and have been constructed using relatively simple and largely standard forms of dynamic systems without sufficient structure and effective learning methods free from unknown approximation errors especially during the inference step. This latter deficiency can likely be overcome by the improved variational learning methods just discussed.

Functionally speaking, speech recognition is a conversion process from the acoustic data sequence of speech into a word or another linguistic-symbol sequence. Technically, this conversion process requires a number of subprocesses including the use of discrete time stamps, often called frames, to characterize the speech waveform data or acoustic features, and the use of categorical labels (e.g., words, phones, etc.)

to index the acoustic data sequence. The fundamental issues in speech recognition lie in the nature of such labels and data. It is important to clearly understand the unique attributes of speech recognition, in terms of both input data and output labels. From the output viewpoint, ASR produces sentences that consist of a variable number of words. Thus, at least in principle, the number of possible classes (sentences) for the classification is so large that it is virtually impossible to construct models for complete sentences without the use of structure. From the input viewpoint, the acoustic data are also a sequence with a variable length, and typically, the length of data input is vastly different from that of label output, giving rise to the special problem of segmentation or alignment that the "static" classification problems in machine learning do not encounter. Combining the input and output viewpoints, we state the fundamental problem of speech recognition as a structured sequence classification task, where a (relatively long) sequence of acoustic data is used to infer a (relatively short) sequence of the linguistic units such as words. For this type of structured pattern recognition, both the standard HMM and its variants discussed in this chapter have captured some major attributes of the speech problem, especially in the temporal modeling aspect, accounting for their practical success in speech recognition to some degree. However, other key attributes of the problem have been poorly captured by the many types of models discussed in this chapter. Most of the remaining chapters in this book will be devoted to addressing this deficiency.

As a summary of this section, we bridged the HMM as a generative statistical model to practical speech problems including its modeling and classification/recognition. We pointed out the weaknesses of the standard HMM as a generative model for characterizing temporal properties of speech features, motivating its extensions to several variants where the temporal independence of speech data conditioned on each HMM state is replaced by more realistic, temporally correlated dynamic systems with latent structure. The state-space formulation of nonlinear dynamic system models provides an intriguing mechanism to connect to the recurrent neural networks, which we will discuss in great detail later in Chap. 13.

References

1. Acero, A., Deng, L., Kristjansson, T.T., Zhang, J.: HMM adaptation using vector taylor series for noisy speech recognition. In: Proceedings of Annual Conference of International Speech Communication Association (INTERSPEECH), pp. 869–872 (2000)
2. Bahl, L., Brown, P., de Souza, P., Mercer, R.: Maximum mutual information estimation of HMM parameters for speech recognition. In: Proceedings of International Conference on Acoustics, Speech and Signal Processing (ICASSP), pp. 49–52 (1986)
3. Baker, J.: Stochastic modeling for automatic speech recognition. In: Reddy, D. (ed.) Speech Recognition. Academic, New York (1976)
4. Baker, J., Deng, L., Glass, J., Khudanpur, S., Lee, C.H., Morgan, N., O'Shgughnessy, D.: Research developments and directions in speech recognition and understanding, part i. IEEE Signal Process. Mag. 26(3), 75–80 (2009)
5. Baker, J., Deng, L., Glass, J., Khudanpur, S., Lee, C.H., Morgan, N., O'Shgughnessy, D.: Updated minds report on speech recognition and understanding (research developments and

directions in speech recognition and understanding, part ii). IEEE Signal Process. Mag. **26**(4), 78–85 (2009)

6. Baum, L., Petrie, T.: Statistical inference for probabilistic functions of finite state Markov chains. Ann. Math. Statist. **37**(6), 1554–1563 (1966)
7. Bellman, R.: Dynamic Programming. Princeton University Press, Princeton (1957)
8. Bengio, Y.: Estimating or propagating gradients through stochastic neurons. CoRR (2013)
9. Biem, A., Katagiri, S., McDermott, E., Juang, B.H.: An application of discriminative feature extraction to filter-bank-based speech recognition. IEEE Trans. Speech Audio Process. **9**, 96–110 (2001)
10. Bilmes, J.: A gentle tutorial of the EM algorithm and its application to parameter estimation for Gaussian mixture and hidden Markov models. Technical Report TR-97-021, ICSI (1997)
11. Bilmes, J.: Buried Markov models: a graphical modeling approach to automatic speech recognition. Comput. Speech Lang. **17**, 213–231 (2003)
12. Bilmes, J.: What HMMs can do. IEICE Trans. Inf. Syst. **E89-D**(3), 869–891 (2006)
13. Bilmes, J.: Dynamic graphical models. IEEE Signal Process. Mag. **33**, 29–42 (2010)
14. Bilmes, J., Bartels, C.: Graphical model architectures for speech recognition. IEEE Signal Process. Mag. **22**, 89–100 (2005)
15. Bridle, J., Deng, L., Picone, J., Richards, H., Ma, J., Kamm, T., Schuster, M., Pike, S., Reagan, R.: An investigation fo segmental hidden dynamic models of speech coarticulation for automatic speech recognition. Final Report for 1998 Workshop on Langauge Engineering, CLSP, Johns Hopkins (1998)
16. Chen, X., Eversole, A., Li, G., Yu, D., Seide, F.: Pipelined back-propagation for context-dependent deep neural networks. In: Proceedings of Annual Conference of International Speech Communication Association (INTERSPEECH) (2012)
17. Chengalvarayan, R., Deng, L.: HMM-based speech recognition using state-dependent, discriminatively derived transforms on mel-warped DFT features. IEEE Trans. Speech Audio Process. **5**, 243–256 (1997)
18. Chengalvarayan, R., Deng, L.: Speech trajectory discrimination using the minimum classification error learning. IEEE Trans. Speech Audio Process. **6**, 505–515 (1998)
19. Dahl, G., Yu, D., Deng, L., Acero, A.: Large vocabulary continuous speech recognition with context-dependent DBN-HMMs. In: Proceedings of International Conference on Acoustics, Speech and Signal Processing (ICASSP) (2011)
20. Dahl, G., Yu, D., Deng, L., Acero, A.: Context-dependent pre-trained deep neural networks for large-vocabulary speech recognition. IEEE Trans. Audio, Speech Lang. Process. **20**(1), 30–42 (2012)
21. Rezende, D.J., Mohamed, S., Wierstra, D.: Stochastic backpropagation and approximate inference in deep generative models. In: Proceedings of International Conference on Machine Learning (ICML) (2014)
22. Dempster, A.P., Laird, N.M., Rubin, D.B.: Maximum-likelihood from incomplete data via the EM algorithm. J. Roy. Stat. Soc. Ser. B. **39** (1977)
23. Deng, L.: A generalized hidden Markov model with state-conditioned trend functions of time for the speech signal. Signal Process. **27**(1), 65–78 (1992)
24. Deng, L.: A stochastic model of speech incorporating hierarchical nonstationarity. IEEE Trans. Acoust. Speech Signal Process. **1**(4), 471–475 (1993)
25. Deng, L.: A dynamic, feature-based approach to the interface between phonology and phonetics for speech modeling and recognition. Speech Commun. **24**(4), 299–323 (1998)
26. Deng, L.: Articulatory features and associated production models in statistical speech recognition. In: Computational Models of Speech Pattern Processing, pp. 214–224. Springer, New York (1999)
27. Deng, L.: Computational models for speech production. In: Computational Models of Speech Pattern Processing, pp. 199–213. Springer, New York (1999)
28. Deng, L.: Switching dynamic system models for speech articulation and acoustics. In: Mathematical Foundations of Speech and Language Processing, pp. 115–134. Springer, New York (2003)

29. Deng, L.: Dynamic Speech Models—Theory, Algorithm, and Applications. Morgan and Claypool (2006)
30. Deng, L.: Front-End, Back-End, and hybrid techniques to noise-robust speech recognition. Chapter 4 in Book: Robust Speech Recognition of Uncertain Data. Springer (2011)
31. Deng, L., Acero, A., Plumpe, M., Huang, X.: Large vocabulary speech recognition under adverse acoustic environment. In: Proceedings of International Conference on Spoken Language Processing (ICSLP), pp. 806–809 (2000)
32. Deng, L., Aksmanovic, M., Sun, D., Wu, J.: Speech recognition using hidden Markov models with polynomial regression functions as non-stationary states. IEEE Trans. Acoust. Speech Signal Process. 2(4), 101–119 (1994)
33. Deng, L., Attias, H., Lee, L., Acero, A.: Adaptive kalman smoothing for tracking vocal tract resonances using a continuous-valued hidden dynamic model. IEEE Trans. Audio, Speech Lang. Process. 15, 13–23 (2007)
34. Deng, L., Bazzi, I., Acero, A.: Tracking vocal tract resonances using an analytical nonlinear predictor and a target-guided temporal constraint. In: Proceedings of Annual Conference of International Speech Communication Association (INTERSPEECH) (2003)
35. Deng, L., Dang, J.: Speech analysis: the production-perception perspective. In: Advances in Chinese Spoken Language Processing. World Scientific Publishing, Singapore (2007)
36. Deng, L., Droppo, J., Acero, A.: Recursive estimation of nonstationary noise using iterative stochastic approximation for robust speech recognition. IEEE Trans. Speech Audio Process. 11, 568–580 (2003)
37. Deng, L., Droppo, J., Acero, A.: A Bayesian approach to speech feature enhancement using the dynamic cepstral prior. In: Proceedings of International Conference on Acoustics, Speech and Signal Processing (ICASSP), vol. 1, pp. I-829–I-832 (2002)
38. Deng, L., Droppo, J., Acero, A.: Enhancement of log mel power spectra of speech using a phase-sensitive model of the acoustic environment and sequential estimation of the corrupting noise. IEEE Trans. Speech Audio Process. 12(2), 133–143 (2004)
39. Deng, L., Kenny, P., Lennig, M., Gupta, V., Seitz, F., Mermelsten, P.: Phonemic hidden Markov models with continuous mixture output densities for large vocabulary word recognition. IEEE Trans. Acoust. Speech Signal Process. 39(7), 1677–1681 (1991)
40. Deng, L., Lennig, M., Seitz, F., Mermelstein, P.: Large vocabulary word recognition using context-dependent allophonic hidden Markov models. Comput. Speech Lang. 4, 345–357 (1991)
41. Deng, L., Li, X.: Machine learning paradigms in speech recognition: an overview. IEEE Trans. Audio, Speech Lang. Process. 21(5), 1060–1089 (2013)
42. Deng, L., Mark, J.: Parameter estimation for Markov modulated poisson processes via the em algorithm with time discretization. In: Telecommunication Systems (1993)
43. Deng, L., O'Shaughnessy, D.: Speech Processing—A Dynamic and Optimization-Oriented Approach. Marcel Dekker Inc, New York (2003)
44. Deng, L., Ramsay, G., Sun, D.: Production models as a structural basis for automatic speech recognition. Speech Commun. 33(2–3), 93–111 (1997)
45. Deng, L., Rathinavelu, C.: A Markov model containing state-conditioned second-order nonstationarity: application to speech recognition. Comput. Speech Lang. 9(1), 63–86 (1995)
46. Deng, L., Sameti, H.: Transitional speech units and their representation by regressive Markov states: applications to speech recognition. IEEE Trans. Speech Audio Process. 4(4), 301–306 (1996)
47. Deng, L., Sun, D.: A statistical approach to automatic speech recognition using the atomic speech units constructed from overlapping articulatory features. J. Acoust. Soc. Am. 85, 2702–2719 (1994)
48. Deng, L., Wang, K., Acero, A., Hon, H., Droppo, J., Boulis, C., Wang, Y., Jacoby, D., Mahajan, M., Chelba, C., Huang, X.: Distributed speech processing in mipad's multimodal user interface. IEEE Trans. Audio, Speech Lang. Process. 20(9), 2409–2419 (2012)
49. Deng, L., Wu, J., Droppo, J., Acero, A.: Analysis and comparisons of two speech feature extraction/compensation algorithms (2005)

50. Deng, L., Yu, D.: Use of differential cepstra as acoustic features in hidden trajectory modelling for phonetic recognition. In: Proceedings of International Conference on Acoustics, Speech and Signal Processing (ICASSP), pp. 445–448 (2007)
51. Deng, L., Yu, D., Acero, A.: A bidirectional target filtering model of speech coarticulation: two-stage implementation for phonetic recognition. IEEE Trans. Speech Audio Process. **14**, 256–265 (2006)
52. Deng, L., Yu, D., Acero, A.: Structured speech modeling. IEEE Trans. Speech Audio Process. **14**, 1492–1504 (2006)
53. Droppo, J., Acero, A.: Noise robust speech recognition with a switching linear dynamic model. In: Proceedings of International Conference on Acoustics, Speech and Signal Processing (ICASSP), vol. 1, pp. I-953–I-956 (2004)
54. Fox, E., Sudderth, E., Jordan, M., Willsky, A.: Bayesian nonparametric methods for learning Markov switching processes. IEEE Signal Process. Mag. **27**(6), 43–54 (2010)
55. Frey, B., Deng, L., Acero, A., Kristjansson, T.: Algonquin: iterating laplaces method to remove multiple types of acoustic distortion for robust speech recognition. In: Proceedings of European Conference on Speech Communication and Technology (EUROSPEECH) (2001)
56. Fu, Q., Zhao, Y., Juang, B.H.: Automatic speech recognition based on non-uniform error criteria. IEEE Trans. Audio, Speech Lang. Process. **20**(3), 780–793 (2012)
57. Gales, M., Watanabe, S., Fosler-Lussier, E.: Structured discriminative models for speech recognition. IEEE Signal Process. Mag. **29**, 70–81 (2012)
58. Gales, M., Young, S.: Robust continuous speech recognition using parallel model combination. IEEE Trans. Speech Audio Process. **4**(5), 352–359 (1996)
59. Gao, Y., Bakis, R., Huang, J., Xiang, B.: Multistage coarticulation model combining articulatory, formant and cepstral features. In: Proceedings of International Conference on Spoken Language Processing (ICSLP), pp. 25–28. Beijing, China (2000)
60. Gemmeke, J., Virtanen, T., Hurmalainen, A.: Exemplar-based sparse representations for noise robust automatic speech recognition. IEEE Trans. Audio, Speech Lang. Process. **19**(7), 2067–2080 (2011)
61. Ghahramani, Z., Hinton, G.E.: Variational learning for switching state-space models. Neural Comput. **12**, 831–864 (2000)
62. Gong, Y., Illina, I., Haton, J.P.: Modeling long term variability information in mixture stochastic trajectory framework. In: Proceedings of International Conference on Spoken Language Processing (1996)
63. He, X., Deng, L.: Discriminative Learning for Speech Recognition: Theory and Practice. Morgan and Claypool (2008)
64. He, X., Deng, L.: Speech recognition, machine translation, and speech translation—a unified discriminative learning paradigm. IEEE Signal Process. Mag. **27**, 126–133 (2011)
65. He, X., Deng, L., Chou, W.: Discriminative learning in sequential pattern recognition—a unifying review for optimization-oriented speech recognition. IEEE Signal Process. Mag. **25**(5), 14–36 (2008)
66. Heigold, G., Ney, H., Schluter, R.: Investigations on an EM-style optimization algorithm for discriminative training of HMMs. IEEE Trans. Audio, Speech Lang. Process. **21**(12), 2616–2626 (2013)
67. Heigold, G., Wiesler, S., Nubbaum-Thom, M., Lehnen, P., Schluter, R., Ney, H.: Discriminative HMMs. log-linear models, and CRFs: what is the difference? In: Proceedings of International Conference on Acoustics, Speech and Signal Processing (ICASSP) (2010)
68. Hoffman, M.D., Blei, D.M., Wang, C., Paisley, J.: Stochastic variational inference (2013)
69. Holmes, W., Russell, M.: Probabilistic-trajectory segmental HMMs. Comput. Speech Lang. **13**, 3–37 (1999)
70. Huang, X., Acero, A., Hon, H.W., et al.: Spoken Language Processing, vol. 18. Prentice Hall, Englewood Cliffs (2001)
71. Huang, X., Deng, L.: An overview of modern speech recognition. In: Indurkhya, N., Damerau, F.J. (eds.) Handbook of Natural Language Processing, 2nd edn. CRC Press, Taylor and Francis Group, Boca Raton, FL (2010). ISBN 978-1420085921

72. Jelinek, F.: Continuous speech recognition by statistical methods. Proc. IEEE **64**(4), 532–557 (1976)
73. Juang, B.H., Hou, W., Lee, C.H.: Minimum classification error rate methods for speech recognition. IEEE Trans. Speech Audio Process. **5**(3), 257–265 (1997)
74. Juang, B.H., Levinson, S.E., Sondhi, M.M.: Maximum likelihood estimation for mixture multivariate stochastic observations of Markov chains. IEEE Int. Symp. Inf. Theory **32**(2), 307–309 (1986)
75. Kalinli, O., Seltzer, M.L., Droppo, J., Acero, A.: Noise adaptive training for robust automatic speech recognition. IEEE Trans. Audio, Speech Lang. Process. **18**(8), 1889–1901 (2010)
76. Kello, C.T., Plaut, D.C.: A neural network model of the articulatory-acoustic forward mapping trained on recordings of articulatory parameters. J. Acoust. Soc. Am. **116**(4), 2354–2364 (2004)
77. King, S., Frankel, J., Livescu, K., McDermott, E., Richmond, K., Wester, M.: J. Acoust. Soc. Am. **121**, 723–742 (2007)
78. Kingma, D., Welling, M.: Efficient gradient-based inference through transformations between bayes nets and neural nets. In: Proceedings of International Conference on Machine Learning (ICML) (2014)
79. Lee, L., Attias, H., Deng, L.: Variational inference and learning for segmental switching state space models of hidden speech dynamics. In: Proceedings of International Conference on Acoustics, Speech and Signal Processing (ICASSP), vol. 1, pp. I-872–I-875 (2003)
80. Lee, L.J., Fieguth, P., Deng, L.: A functional articulatory dynamic model for speech production. In: Proceedings of International Conference on Acoustics, Speech and Signal Processing (ICASSP), vol. 2, pp. 797–800. Salt Lake City, Utah (2001)
81. Li, J., Deng, L., Yu, D., Gong, Y., Acero, A.: High-performance hmm adaptation with joint compensation of additive and convolutive distortions via vector taylor series. In: Proceedings of IEEE Workshop on Automfatic Speech Recognition and Understanding (ASRU), pp. 65–70 (2007)
82. Li, J., Deng, L., Yu, D., Gong, Y., Acero, A.: HMM adaptation using a phase-sensitive acoustic distortion model for environment-robust speech recognition. In: Proceedings of International Conference on Acoustics, Speech and Signal Processing (ICASSP), pp. 4069–4072 (2008)
83. Li, J., Deng, L., Yu, D., Gong, Y., Acero, A.: A unified framework of HMM adaptation with joint compensation of additive and convolutive distortions. Comput. Speech Lang. **23**(3), 389–405 (2009)
84. Liu, F.H., Stern, R.M., Huang, X., Acero, A.: Efficient cepstral normalization for robust speech recognition. In: Proceedings of ACL Workshop on Human Language Technologies (ACL-HLT), pp. 69–74 (1993)
85. Liu, S., Sim, K.: Temporally varying weight regression: a semi-parametric trajectory model for automatic speech recognition. IEEE Trans. Audio, Speech Lang. Process. **22**(1), 151–160 (2014)
86. Livescu, K., Fosler-Lussier, E., Metze, F.: Subword modeling for automatic speech recognition: past, present, and emerging approaches. IEEE Signal Process. Mag. **29**(6), 44–57 (2012)
87. Ma, J., Deng, L.: A path-stack algorithm for optimizing dynamic regimes in a statistical hidden dynamic model of speech. Comput. Speech Lang. **14**, 101–104 (2000)
88. Ma, J., Deng, L.: Efficient decoding strategies for conversational speech recognition using a constrained nonlinear state-space model. IEEE Trans. Audio, Speech Lang. Process. **11**(6), 590–602 (2004)
89. Ma, J., Deng, L.: Target-directed mixture dynamic models for spontaneous speech recognition. IEEE Trans. Audio Speech Process. **12**(1), 47–58 (2004)
90. Macherey, W., Ney, H.: A comparative study on maximum entropy and discriminative training for acoustic modeling in automatic speech recognition. In: Proceedings of Eurospeech, pp. 493–496 (2003)
91. Mak, B., Tam, Y., Li, P.: Discriminative auditory-based features for robust speech recognition. IEEE Trans. Speech Audio Process. **12**, 28–36 (2004)

92. Mesot, B., Barber, D.: Switching linear dynamical systems for noise robust speech recognition. IEEE Trans. Audio, Speech Lang. Process. **15**(6), 1850–1858 (2007)
93. Mnih, A., Gregor, K.: Neural variational inference and learning in belief networks. In: Proceedings of International Conference on Machine Learning (ICML) (2014)
94. Moon, T.K.: The expectation-maximization algorithm. IEEE Signal Process. Mag. **13**(6), 47–60 (1996)
95. Ostendorf, M., Digalakis, V., Kimball, O.: From HMM's to segment models: a unified view of stochastic modeling for speech recognition. IEEE Trans. Audio Speech Process. **4**(5) (1996)
96. Ostendorf, M., Kannan, A., Kimball, O., Rohlicek, J.: Continuous word recognition based on the stochastic segment model. In: Proceedings of the DARPA Workshop CSR (1992)
97. Pavlovic, V., Frey, B., Huang, T.: Variational learning in mixed-state dynamic graphical models. In: UAI, pp. 522–530. Stockholm (1999)
98. Picone, J., Pike, S., Regan, R., Kamm, T., Bridle, J., Deng, L., Ma, Z., Richards, H., Schuster, M.: Initial evaluation of hidden dynamic models on conversational speech. In: Proceedings of International Conference on Acoustics, Speech and Signal Processing (ICASSP) (1999)
99. Povey, D., Kanevsky, D., Kingsbury, B., Ramabhadran, B., Saon, G., Visweswariah, K.: Boosted MMI for model and feature-space discriminative training. In: Proceedings of International Conference on Acoustics, Speech and Signal Processing (ICASSP), pp. 4057–4060 (2008)
100. Povey, D., Kingsbury, B., Mangu, L., Saon, G., Soltau, H., Zweig, G.: FMPE: discriminatively trained features for speech recognition. In: Proceedings of International Conference on Acoustics, Speech and Signal Processing (ICASSP), vol. 1, pp. 961–964 (2005)
101. Povey, D., Woodland, P.C.: Minimum phone error and I-smoothing for improved discriminative training. In: Proceedings of International Conference on Acoustics, Speech and Signal Processing (ICASSP), vol. 1, pp. I–105 (2002)
102. Povey, D., Woodland, P.C.: Minimum phone error and i-smoothing for improved discriminative training. In: Proceedings of International Conference on Acoustics, Speech and Signal Processing (ICASSP), pp. 105–108 (2002)
103. Rabiner, L.: A tutorial on hidden Markov models and selected applications in speech recognition. Proc. IEEE **77**(2), 257–286 (1989)
104. Rabiner, L., Juang, B.H.: An introduction to hidden Markov models. IEEE ASSP Mag. **3**(1), 4–16 (1986)
105. Rabiner, L., Juang, B.H.: Fundamentals of Speech Recognition. Prentice-Hall, Upper Saddle River (1993)
106. Rosti, A., Gales, M.: Rao-blackwellised gibbs sampling for switching linear dynamical systems. In: Proceedings of International Conference on Acoustics, Speech and Signal Processing (ICASSP), vol. 1, pp. I-809–I-812 (2004)
107. Russell, M., Jackson, P.: A multiple-level linear/linear segmental HMM with a formant-based intermediate layer. Comput. Speech Lang. **19**, 205–225 (2005)
108. Sakoe, H., Chiba, S.: Dynamic programming algorithm optimization for spoken word recognition. In: Readings in Speech Recognition, pp. 159–165. Morgan Kaufmann Publishers Inc, San Francisco (1990)
109. Schlueter, R., Macherey, W., Mueller, B., Ney, H.: Comparison of discriminative training criteria and optimization methods for speech recognition. Speech Commun. **31**, 287–310 (2001)
110. Seide, F., Li, G., Yu, D.: Conversational speech transcription using context-dependent deep neural networks. In: Proceedings of Annual Conference of International Speech Communication Association (INTERSPEECH), pp. 437–440 (2011)
111. Sun, J., Deng, L.: An overlapping-feature based phonological model incorporating linguistic constraints: applications to speech recognition. J. Acoust. Soc. Am. **111**, 1086–1101 (2002)
112. Suzuki, J., Fujino, A., Isozaki, H.: Semi-supervised structured output learning based on a hybrid generative and discriminative approach. In: Proceedings of EMNLP-CoNLL (2007)

113. Wang, Y., Gales, M.J.: Speaker and noise factorization for robust speech recognition. IEEE Trans. Audio, Speech Lang. Process. **20**(7), 2149–2158 (2012)
114. Woodland, P.C., Povey, D.: Large scale discriminative training of hidden Markov models for speech recognition. Comput. Speech Lang. (2002)
115. Wright, S., Kanevsky, D., Deng, L., He, X., Heigold, G., Li, H.: Optimization algorithms and applications for speech and language processing. IIEEE Trans. Audio, Speech Lang. Process. **21**(11), 2231–2243 (2013)
116. Xing, E., Jordan, M., Russell, S.: A generalized mean field algorithm for variational inference in exponential families. In: Proceedings of Uncertainty in Artificial Intelligence (2003)
117. Yu, D., Deng, L.: Speaker-adaptive learning of resonance targets in a hidden trajectory model of speech coarticulation. Comput. Speech Lang. **27**, 72–87 (2007)
118. Yu, D., Deng, L., Acero, A.: A lattice search technique for a long-contextual-span hidden trajectory model of speech. Speech Commun. **48**, 1214–1226 (2006)
119. Yu, D., Deng, L., Dahl, G.: Roles of pre-training and fine-tuning in context-dependent DBN-HMMs for real-world speech recognition. In: NIPS Workshop on Deep Learning and Unsupervised Feature Learning (2010)
120. Yu, D., Deng, L., Gong, Y., Acero, A.: A novel framework and training algorithm for variable-parameter hidden Markov models. IEEE Trans. Audio, Speech Lang. Process. **17**(7), 1348–1360 (2009)
121. Yu, D., Deng, L., He, X., Acero, A.: Use of incrementally regulated discriminative margins in MCE training for speech recognition. In: Proceedings of Annual Conference of International Speech Communication Association (INTERSPEECH) (2006)
122. Yu, D., Deng, L., He, X., Acero, A.: Use of incrementally regulated discriminative margins in mce training for speech recognition. In: Proceedings of International Conference on Spoken Language Processing (ICSLP), pp. 2418–2421 (2006)
123. Yu, D., Deng, L., He, X., Acero, A.: Large-margin minimum classification error training: a theoretical risk minimization perspective. Comput. Speech Lang. **22**, 415–429 (2008)
124. Zen, H., Tokuda, K., Kitamura, T.: An introduction of trajectory model into HMM-based speech synthesis. In: Proceedings of ISCA SSW5, pp. 191–196 (2004)
125. Zhang, B., Matsoukas, S., Schwartz, R.: Discriminatively trained region dependent feature transforms for speech recognition. In: Proceedings of International Conference on Acoustics, Speech and Signal Processing (ICASSP), vol. 1, pp. I–I (2006)
126. Zhang, L., Renals, S.: Acoustic-articulatory modelling with the trajectory HMM. IEEE Signal Process. Lett. **15**, 245–248 (2008)
127. Zhang, S., Gales, M.: Structured SVMs for automatic speech recognition. IEEE Trans. Audio, Speech Lang. Process. **21**(3), 544–555 (2013)
128. Zhou, J.L., Seide, F., Deng, L.: Coarticulation modeling by embedding a target-directed hidden trajectory model into HMM—model and training. In: Proceedings of International Conference on Acoustics, Speech and Signal Processing (ICASSP), vol. 1, pp. 744–747. Hongkong (2003)

Part II
Deep Neural Networks

Chapter 4
Deep Neural Networks

Abstract In this chapter, we introduce deep neural networks (DNNs)—multilayer perceptrons with many hidden layers. DNNs play an important role in the modern speech recognition systems, and are the focus of the rest of the book. We depict the architecture of DNNs, describe the popular activation functions and training criteria, illustrate the famous backpropagation algorithm for learning DNN model parameters, and introduce practical tricks that make the training process robust.

4.1 The Deep Neural Network Architecture

A deep neural network (DNN)[1] is a conventional multilayer perceptron (MLP) with many (often more than two) hidden layers. See a comprehensive review of the DNN in use as an acoustic model for speech recognition in [11]. Figure 4.1 depicts a DNN with a total of five layers that include an input layer, three hidden layers and an output layer. For the sake of notation simplicity, we denote the input layer as layer 0 and the output layer as layer L for an $L + 1$-layer DNN.

In the first L layers

$$\mathbf{v}^\ell = f\left(\mathbf{z}^\ell\right) = f\left(\mathbf{W}^\ell \mathbf{v}^{\ell-1} + \mathbf{b}^\ell\right), \text{ for } 0 < \ell < L, \tag{4.1}$$

where $\mathbf{z}^\ell = \mathbf{W}^\ell \mathbf{v}^{\ell-1} + \mathbf{b}^\ell \in \mathbb{R}^{N_\ell \times 1}$, $\mathbf{v}^\ell \in \mathbb{R}^{N_\ell \times 1}$, $\mathbf{W}^\ell \in \mathbb{R}^{N_\ell \times N_{\ell-1}}$, $\mathbf{b}^\ell \in \mathbb{R}^{N_\ell \times 1}$, and $N_\ell \in \mathbb{R}$ are, respectively, the excitation vector, the activation vector, the weight matrix, the bias vector, and the number of neurons at layer ℓ. $\mathbf{v}^0 = \mathbf{o} \in \mathbb{R}^{N_0 \times 1}$ is the observation (or feature) vector, $N_0 = D$ is the feature dimension, and $f(\cdot) : \mathbb{R}^{N_\ell \times 1} \rightarrow \mathbb{R}^{N_\ell \times 1}$ is the activation function applied to the excitation vector element-wise. In most applications, the sigmoid function

---

[1] The term deep neural network first appeared in [21] in the context of speech recognition, but was coined in [5] which converted the term deep belief network in the earlier studies into the more appropriate term of deep neural network [4, 17, 24]. The term deep neural network was originally introduced to mean multilayer perceptrons with many hidden layers, but was later extended to mean any neural network with a deep structure.

© Springer-Verlag London 2015

D. Yu and L. Deng, *Automatic Speech Recognition*,

Signals and Communication Technology, DOI 10.1007/978-1-4471-5779-3_4

Fig. 4.1 An example deep
neural network with an input
layer, three hidden layers, and
an output layer

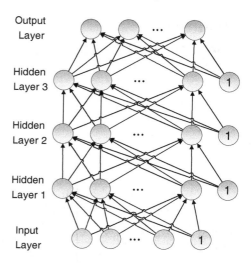

$$\sigma\left(z\right) = \frac{1}{1 + e^{-z}} \tag{4.2}$$

or the hyperbolic tangent function

$$\tanh\left(z\right) = \frac{e^z - e^{-z}}{e^z + e^{-z}} \tag{4.3}$$

is used as the activation function. Since the tanh (z) function is a rescaled version of
the sigmoid function, these two activation functions have the same modeling power.
However, the output range of $\sigma\left(z\right)$ is $(0, 1)$, which encourages sparse but, at the same
time, asymmetric activation values. On the other hand, the output range of tanh (z)
is $(-1, +1)$, and thus the activation value is symmetric, which was believed to help
the model training [14]. Another popular activation function is the rectified linear
unit (ReLU) function

$$\text{ReLU}\left(z\right) = \max\left(0, z\right), \tag{4.4}$$

which enforces sparse activations[2] [7] and has very simple gradient as we will discus
in Sect. 4.2. Since sigmoid function is still the most popular activation functions used
by the practitioners, in all the following discussions we assume that the sigmoid
activation function is used unless otherwise noted.

The output layer needs to be chosen based on the tasks in hand. For the regression
tasks, a linear layer

$$\mathbf{v}^L = \mathbf{z}^L = \mathbf{W}^L \mathbf{v}^{L-1} + \mathbf{b}^L \tag{4.5}$$

---

[2] The output of the sigmoid function can be very close to 0 but cannot reach 0, while the output of
the ReLU function can be exactly 0.

is typically used to generate the output vector $\mathbf{v}^L \in \mathbb{R}^{N_L}$, where N_L is the output dimension.

For the multi-class classification tasks each output neuron represents a class $i \in \{1, \ldots, C\}$, where $C = N_L$ is the number of classes. The value of the ith output neuron v_i^L represents the probability $P_{\text{dnn}}(i|\mathbf{o})$ that the observation vector $\mathbf{o}$ belongs to class i. To serve as a valid multinomial probability distribution, the output vector $\mathbf{v}^L$ should satisfy the requirements $v_i^L \geq 0$ and $\sum_{i=1}^C v_i^L = 1$. This can be done by normalizing the excitation with a softmax function

$$v_i^L = P_{\text{dnn}}(i|\mathbf{o}) = \text{softmax}_i\left(\mathbf{z}^L\right) = \frac{e^{z_i^L}}{\sum_{j=1}^C e^{z_j^L}}, \qquad (4.6)$$

where z_i^L is the ith element of the excitation vector $\mathbf{z}^L$.

Given an observation vector $\mathbf{o}$, the output of the DNN specified by the model parameters $\{\mathbf{W}, \mathbf{b}\} = \left\{\mathbf{W}^\ell, \mathbf{b}^\ell | 0 < \ell \leq L\right\}$ can be calculated by computing the activation vectors with Eq. (4.1) layer by layer from layer 1 to layer $L - 1$ and with Eq. (4.5) for the regression tasks and Eq. (4.6) for the classification tasks to calculate the output of the DNN. This process is often called forward computation and is summarized in Algorithm 4.1.

Algorithm 4.1 DNN Forward Computation

1: **procedure** FORWARDCOMPUTATION($\mathbf{O}$)
                ▷ Each column of $\mathbf{O}$ is an observation vector
2:      $\mathbf{V}^0 \leftarrow \mathbf{O}$
3:      **for** $\ell \leftarrow 1; \ell < L; \ell \leftarrow \ell + 1$ **do**   ▷ L is the total number of layers
4:          $\mathbf{Z}^\ell \leftarrow \mathbf{W}^\ell \mathbf{V}^{\ell-1} + \mathbf{B}^\ell$ ▷ Each column of $\mathbf{B}^\ell$ is $\mathbf{b}^\ell$
5:          $\mathbf{V}^\ell \leftarrow f\left(\mathbf{Z}^\ell\right)$ ▷ $f(.)$ can be sigmoid, tanh, ReLU, or other functions
6:      **end for**
7:      $\mathbf{Z}^L \leftarrow \mathbf{W}^L \mathbf{V}^{L-1} + \mathbf{B}^L$
8:      **if** regression **then**   ▷ regression task
9:          $\mathbf{V}^L \leftarrow \mathbf{Z}^L$
10:     **else**   ▷ classification task
11:         $\mathbf{V}^L \leftarrow \text{softmax}\left(\mathbf{Z}^L\right)$ ▷ Apply softmax column-wise
12:     **end if**
13:     Return $\mathbf{V}^L$
14: **end procedure**

4.2 Parameter Estimation with Error Backpropagation

It has been known since 1980s that an MLP with a sufficiently large hidden layer is a universal approximator [12]. In other words, such an MLP can approximate any mapping $g : \mathbb{R}^D \rightarrow \mathbb{R}^C$ from the input space $\mathbb{R}^D$ to the output space $\mathbb{R}^C$. It is obvious that a DNN, which is an MLP with more than two hidden layers, can also serve as a universal approximator.

The model parameters $\{\mathbf{W}, \mathbf{b}\}$ in a DNN, however, are unknown and need to be estimated from training samples $\mathbb{S} = \{(\mathbf{o}^m, \mathbf{y}^m) | 0 \leq m < M\}$ for each task, where M is the number of training samples, $\mathbf{o}^m$ is the mth observation vector and $\mathbf{y}^m$ is the corresponding desired output vector. This process is often called the training process or the parameter estimation process, which can be specified by a training criterion and a learning algorithm.

4.2.1 Training Criteria

The training criterion should be easy to evaluate and be highly correlated to the final goal of the task so that the improvement in the training criterion would lead to the improvement in the final evaluation score. Ideally, the model parameters should be trained to minimize the expected loss

$$J_{\text{EL}} = \mathbb{E}\left(J\left(\mathbf{W}, \mathbf{b}; \mathbf{o}, \mathbf{y}\right)\right) = \int_{\mathbf{o}} J\left(\mathbf{W}, \mathbf{b}; \mathbf{o}, \mathbf{y}\right) p\left(\mathbf{o}\right) d\left(\mathbf{o}\right), \qquad (4.7)$$

where $J\left(\mathbf{W}, \mathbf{b}; \mathbf{o}, \mathbf{y}\right)$ is the loss function given the model parameters $\{\mathbf{W}, \mathbf{b}\}$, the observation $\mathbf{o}$, and the corresponding output vector $\mathbf{y}$, and $p\left(\mathbf{o}\right)$ is the probability density function of observation $\mathbf{o}$. Unfortunately, $p\left(\mathbf{o}\right)$ is typically unknown and needs to be estimated from the training set, and $J\left(\mathbf{W}, \mathbf{b}; \mathbf{o}, \mathbf{y}\right)$ is not well-defined (the desired output vector is unknown) for samples unseen in the training set. For this reason, the DNN model parameters are often trained to optimize the empirical criteria.

There are two popular empirical training criteria in DNN model learning. For the regression tasks, the mean square error (MSE) criterion

$$J_{\text{MSE}}\left(\mathbf{W}, \mathbf{b}; \mathbb{S}\right) = \frac{1}{M} \sum_{m=1}^{M} J_{\text{MSE}}\left(\mathbf{W}, \mathbf{b}; \mathbf{o}^m, \mathbf{y}^m\right) \qquad (4.8)$$

is typically used, where

$$J_{\text{MSE}}\left(\mathbf{W}, \mathbf{b}; \mathbf{o}, \mathbf{y}\right) = \frac{1}{2}\left\|\mathbf{v}^L - \mathbf{y}\right\|^2 = \frac{1}{2}\left(\mathbf{v}^L - \mathbf{y}\right)^{\text{T}}\left(\mathbf{v}^L - \mathbf{y}\right). \qquad (4.9)$$

For the classification tasks, $\mathbf{y}$ is a probability distribution and the cross-entropy (CE) criterion

$$J_{\text{CE}}\left(\mathbf{W}, \mathbf{b}; \mathbb{S}\right) = \frac{1}{M} \sum_{m=1}^{M} J_{\text{CE}}\left(\mathbf{W}, \mathbf{b}; \mathbf{o}^m, \mathbf{y}^m\right) \qquad (4.10)$$

is often used, where

$$J_{CE}\left(\mathbf{W}, \mathbf{b}; \mathbf{o}, \mathbf{y}\right) = -\sum_{i=1}^{C} y_i \log v_i^L, \tag{4.11}$$

$y_i = P_{emp}\left(i|\mathbf{o}\right)$ is the empirical (observed in the training set) probability that the observation $\mathbf{o}$ belongs to class i, and $v_i^L = P_{dnn}\left(i|\mathbf{o}\right)$ is the same probability estimated from the DNN. Minimizing the cross-entropy criterion is equivalent to minimizing the Kullback-Leibler divergence (KLD) between the empirical probability distribution and the probability distribution estimated from the DNN. In most cases, a hard class label is used, or $y_i = \mathbb{I}\left(c = i\right)$, where

$$\mathbb{I}\left(x\right) = \begin{cases} 1, & \text{if } x \text{ is true} \\ 0, & \text{otherwise} \end{cases} \tag{4.12}$$

is the indicator function, and c is the class label in the training set for observation $\mathbf{o}$. In these cases, the CE criterion specified by Eq. (4.11) is reduced to the negative log-likelihood (NLL)

$$J_{NLL}\left(\mathbf{W}, \mathbf{b}; \mathbf{o}, \mathbf{y}\right) = -\log v_c^L. \tag{4.13}$$

4.2.2 Training Algorithms

Given the training criterion, the model parameters $\{\mathbf{W}, \mathbf{b}\}$ can be learned with the famous error backpropagation (BP) algorithm [19], which can be derived from the chain rule used for gradient computation.[3]

In its simplest form, the model parameters can be improved based on the first-order gradient information as

$$\mathbf{W}_{t+1}^\ell \leftarrow \mathbf{W}_t^\ell - \varepsilon \triangle \mathbf{W}_t^\ell \tag{4.14}$$

and

$$\mathbf{b}_{t+1}^\ell \leftarrow \mathbf{b}_t^\ell - \varepsilon \triangle \mathbf{b}_t^\ell, \tag{4.15}$$

where $\mathbf{W}_t^\ell$ and $\mathbf{b}_t^\ell$ are the weight matrix and the bias vector at layer ℓ after the tth update,

$$\triangle \mathbf{W}_t^\ell = \frac{1}{M_b} \sum_{m=1}^{M_b} \nabla_{\mathbf{W}_t^\ell} J\left(\mathbf{W}, \mathbf{b}; \mathbf{o}^m, \mathbf{y}^m\right) \tag{4.16}$$

---

[3] Although the name backpropagation was coined in 1986 [19] the algorithm itself can be traced back at least to 1969 [3] as a multistage dynamic system optimization method.

and

$$\triangle \mathbf{b}_t^\ell = \frac{1}{M_b} \sum_{m=1}^{M_b} \nabla_{\mathbf{b}_t^\ell} J \left(\mathbf{W}, \mathbf{b}; \mathbf{o}^m, \mathbf{y}^m \right) \tag{4.17}$$

are, respectively, the average weight matrix gradient and the average bias vector gradient at iteration t estimated from the training batch consisted of M_b samples, ε is the learning rate, and $\nabla_{\mathbf{x}} J$ is the gradient of J with regard to $\mathbf{x}$.

The gradient of the training criterion with regard to the top-layer weight matrix and bias vector depends on the criterion chosen. For the regression tasks, where the MSE training criterion (Eq. 4.9) and the linear output layer (Eq. 4.5) are used, the gradient with regard to the output layer weight matrix is

$$
\begin{aligned}
\nabla_{\mathbf{W}_t^L} J_{\text{MSE}} \left(\mathbf{W}, \mathbf{b}; \mathbf{o}, \mathbf{y} \right) &= \nabla_{\mathbf{z}_t^L} J_{\text{MSE}} \left(\mathbf{W}, \mathbf{b}; \mathbf{o}, \mathbf{y} \right) \frac{\partial \mathbf{z}_t^L}{\partial \mathbf{W}_t^L} \\
&= \mathbf{e}_t^L \frac{\partial \left(\mathbf{W}_t^L \mathbf{v}_t^{L-1} + \mathbf{b}_t^L \right)}{\partial \mathbf{W}_t^L} \\
&= \mathbf{e}_t^L \left(\mathbf{v}_t^{L-1} \right)^{\text{T}} \\
&= \left(\mathbf{v}_t^L - \mathbf{y} \right) \left(\mathbf{v}_t^{L-1} \right)^{\text{T}},
\end{aligned}
\tag{4.18}
$$

where we have defined the error signal at the output layer as

$$
\begin{aligned}
\mathbf{e}_t^L &\triangleq \nabla_{\mathbf{z}_t^L} J_{\text{MSE}} \left(\mathbf{W}, \mathbf{b}; \mathbf{o}, \mathbf{y} \right) \\
&= \frac{1}{2} \frac{\partial \left(\mathbf{z}_t^L - \mathbf{y} \right)^T \left(\mathbf{z}_t^L - \mathbf{y} \right)}{\partial \mathbf{z}_t^L} \\
&= \left(\mathbf{v}_t^L - \mathbf{y} \right).
\end{aligned}
\tag{4.19}
$$

Similarly,

$$\nabla_{\mathbf{b}_t^L} J_{\text{MSE}} \left(\mathbf{W}, \mathbf{b}; \mathbf{o}, \mathbf{y} \right) = \left(\mathbf{v}_t^L - \mathbf{y} \right). \tag{4.20}$$

For the classification tasks, where the CE training criterion (Eq. 4.11) and the softmax output layer (Eq. 4.6) are used, the gradient with regard to the output layer weight matrix can be derived as

$$
\begin{aligned}
\nabla_{\mathbf{W}_t^L} J_{\text{CE}} \left(\mathbf{W}, \mathbf{b}; \mathbf{o}, \mathbf{y} \right) &= \nabla_{\mathbf{z}_t^L} J_{\text{CE}} \left(\mathbf{W}, \mathbf{b}; \mathbf{o}, \mathbf{y} \right) \frac{\partial \mathbf{z}_t^L}{\partial \mathbf{W}_t^L} \\
&= \mathbf{e}_t^L \frac{\partial \left(\mathbf{W}_t^L \mathbf{v}_t^{L-1} + \mathbf{b}_t^L \right)}{\partial \mathbf{W}_t^L}
\end{aligned}
$$

$$= \mathbf{e}_t^L \left(\mathbf{v}_t^{L-1}\right)^{\mathsf{T}}$$
$$= \left(\mathbf{v}_t^L - \mathbf{y}\right) \left(\mathbf{v}_t^{L-1}\right)^{\mathsf{T}}, \tag{4.21}$$

where we have similarly defined the error signal at the output layer as

$$\mathbf{e}_t^L \triangleq \nabla_{\mathbf{z}_t^L} J_{\text{CE}} \left(\mathbf{W}, \mathbf{b}; \mathbf{o}, \mathbf{y}\right)$$

$$= -\frac{\partial \sum_{i=1}^C y_i \log \text{softmax}_i \left(\mathbf{z}_t^L\right)}{\partial \mathbf{z}_t^L}$$

$$= \frac{\partial \sum_{i=1}^C y_i \log \sum_{j=1}^C e^{z_j^L}}{\partial \mathbf{z}_t^L} - \frac{\partial \sum_{i=1}^C y_i \log e^{z_i^L}}{\partial \mathbf{z}_t^L}$$

$$= \frac{\partial \log \sum_{j=1}^C e^{z_j^L}}{\partial \mathbf{z}_t^L} - \frac{\partial \sum_{i=1}^C y_i z_i^L}{\partial \mathbf{z}_t^L}$$

$$= \begin{bmatrix} \dfrac{e^{z_1^L}}{\sum_{j=1}^C e^{z_j^L}} \\ \vdots \\ \dfrac{e^{z_i^L}}{\sum_{j=1}^C e^{z_j^L}} \\ \vdots \\ \dfrac{e^{z_C^L}}{\sum_{j=1}^C e^{z_j^L}} \end{bmatrix} - \begin{bmatrix} y_1 \\ \vdots \\ y_i \\ \vdots \\ y_C \end{bmatrix}$$

$$= \left(\mathbf{v}_t^L - \mathbf{y}\right).$$

Similarly,

$$\nabla_{\mathbf{b}_t^L} J_{\text{CE}} \left(\mathbf{W}, \mathbf{b}; \mathbf{o}, \mathbf{y}\right) = \left(\mathbf{v}_t^L - \mathbf{y}\right). \tag{4.22}$$

Note that $\nabla_{\mathbf{W}_t^L} J_{\text{CE}} \left(\mathbf{W}, \mathbf{b}; \mathbf{o}, \mathbf{y}\right)$ (Eq. 4.21) appears to have the same form as $\nabla_{\mathbf{W}_t^L} J_{\text{MSE}} \left(\mathbf{W}, \mathbf{b}; \mathbf{o}, \mathbf{y}\right)$ (Eq. 4.18). However, they are actually different since in the regression case $\mathbf{v}_t^L = \mathbf{z}_t^L$ and in the classification case $\mathbf{v}_t^L = \text{softmax} \left(\mathbf{z}_t^L\right)$.

Note also that for $0 < \ell < L$,

$$\nabla_{\mathbf{W}_t^\ell} J \left(\mathbf{W}, \mathbf{b}; \mathbf{o}, \mathbf{y}\right) = \nabla_{\mathbf{v}_t^\ell} J \left(\mathbf{W}, \mathbf{b}; \mathbf{o}, \mathbf{y}\right) \frac{\partial \mathbf{v}_t^\ell}{\partial \mathbf{W}_t^\ell}$$

$$= \text{diag} \left(f' \left(\mathbf{z}_t^\ell\right)\right) \mathbf{e}_t^\ell \frac{\partial \left(\mathbf{W}_t^\ell \mathbf{v}_t^{\ell-1} + \mathbf{b}_t^\ell\right)}{\partial \mathbf{W}_t^\ell}$$

$$= \text{diag}\left(f'\left(\mathbf{z}_t^\ell\right)\right)\mathbf{e}_t^\ell\left(\mathbf{v}_t^{\ell-1}\right)^{\mathrm{T}}$$

$$= \left[f'\left(\mathbf{z}_t^\ell\right)\bullet\mathbf{e}_t^\ell\right]\left(\mathbf{v}_t^{\ell-1}\right)^{\mathrm{T}}, \tag{4.23}$$

$$\nabla_{\mathbf{b}_t^\ell} J\left(\mathbf{W}, \mathbf{b}; \mathbf{o}, \mathbf{y}\right) = \nabla_{\mathbf{v}_t^\ell} J\left(\mathbf{W}, \mathbf{b}; \mathbf{o}, \mathbf{y}\right)\frac{\partial \mathbf{v}_t^\ell}{\partial \mathbf{b}_t^\ell}$$

$$= \text{diag}\left(f'\left(\mathbf{z}_t^\ell\right)\right)\mathbf{e}_t^\ell\frac{\partial\left(\mathbf{W}_t^\ell\mathbf{v}_t^{\ell-1} + \mathbf{b}_t^\ell\right)}{\partial \mathbf{b}_t^\ell}$$

$$= \text{diag}\left(f'\left(\mathbf{z}_t^\ell\right)\right)\mathbf{e}_t^\ell$$

$$= f'\left(\mathbf{z}_t^\ell\right)\bullet\mathbf{e}_t^\ell, \tag{4.24}$$

where $\mathbf{e}_t^\ell \triangleq \nabla_{\mathbf{v}_t^\ell} J\left(\mathbf{W}, \mathbf{b}; \mathbf{o}, \mathbf{y}\right)$ is the error signal at layer ℓ, $\bullet$ is the element-wise product, $\text{diag}(\mathbf{x})$ is a square matrix whose diagonal equals to $\mathbf{x}$, and $f'\left(\mathbf{z}_t^\ell\right)$ is the element-wise derivative of the activation function. For the sigmoid activation function,

$$\sigma'\left(\mathbf{z}_t^\ell\right) = \left(1 - \sigma\left(\mathbf{z}_t^\ell\right)\right)\bullet\sigma\left(\mathbf{z}_t^\ell\right) = \left(1 - \mathbf{v}_t^\ell\right)\bullet\mathbf{v}_t^\ell. \tag{4.25}$$

Similarly, for the tanh activation function

$$\tanh'\left(z_{t,i}^\ell\right) = 1 - \left[\tanh\left(z_{t,i}^\ell\right)\right]^2 = 1 - \left[v_{t,i}^\ell\right]^2, \tag{4.26}$$

or

$$\tanh'\left(\mathbf{z}_t^\ell\right) = 1 - \mathbf{v}_t^\ell\bullet\mathbf{v}_t^\ell. \tag{4.27}$$

For the ReLU activation function

$$\text{ReLU}'\left(z_{t,i}^\ell\right) = \begin{cases} 1, & \text{if } z_{t,i}^\ell > 0 \\ 0, & \text{otherwise} \end{cases} \tag{4.28}$$

or

$$\text{ReLU}'\left(\mathbf{z}_t^\ell\right) = \max\left(0, \text{sgn}\left(\mathbf{z}_t^\ell\right)\right), \tag{4.29}$$

where $\text{sgn}\left(\mathbf{z}_t^\ell\right)$ is the sign of $\mathbf{z}_t^\ell$ applied element-wise. The error signal can be back-propagated from top to bottom as

$$\mathbf{e}_t^{L-1} = \nabla_{\mathbf{v}_t^{L-1}} J\left(\mathbf{W}, \mathbf{b}; \mathbf{o}, \mathbf{y}\right)$$

$$= \frac{\partial \mathbf{z}_t^L}{\partial \mathbf{v}_t^{L-1}} \nabla_{\mathbf{z}_t^L} J\left(\mathbf{W}, \mathbf{b}; \mathbf{o}, \mathbf{y}\right)$$

$$= \frac{\partial \left(\mathbf{W}_t^L \mathbf{v}_t^{L-1} + \mathbf{b}_t^L\right)}{\partial \mathbf{v}_t^{L-1}} \mathbf{e}_t^L$$

$$= \left(\mathbf{W}_t^L\right)^{\mathrm{T}} \mathbf{e}_t^L. \tag{4.30}$$

For $\ell < L$,

$$\mathbf{e}_t^{\ell-1} = \nabla_{\mathbf{v}_t^{\ell-1}} J\left(\mathbf{W}, \mathbf{b}; \mathbf{o}, \mathbf{y}\right)$$

$$= \frac{\partial \mathbf{v}_t^\ell}{\partial \mathbf{v}_t^{\ell-1}} \nabla_{\mathbf{v}_t^\ell} J\left(\mathbf{W}, \mathbf{b}; \mathbf{o}, \mathbf{y}\right)$$

$$= \frac{\partial \left(\mathbf{W}_t^\ell \mathbf{v}_t^{\ell-1} + \mathbf{b}_t^\ell\right)}{\partial \mathbf{v}_t^{\ell-1}} \mathrm{diag}\left(f'\left(\mathbf{z}_t^\ell\right)\right) \mathbf{e}_t^\ell$$

$$= \left(\mathbf{W}_t^\ell\right)^{\mathrm{T}} \left[f'\left(\mathbf{z}_t^\ell\right) \bullet \mathbf{e}_t^\ell\right]. \tag{4.31}$$

The key steps in the backpropagation algorithm is summarized in Algorithm 4.2.

4.3 Practical Considerations

The basic backpropagation algorithm described in Sect. 4.2 is theoretically simple. However, learning a model efficiently and effectively requires taking consideration many practical issues [1, 14].

4.3.1 Data Preprocessing

Data preprocessing plays an important role in many machine learning algorithms. The two most popular preprocessing techniques are per-sample feature normalization and global feature standardization.

If the mean of each sample reflects a variation that is irrelevant to the problem in hand, this mean should be subtracted from the feature to reduce the variability in the final feature fed to the DNN. For example, subtracting the mean intensity of an image can reduce the variability introduced by the brightness. In the handwriting character recognition tasks, normalizing the image center can reduce the variability caused by the shifted character position. In speech recognition, the cepstral mean

---

Algorithm 4.2 Backpropagation Algorithm

1: **procedure** BACKPROPAGATION($\mathbb{S} = \{(\mathbf{o}^m, \mathbf{y}^m) \mid 0 \leq m < M\}$)
                                               $\triangleright$ $\mathbb{S}$ is the training set with M samples
2:     Randomly initialize $\left\{\mathbf{W}_0^\ell, \mathbf{b}_0^\ell\right\}, 0 < \ell \leq L$ $\triangleright$ L is the total number of layers
3:     **while** Stopping Criterion Not Met **do**
                $\triangleright$ Stop if reached max iterations or the training criterion improvement is small
4:         Randomly select a minibatch $\mathbf{O}, \mathbf{Y}$ with M_b samples.
5:         Call ForwardComputation($\mathbf{O}$)
6:         $\mathbf{E}_t^L \leftarrow \mathbf{V}_t^L - \mathbf{Y}$ $\triangleright$ Each column of $\mathbf{E}_t^L$ is $\mathbf{e}_t^L$
7:         $\mathbf{G}_t^L \leftarrow \mathbf{E}_t^L$
8:         **for** $\ell \leftarrow L; \ell > 0; \ell \leftarrow \ell - 1$ **do**
9:             $\nabla_{\mathbf{W}_t^\ell} \leftarrow \mathbf{G}_t^\ell \left(\mathbf{v}_t^{\ell-1}\right)^{\mathrm{T}}$
10:            $\nabla_{\mathbf{b}_t^\ell} \leftarrow \mathbf{G}_t^\ell$
11:            $\mathbf{W}_{t+1}^\ell \leftarrow \mathbf{W}_t^\ell - \frac{\varepsilon}{M_b} \nabla_{\mathbf{W}_t^\ell}$ $\triangleright$ Update $\mathbf{W}$
12:            $\mathbf{b}_{t+1}^\ell \leftarrow \mathbf{b}_t^\ell - \frac{\varepsilon}{M_b} \nabla_{\mathbf{b}_t^\ell}$ $\triangleright$ Update $\mathbf{b}$
13:            $\mathbf{E}_t^{\ell-1} \leftarrow \left(\mathbf{W}_t^\ell\right)^{\mathrm{T}} \mathbf{G}_t^\ell$ $\triangleright$ Error backpropagation
14:            **if** $\ell > 1$ **then**
15:                $\mathbf{G}_t^{\ell-1} \leftarrow f'\left(\mathbf{Z}_t^{\ell-1}\right) \bullet \mathbf{E}_t^{\ell-1}$
16:            **end if**
17:         **end for**
18:     **end while**
19:     Return $dnn = \left\{\mathbf{W}^\ell, \mathbf{b}^\ell\right\}, 0 < \ell \leq L$
20: **end procedure**

---

normalization (CMN) [15] technique, which subtracts the per-utterance mean of the MFCC features can reduce acoustic channel distortions. Using CMN as an example, the per-sample normalization can be carried out by first estimating the per-utterance mean

$$\bar{\mu}_i = \frac{1}{T} \sum_{t=1}^{T} o_i^t \,, \tag{4.32}$$

for each dimension i, where T is the total number of frames in the utterance, and then subtracting the mean from all frames in the utterance as

$$\bar{o}_i^t = o_i^t - \bar{\mu}_i. \tag{4.33}$$

The goal of global feature standardization is to scale the data along each dimension using a global transformation so that the final data vectors lie in the similar range. For example, in image processing, we often rescale the pixel values in the range of [0, 255] to the range of [0, 1]. For real-valued features such as the MFCC and the log filter-bank feature in speech recognition tasks, each dimension of the feature is often standardized to have zero-mean and unit-variance using a global transformation (e.g., in [5]). The global transformation in both cases is estimated from the training data only and it is then applied to both training and test sets. Given the training set $\mathbb{S} = \{(\mathbf{o}^m, \mathbf{y}^m) \mid 0 \leq m < M\}$ (which may have been normalized per sample), we can calculate the average value

$$\mu_i = \frac{1}{M} \sum_{m=1}^{M} o_i^m \qquad (4.34)$$

and the standard deviation

$$\sigma_i = \sqrt{\frac{1}{M} \sum_{m=1}^{M} \left(o_i^m - \mu_i\right)^2} \qquad (4.35)$$

for each dimension i. All samples in both training and test sets can be standardized as

$$\tilde{o}_i^m = \frac{o_i^m - \mu_i}{\sigma_i}. \qquad (4.36)$$

The global feature standardization described above is useful because the later processing steps often perform better when each dimension is scaled to a similar range of numerical values [14]. For example, in DNN training, by standardizing features we can use the same learning rate across all weight dimensions and still get a good model. Without feature standardization, the energy component or the first MFCC component c_0 would overshadow the other components and dominate the learning process unless learning algorithms (such as AdaGrad [6]), which automatically adjust learning rate for each separate dimension, are used.

4.3.2 Model Initialization

The learning algorithm specified in Sect. 4.2 starts from an initial model. Since the DNN is a highly nonlinear model and the training criterion with regard to the model parameters is nonconvex, the initial model can greatly affect the resulting model.

There are many heuristic tricks in initializing the DNN model. Most of these tricks are based on two considerations. First, the weights should be initialized so that each neuron operates in the linear range of the sigmoid function at the start of the learning. If weights were all very large, many neurons would saturate (close to zero or one) and the gradients would be very small according to Eq. (4.25). When the neurons operate in the linear range, instead, the gradients are large enough (close to the maximum value of 0.25) that learning can proceed effectively. Note that the excitation value depends on both the input values and the weights. When the input features are standardized as described in Sect. 4.3.1, determining the initial weights can become easier. Second, it is important to initialize the parameters randomly. This is because neurons in the DNNs are symmetric and interchangeable. If all the model parameters have identical values, all the hidden layer neurons will have the same output value and detect the same feature patterns in the lower layers. Random initialization serves the purpose of symmetry breaking.

LeCun and Bottou [14] suggested to draw values from a zero-mean Gaussian distribution with standard deviation $\sigma_{\mathbf{W}^{\ell+1}} = \frac{1}{\sqrt{N_\ell}}$ to initialize the weights in layer ℓ as defined in Eq. (4.1), where N_ℓ is the number of connections feeding into the node. For the DNNs used in speech recognition tasks, where each hidden layer has 1,000–2,000 neurons, we have found that initializing the weight matrices by drawing from a Gaussian distribution $\mathcal{N}(w; 0, 0.05)$ or from a uniform distribution in the range of $[-0.05, 0.05]$ performs very well. The bias vectors $\mathbf{b}^\ell$ can be simply initialized to zero.

4.3.3 Weight Decay

As in many machine learning algorithms, overfitting can be a problem especially since the number of parameters in a DNN is huge compared to many other learning machines. Overfitting happens because we are interested in minimizing the expected loss (4.7), but instead we actually minimize the empirical losses defined on the training set.

The simplest way to control overfitting is to regularize the training criterion so that the model parameters will not be tuned to fit the training data too well. The most widely used regularization terms include

$$R_1(\mathbf{W}) = \|\text{vec}(\mathbf{W})\|_1 = \sum_{\ell=1}^{L} \left\|\text{vec}\left(\mathbf{W}^\ell\right)\right\|_1 = \sum_{\ell=1}^{L} \sum_{i=1}^{N_\ell} \sum_{j=1}^{N_{\ell-1}} |\mathbf{W}_{ij}^\ell| \qquad (4.37)$$

that is based on the L_1 norm, and

$$R_2(\mathbf{W}) = \|\text{vec}(\mathbf{W})\|_2^2 = \sum_{\ell=1}^{L} \left\|\text{vec}\left(\mathbf{W}^\ell\right)\right\|_2^2 = \sum_{\ell=1}^{L} \sum_{i=1}^{N_\ell} \sum_{j=1}^{N_{\ell-1}} \left(\mathbf{W}_{ij}^\ell\right)^2 \qquad (4.38)$$

that is based on the L_2 norm, where $\mathbf{W}_{ij}$ is the (i, j)-th value in the matrix $\mathbf{W}$, $\text{vec}\left(\mathbf{W}^\ell\right) \in \mathbb{R}^{[N_\ell \times N_{\ell-1}] \times 1}$ is the vector generated by concatenating all the columns in the matrix $\mathbf{W}^\ell$, and $\left\|\text{vec}\left(\mathbf{W}^\ell\right)\right\|_2$ equals to $\left\|\mathbf{W}^\ell\right\|_F$—the Frobenious norm of the matrix $\mathbf{W}^\ell$. These regularization terms are often called weight decay in the neural network literature.

When regularization terms are included, the training criterion becomes

$$\ddot{J}(\mathbf{W}, \mathbf{b}; \mathbb{S}) = J(\mathbf{W}, \mathbf{b}; \mathbb{S}) + \lambda R(\mathbf{W}) \qquad (4.39)$$

where $J(\mathbf{W}, \mathbf{b}; \mathbb{S})$ is either $J_{\text{MSE}}(\mathbf{W}, \mathbf{b}; \mathbb{S})$ or $J_{\text{CE}}(\mathbf{W}, \mathbf{b}; \mathbb{S})$ that optimizes the empirical loss on the training set $\mathbb{S}$, $R(\mathbf{W})$ is either $R_1(\mathbf{W})$ or $R_2(\mathbf{W})$ described above, and λ is an interpolation weight, which is sometimes called regularization weight. Note that

$$\nabla_{\mathbf{W}_t^\ell} \ddot{J}(\mathbf{W}, \mathbf{b}; \mathbf{o}, \mathbf{y}) = \nabla_{\mathbf{W}_t^\ell} J(\mathbf{W}, \mathbf{b}; \mathbf{o}, \mathbf{y}) + \lambda \nabla_{\mathbf{W}_t^\ell} R(\mathbf{W}) \qquad (4.40)$$

$$\nabla_{\mathbf{b}_t^\ell} \ddot{J}(\mathbf{W}, \mathbf{b}; \mathbf{o}, \mathbf{y}) = \nabla_{\mathbf{b}_t^\ell} J(\mathbf{W}, \mathbf{b}; \mathbf{o}, \mathbf{y}) \qquad (4.41)$$

where

$$\nabla_{\mathbf{W}_t^\ell} R_1(\mathbf{W}) = \mathrm{sgn}\left(\mathbf{W}_t^\ell\right), \qquad (4.42)$$

and

$$\nabla_{\mathbf{W}_t^\ell} R_2(\mathbf{W}) = 2\mathbf{W}_t^\ell. \qquad (4.43)$$

Weight decay often helps when the training set size is small compared to the number of parameters in the DNN. Since there are typically over one million parameters in the weight matrices in typical DNNs used in the speech recognition tasks, the interpolation weight λ should be small (often in the range of 10^{-4}) or even 0 when the training set size is large.

4.3.4 Dropout

Weight decay is one way to control the overfitting. Another popular approach is dropout [10]. The basic idea of dropout is to randomly omit a certain percentage (e.g., α) of the neurons in each hidden layer for each presentation of the samples during training. This means during the training each random combination of the $(1 - \alpha)$ remaining hidden neurons needs to perform well even in the absence of the omitted neurons. This requires each neuron to depend less on other neurons to detect patterns.

Alternatively, dropout can be considered a technique that adds random noise to the training data. This is because each higher-layer neuron gets input from a *random* collection of the lower-layer neurons. The excitation received by each neuron is different even if the same input is fed into the DNN. With dropout, DNNs need to waste some of the weights to remove the effect of the random noise introduced. As such, dropout essentially reduces the capacity of the DNN, and thus can improve generalization of the resulting model.

When a hidden neuron is dropped out, its activation is set to 0 and so no error signal will pass through it. This means that other than the random dropout operation, no other changes to the training algorithm are needed to implement this feature. At the test time, however, instead of using a random combination of the neurons at each hidden layer, we use the average of all the possible combinations. This can be easily accomplished by discounting all the weights involved in dropout training by $(1 - \alpha)$, and then using the resulting model as a normal DNN (i.e., without dropout). Thus, dropout can also be interpreted as an efficient way of performing (geometric) model averaging (similar to bagging) in the DNN framework.

A slightly different implementation is to divide each activation by $(1-\alpha)$ before neurons are dropped in the training. By doing so the weights are automatically discounted by $(1-\alpha)$, and thus no weight compensation is needed when the model is used for testing. Another benefit of this approach is that we may apply different dropout rates at different epoch in the training. Our experience indicates that a dropout rate of 0.1–0.2 often helps to improve the recognition rate. An intelligently designed training schedule that uses a large dropout rate (e.g., 0.5) initially and reduces the dropout rate gradually can further improve the performance. This is because the model trained with a larger dropout rate can be considered as the seed model of that trained with a smaller dropout rate. Since the objective function associated with the larger dropout rate is smoother, it is less likely to trap into a very bad local optimum.

During the dropout training, we need to repeatedly sample a random subset of activations at each layer. This would slow down the training significantly. For this reason, the speed of the random number generation and sampling code is critical to reducing the training time. Alternatively, a fast dropout training algorithm proposed in [23] can be used. The key idea of this algorithm is to sample from or integrate a Gaussian approximation, instead of doing Monte Carlo sampling. This approximation, justified by the central limit theorem and empirical evidence, provides significant speedup and more stability. This technique can be extended to integrate out other types of noise and transformations.

4.3.5 Batch Size Selection

The parameter update formulas Eqs. (4.14) and (4.15) require the calculation of the empirical gradient estimated from a batch of training samples. The choice of the batch size will affect both the convergence speed and the resulting model.

The simplest and obvious choice of the batch is the whole training set. If our only goal is to minimize the empirical loss on the training set, the gradient estimated from the whole training set is the true gradient (i.e., the variance is zero). Even if our goal is to optimize the expected loss the gradient estimated from the whole training set still has smaller variance than that estimated from any subset of the training data. This approach, often referred to as batch training, has several advantages: First, the convergence property of the batch training is well-known. Second, many accelerating techniques such as conjugate gradient [9] and L-BFGS [16] work best in batch training. Third, batch training can be easily parallelized across computers. Unfortunately, batch training requires a complete pass through the entire dataset before model parameters are updated, and is thus not efficient in many large scale problems even though embarrassing parallelization is possible.

Alternatively, we can use the stochastic gradient descent (SGD) [2] technique, which is sometimes referred to as online leaning in the machine learning literature. SGD updates the model parameters based on the gradients estimated from a single training sample. If the sample is i.i.d. (independently and identically distributed), which is easy to guarantee if we draw the sample from the training set following the uniform distribution, we can show that

$$\mathbb{E}\left(\nabla J_t\left(\mathbf{W}, \mathbf{b}; \mathbf{o}, \mathbf{y}\right)\right) = \frac{1}{M} \sum_{m=1}^{M} \nabla J\left(\mathbf{W}, \mathbf{b}; \mathbf{o}^m, \mathbf{y}^m\right). \qquad (4.44)$$

In other words, the gradient estimated from the single sample is an unbiased estimation of the gradient on the whole training set. However, the variance of the estimation is

$$\mathbb{V}\left(\nabla J_t\left(\mathbf{W}, \mathbf{b}; \mathbf{o}, \mathbf{y}\right)\right) = \mathbb{E}\left[\left(\mathbf{x} - \mathbb{E}\left(\mathbf{x}\right)\right)\left(\mathbf{x} - \mathbb{E}\left(\mathbf{x}\right)\right)^{\mathrm{T}}\right]$$

$$= \mathbb{E}\left(\mathbf{x}\mathbf{x}^{\mathrm{T}}\right) - \mathbb{E}\left(\mathbf{x}\right)\mathbb{E}\left(\mathbf{x}\right)^{T}$$

$$= \frac{1}{M} \sum_{m=1}^{M} \mathbf{x}_m \mathbf{x}_m^{\mathrm{T}} - \mathbb{E}\left(\mathbf{x}\right)\mathbb{E}\left(\mathbf{x}\right)^{T}, \qquad (4.45)$$

which is nonzero unless all samples are identical; i.e., $\nabla J_t\left(\mathbf{W}, \mathbf{b}; \mathbf{o}, \mathbf{y}\right) = \mathbb{E}\left(\nabla J_t\left(\mathbf{W}, \mathbf{b}; \mathbf{o}, \mathbf{y}\right)\right)$, where for simplicity we have defined $\mathbf{x} \triangleq \nabla J_t\left(\mathbf{W}, \mathbf{b}; \mathbf{o}, \mathbf{y}\right)$. Because this estimate of the gradient is noisy, the model parameters may not move precisely down the gradient at each iteration. This seemingly drawback, however, is actually an important advantage of the SGD algorithm over the batch learning algorithm. This is because DNNs are highly nonlinear and nonconvex. The objective function contains many local optima, many of which are very poor. Batch learning will find the minimum of whatever basin the model parameters are initially in and results in a model that is highly dependent on the initial model. The SGD algorithm, however, due to the noisy gradient estimation, can jump out of the poor local optima and enter a better basin. This property is similar to the simulated annealing [13], which allows the model parameters to move in a direction that is inferior locally but superior globally.

SGD is also often much faster than batch learning especially on large datasets. This is due to two reasons. First, typically there are many similar, sometimes redundant, samples in the large datasets. Estimating the gradient by going through the whole dataset is thus waste of computation. Second, and more importantly, in the SGD training, one can make quick updates after seeing each sample. The new gradient is estimated based on the new model instead of the old model and so can move more quickly toward finding the best model.

The SGD algorithm, however, is difficult to parallelize even on the same computer. In addition, it cannot fully converge to the local minimum due to the noisy estimation of the gradient. Instead, it will fluctuate around the minimum. The size of the fluctuation depends on the learning rate and the amplitude of the gradient estimation variance. Although this fluctuation can sometimes reduce overfitting, it is not desirable in many cases.

A compromise between batch learning and the SGD algorithm is minibatch training, which estimates the gradient based on a small batch of randomly drawn training samples. It is easy to show that the gradient estimated from the minibatch is also unbiased and the variance of the estimation is smaller than that in the SGD algorithm. Minibatch training allows us to easily parallelize within the minibatch, and thus can converge faster than SGD. Since we prefer large gradient estimation variance in the early stage of the training to quickly jump out of poor local optima and smaller variance at the later stage to settle down to the minimum, we can choose smaller minibatch size initially and large ones in the later stage. In speech recognition tasks, we have found that a better model can be learned if we use 64–256 samples in early stages and 1,024–8,096 samples in later stages. An even smaller batch size is preferred at the very initial stage when a deeper network is to be trained. The minibatch size may be automatically determined based on the gradient estimation variance. Alternatively, the batch size can be determined by searching on a small subset of samples in each epoch [8, 20].

4.3.6 Sample Randomization

Sample randomization is irrelevant to the batch training since all samples are used to estimate the gradient. However, it is very important for SGD and minibatch training. This is because to get an unbiased estimate of the gradient the samples have to be IID. Heuristically, if successive samples are not randomly drawn from the training set (e.g., all belong to the same speaker), the model parameters will likely to move along the similar direction for too long.

If the whole training set can be loaded into the memory, sample randomization can be easily done by permuting an index array. Samples can then be drawn one by one according to the permuted index array. Since the index array is typically much smaller than the features, this would cost less than permuting the feature vectors themselves, especially if each data pass requires a different randomization order. This trick also guarantees that each sample will be presented to the training algorithm once for each data pass, and thus will not affect the data distribution. This property will guarantee that the model learned is consistent.

If the training set is huge, which is typically the case in speech recognition, we cannot load the whole training set into memory. In that case, we can load a large chunk (typically 24–48 h of speech or 8.6–17.2 M samples) of the data into memory each time using a rolling window and randomize inside the window. If the training data are from different sources (e.g., different languages), randomizing the utterance list files before feeding them into the DNN training tool also will help.

4.3.7 Momentum

It is well-known that the convergence speed can be improved if the model update is based on all the previous gradients (more global view) instead of only the current one (local view). Nesterov's accelerated gradient algorithm [18], which is proved to be optimal for the convex condition, is an example of this trick. In the DNN training, this is typically achieved with a simple technique named momentum. When momentum is applied, Eqs. (4.16) and (4.17) are replaced with

$$\triangle \mathbf{W}_t^\ell = \rho \triangle \mathbf{W}_{t-1}^\ell + (1 - \rho) \frac{1}{M_b} \sum_{m=1}^{M_b} \nabla_{\mathbf{W}_t^\ell} \ddot{J} \left(\mathbf{W}, \mathbf{b}; \mathbf{o}^m, \mathbf{y}^m \right) \tag{4.46}$$

and

$$\triangle \mathbf{b}_t^\ell = \rho \triangle \mathbf{b}_{t-1}^\ell + (1 - \rho) \frac{1}{M_b} \sum_{m=1}^{M_b} \nabla_{\mathbf{b}_t^\ell} \ddot{J} \left(\mathbf{W}, \mathbf{b}; \mathbf{o}^m, \mathbf{y}^m \right) \tag{4.47}$$

where ρ is the momentum factor, which typically takes value of 0.9–0.99 when SGD or minibatch training is used.[4] Momentum smoothes the parameter update and reduces the variance of the gradient estimation. Practically, it can reduce the oscillation problems commonly seen in the regular backpropagation algorithm when the error surface has a very narrow minimum, and thus speed up the training.

The above definition of momentum works great when the minibatch size is the same. However, sometimes we may want to use variable minibatch sizes. For example, we may want to use smaller minibatch initially, and then larger minibatch later as discussed in Sect. 4.3.5. In the sequence-discriminative training, which we will discuss in Chap. 8, each minibatch may have a different size since each utterance is of different length. Under these conditions, the above definition of momentum is no longer valid. Since the momentum can be considered as a finite impulse response (FIR) filter, we can define the momentum at the sample level as ρ_s and derive the momentum for different minibatch size M_b as

$$\rho = \exp \left(M_b \rho_s \right) . \tag{4.48}$$

4.3.8 Learning Rate and Stopping Criterion

One of the difficulties in training a DNN is selecting an appropriate learning strategy. It has been shown in theory that when the learning rate is set to

---

[4] In practice, we have found out that we may achieve slightly better result if we only use momentum after the first epoch.

$$\varepsilon = \frac{c}{t}, \tag{4.49}$$

where t is the number of samples presented and c is a constant, SGD converges asymptotically [2]. In practice, however, this annealing scheme converges very slowly since the learning rate will quickly become very small.

Note that it is the combination of the learning rate and the batch size that affects the learning behavior. As we have discussed in Sect. 4.3.5 that a smaller batch size should be used in the first several data passes and a larger batch size should be used in the later data passes. Since the batch size is variable, we can define a per-sample learning rate

$$\varepsilon_s = \frac{\varepsilon}{M_b}, \tag{4.50}$$

and change the model parameter update formulas to

$$\mathbf{W}^\ell_{t+1} \leftarrow \mathbf{W}^\ell_{\mathbf{t}} - \varepsilon_s \triangle \widetilde{\mathbf{W}}^\ell_t \tag{4.51}$$

$$\mathbf{b}^\ell_{t+1} \leftarrow \mathbf{b}^\ell_{\mathbf{t}} - \varepsilon_s \triangle \widetilde{\mathbf{b}}^\ell_t \tag{4.52}$$

where

$$\triangle \widetilde{\mathbf{W}}^\ell_t = \rho \triangle \mathbf{W}^\ell_{t-1} + (1 - \rho) \sum_{m=1}^{M_b} \nabla_{\mathbf{W}^\ell_t} \ddot{J}\left(\mathbf{W}, \mathbf{b}; \mathbf{o}^m, \mathbf{y}^m\right) \tag{4.53}$$

and

$$\triangle \widetilde{\mathbf{b}}^\ell_t = \rho \triangle \mathbf{b}^\ell_{t-1} + (1 - \rho) \sum_{m=1}^{M_b} \nabla_{\mathbf{b}^\ell_t} \ddot{J}\left(\mathbf{W}, \mathbf{b}; \mathbf{o}^m, \mathbf{y}^m\right). \tag{4.54}$$

This change also reduces one matrix division compared to using the original definition of the learning rate.

With this new update formulas, we can determine the learning strategy empirically. We first decide on a batch size and a large learning rate. We then run the training for hundreds of minibatchs, which typically takes several minutes on multi-core CPUs or GPUs. We monitor the training criterion on these minibatchs and reduce the batch size, learning rate, or both so that the product of $\varepsilon_s M_b$ is halved until the training criterion is obviously improved. We then divide this learning rate by two and use it as the initial learning rate. We run through a large subset of the training set and increase $\varepsilon_s M_b$ by four to eight folds. Note that since at this stage the model parameters are already adjusted to a relatively good location, increasing $\varepsilon_s M_b$ will not lead to diverge but improve the training speed. This procedure can be automated as in [8].

We found two strategies useful in determining the rest learning scheme. The first strategy is to double the batch size and reduce the learning rate by fourfolds whenever the training criterion fluctuates as measured on a large training subset or on a development set, and stop the training when the learning rate is smaller than a threshold or the preset number of data passes is reached. The second strategy is to reduce the learning rate to a very small number after the training criterion fluctuates and stop the training when the fluctuation happens again on either the training or the development set. For the speech recognition tasks in which the models are learned from scratch, we found that ε_s of $0.8e^{-4}$ and $0.3e^{-3}$ for deep and shallow networks, respectively, for the first phase, $1.25e^{-2}$ for the second phase, and $0.8e^{-6}$ for the third phase worked very well in practice. The hyper-parameter searching can also be automated using random search techniques [1] or Bayesian optimization techniques [22].

4.3.9 Network Architecture

Network architecture can be considered as another hyper "parameter" that needs to be determined. Since each layer can be considered as a feature extractor of the previous layer, the number of neurons at each layer should be large enough to capture the essential patterns. This is especially important at several lower layers since the first-layer features are more variable and it requires more neurons to model the patterns than other layers. However, if the layer size is too large, it is easier to overfit to the training data. Basically, the wide, shallow models are easier to overfit and the deep, narrow models are easier to underfit. Actually, if one of the layers is small (often called bottleneck) the performance will be significantly deteriorated especially when the bottleneck layer is close to the input layer. If each layer has the same number of neurons, adding more layers may also convert the model from overfitting to underfiting. This is because additional layers impose additional constraints to the model parameters. Given this observation, we can optimize the number of neurons at each layer on a one-hidden-layer network first. We then stack more layers with the same hidden layer size. In the speech recognition tasks, we have found that 5–7 layer DNNs with 1,000–3,000 neurons at each layer worked very well. It is often easier to find a good configuration on a wide and deep model than a narrow and shallow one. This is because there are many good local optima that perform similarly on a wide and deep model.

4.3.10 Reproducibility and Restartability

In the DNN training, the model parameters are initialized randomly and the training samples are also fed into the trainer in random orders. This inevitably will raise the concern about reproducibility of the results of the training. If our goal is to compare two algorithms or models, we can run the experiments multiple times, each with a new random seed, and report the average result and the standard deviation. Under many other conditions, however, we want to get exactly the same model and test result when we run the trainer twice. This can be achieved by always using the same random seed when generating the initial model and permuting the training samples.

When the training set is large, it is often desirable to stop the training in the middle and continue the training from the last check-point. Some mechanisms need to be built into the training tool to guarantee that restarting from a check-point will generate the exact same result if the training never interrupted. A simple trick is to save all the necessary information, including model parameters, current random number, parameter gradient, momentum, and so on in the check-point file. Another working approach that requires saving less data is to reset all learning parameters after each check-point.

References

1. Bengio, Y.: Practical recommendations for gradient-based training of deep architectures. In: Neural Networks: Tricks of the Trade, pp. 437–478. Springer (2012)
2. Bottou, L.: Online learning and stochastic approximations. On-line Learn. Neural Netw. **17**, 9 (1998)
3. Bryson, E.A., Ho, Y.C.: Applied Optimal Control: Optimization, Estimation, and Control. Blaisdell Publishing Company, US (1969)
4. Dahl, G.E., Yu, D., Deng, L., Acero, A.: Large vocabulary continuous speech recognition with context-dependent DBN-HMMs. In: Proceedings of International Conference on Acoustics, Speech and Signal Processing (ICASSP), pp. 4688–4691 (2011)
5. Dahl, G.E., Yu, D., Deng, L., Acero, A.: Context-dependent pre-trained deep neural networks for large-vocabulary speech recognition. IEEE Trans. Audio, Speech Lang. Process. **20**(1), 30–42 (2012)
6. Duchi, J., Hazan, E., Singer, Y.: Adaptive subgradient methods for online learning and stochastic optimization. J. Mach. Learn. Res. (JMLR) 2121–2159 (2011)
7. Glorot, X., Bordes, A., Bengio, Y.: Deep sparse rectifier networks, pp. 315–323 (2011)
8. Guenter, B., Yu, D., Eversole, A., Kuchaiev, O., Seltzer, M.L.: "Stochastic gradient descent algorithm in the computational network toolkit", OPT2013: NIPS 2013 Workshop on Optimization for Machine Learning (2013)
9. Hestenes, M.R., Stiefel, E.: Methods of conjugate gradients for solving linear systems (1952)
10. Hinton, G.E., Srivastava, N., Krizhevsky, A., Sutskever, I., Salakhutdinov, R.R.: Improving neural networks by preventing co-adaptation of feature detectors. arXiv preprint arXiv:1207.0580 (2012)
11. Hinton, G., Deng, L., Yu, D., Dahl, G.E., Mohamed, A.R., Jaitly, N., Senior, A., Vanhoucke, V., Nguyen, P., Sainath, T.N., et al.: Deep neural networks for acoustic modeling in speech recognition: the shared views of four research groups. IEEE Signal Process. Mag. **29**(6), 82–97 (2012)
12. Hornik, K., Stinchcombe, M., White, H.: Multilayer feedforward networks are universal approximators. Neural Netw. **2**(5), 359–366 (1989)
13. Kirkpatrick, S., Gelatt Jr, D., Vecchi, M.P.: Optimization by simmulated annealing. Science **220**(4598), 671–680 (1983)
14. LeCun, Y., Bottou, L., Orr, G.B., Müller, K.R.: Efficient backprop. Neural Networks: Tricks of The Trade, pp. 9–50. Springer, Berlin (1998)
15. Liu, F.H., Stern, R.M., Huang, X., Acero, A.: Efficient cepstral normalization for robust speech recognition. In: Proceedings of ACL Workshop on Human Language Technologies (ACL-HLT), pp. 69–74 (1993)

16. Liu, D.C., Nocedal, J.: On the limited memory BFGS method for large scale optimization. Math. Program. **45**(1–3), 503–528 (1989)
17. Mohamed, A., Dahl, G.E., Hinton, G.E.: Deep belief networks for phone recognition. In: NIPS Workshop on Deep Learning for Speech Recognition and Related Applications (2009)
18. Nesterov, Y.: A method of solving a convex programming problem with convergence rate O (1/k2). Sov. Math. Dokl. **27**, 372–376 (1983)
19. Rumelhart, D.E., Hintont, G.E., Williams, R.J.: Learning representations by back-propagating errors. Nature **323**(6088), 533–536 (1986)
20. Seide, F., Fu, H., Droppo, J., Li, G., Yu, D.: On parallelizability of stochastic gradient descent for speech dnns. In: Proceedings of International Conference on Acoustics, Speech and Signal Processing (ICASSP) (2014)
21. Seide, F., Li, G., Yu, D.: Conversational speech transcription using context-dependent deep neural networks. In: Proceedings of Annual Conference of International Speech Communication Association (INTERSPEECH), pp. 437–440 (2011)
22. Snoek, J., Larochelle, H., Adams, R.P.: Practical Bayesian optimization of machine learning algorithms. arXiv preprint arXiv:1206.2944 (2012)
23. Wang, S., Manning, C.: Fast dropout training. In: Proceedings of the 30th International Conference on Machine Learning (ICML-13), pp. 118–126 (2013)
24. Yu, D., Deng, L., Dahl, G.: Roles of pre-training and fine-tuning in context-dependent DBN-HMMs for real-world speech recognition. In: Proceedings of Neural Information Processing Systems (NIPS) Workshop on Deep Learning and Unsupervised Feature Learning (2010)

Chapter 5
Advanced Model Initialization Techniques

Abstract In this chapter, we introduce several advanced deep neural network (DNN) model initialization or pretraining techniques. These techniques have played important roles in the early days of deep learning research and continue to be useful under many conditions. We focus our presentation of pretraining DNNs on the following topics: the restricted Boltzmann machine (RBM), which by itself is an interesting generative model, the deep belief network (DBN), the denoising autoencoder, and the discriminative pretraining.

5.1 Restricted Boltzmann Machines

The restricted Boltzmann machine (RBM) [20] is a stochastic generative neural network. As its name indicates, it is a variant of the Boltzmann machine. It is essentially an undirected graphical model constructed of a layer of stochastic visible neurons and a layer of stochastic hidden neurons. The visible and hidden neurons form a bipartite graph with no visible–visible or hidden–hidden connections as shown in Fig. 5.1. The hidden neurons usually take binary values and follow Bernoulli distributions. The visible neurons may take binary or real values depending on the input types.

An RBM assigns an energy to every configuration of visible vector $\mathbf{v}$ and hidden vector $\mathbf{h}$. For the Bernoulli-Bernoulli RBM, in which $\mathbf{v} \in \{0, 1\}^{N_v \times 1}$ and $\mathbf{h} \in \{0, 1\}^{N_h \times 1}$, the energy is

$$E(\mathbf{v}, \mathbf{h}) = -\mathbf{a}^{\mathrm{T}}\mathbf{v} - \mathbf{b}^{\mathrm{T}}\mathbf{h} - \mathbf{h}^{\mathrm{T}}\mathbf{W}\mathbf{v}, \tag{5.1}$$

where N_v and N_h are the number of visible and hidden neurons, respectively, $\mathbf{W} \in \mathbb{R}^{N_h \times N_v}$ is the weight matrix connecting visible and hidden neurons, and $\mathbf{a} \in \mathbb{R}^{N_v \times 1}$ and $\mathbf{b} \in \mathbb{R}^{N_h \times 1}$ are, respectively, the visible and hidden layer bias vectors. If the visible neurons take real values, $\mathbf{v} \in \mathbb{R}^{N_v \times 1}$, the RBM, which is often called Gaussian-Bernoulli RBM, assigns the energy

$$E(\mathbf{v}, \mathbf{h}) = \frac{1}{2}(\mathbf{v} - \mathbf{a})^{\mathrm{T}}(\mathbf{v} - \mathbf{a}) - \mathbf{b}^{\mathrm{T}}\mathbf{h} - \mathbf{h}^{\mathrm{T}}\mathbf{W}\mathbf{v} \tag{5.2}$$

to each configuration $(\mathbf{v}, \mathbf{h})$.

© Springer-Verlag London 2015
D. Yu and L. Deng, *Automatic Speech Recognition*,
Signals and Communication Technology, DOI 10.1007/978-1-4471-5779-3_5

Fig. 5.1 An example of
restricted Boltzmann
machines

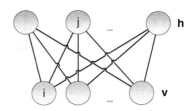

Each configuration is also associated with a probability

$$P(\mathbf{v}, \mathbf{h}) = \frac{e^{-E(\mathbf{v}, \mathbf{h})}}{Z} \tag{5.3}$$

that is defined upon the energy, where $Z = \sum_{\mathbf{v}, \mathbf{h}} e^{-E(\mathbf{v}, \mathbf{h})}$ is the normalization factor known as the partition function.

In the RBM, the posterior probabilities $P(\mathbf{v}|\mathbf{h})$ and $P(\mathbf{h}|\mathbf{v})$ can be efficiently calculated thanks to the lack of direct connections within visible and hidden layers. For example, for the Bernoulli-Bernoulli RBM,

$$
\begin{aligned}
P(\mathbf{h}|\mathbf{v}) &= \frac{e^{-E(\mathbf{v}, \mathbf{h})}}{\sum_{\tilde{\mathbf{h}}} e^{-E(\mathbf{v}, \tilde{\mathbf{h}})}} \\
&= \frac{e^{\mathbf{a}^{\mathsf{T}}\mathbf{v} + \mathbf{b}^{\mathsf{T}}\mathbf{h} + \mathbf{h}^{\mathsf{T}}\mathbf{W}\mathbf{v}}}{\sum_{\tilde{\mathbf{h}}} e^{\mathbf{a}^{\mathsf{T}}\mathbf{v} + \mathbf{b}^{\mathsf{T}}\tilde{\mathbf{h}} + \tilde{\mathbf{h}}^{\mathsf{T}}\mathbf{W}\mathbf{v}}} \\
&= \frac{\prod_i e^{b_i h_i + h_i \mathbf{W}_{i,*}\mathbf{v}}}{\sum_{\tilde{h}_1} \cdots \sum_{\tilde{h}_N} \prod_i e^{b_i \tilde{h}_i + \tilde{h}_i \mathbf{W}_{i,*}v}} \\
&= \frac{\prod_i e^{b_i h_i + h_i \mathbf{W}_{i,*}\mathbf{v}}}{\prod_i \sum_{\tilde{h}_i} e^{b_i \tilde{h}_i + \tilde{h}_i \mathbf{W}_{i,*}v}} \\
&= \prod_i \frac{e^{b_i h_i + h_i \mathbf{W}_{i,*}\mathbf{v}}}{\sum_{\tilde{h}_i} e^{b_i \tilde{h}_i + \tilde{h}_i \mathbf{W}_{i,*}v}} \\
&= \prod_i P(h_i|\mathbf{v}), \tag{5.4}
\end{aligned}
$$

where $\mathbf{W}_{i,*}$ denotes the ith row of $\mathbf{W}$. Equation (5.4) indicates that the hidden neurons are conditionally independent given the visible vector. Since $h_i \in \{0, 1\}$ takes binary values,

$$P(h_i = 1|\mathbf{v}) = \frac{e^{b_i 1 + 1\mathbf{W}_{i,*}\mathbf{v}}}{e^{b_i 1 + 1\mathbf{W}_{i,*}\mathbf{v}} + e^{b_i 0 + 0\mathbf{W}_{i,*}\mathbf{v}}} = \sigma(b_i + \mathbf{W}_{i,*}v) \tag{5.5}$$

or

$$P(\mathbf{h} = \mathbf{1}|\mathbf{v}) = \sigma(\mathbf{W}\mathbf{v} + \mathbf{b}), \tag{5.6}$$

where $\sigma(x) = (1 + e^{-x})^{-1}$ is the element-wise logistic sigmoid function. For the binary visible neuron case, a completely symmetric derivation lets us obtain

$$P(\mathbf{v} = \mathbf{1}|\mathbf{h}) = \sigma(\mathbf{W}^{\mathrm{T}}\mathbf{h} + \mathbf{a}). \tag{5.7}$$

For the Gaussian visible neurons, the conditional probability $P(\mathbf{h} = \mathbf{1}|\mathbf{v})$ is the same as Eq. (5.6); however, $P(\mathbf{v}|\mathbf{h})$ is estimated as

$$P(\mathbf{v}|\mathbf{h}) = \mathcal{N}(\mathbf{v}; \mathbf{W}^{\mathrm{T}}\mathbf{h} + \mathbf{a}, I). \tag{5.8}$$

where I is the appropriate identity covariance matrix.

Note that Eq. (5.6) has the same form as Eq. (4.1) no matter whether binary or real-valued inputs are used. This allows us to use the weights of an RBM to initialize a feedforward neural network with sigmoidal hidden units because we can equate the inference for RBM hidden units with forward computation in a deep neural network (DNN).

5.1.1 Properties of RBMs

An RBM can be used to learn a probability distribution over its set of inputs. Before writing an expression for the probability assigned by an RBM to some visible vector $\mathbf{v}$, it is convenient to define a quantity known as the free energy:

$$F(\mathbf{v}) = -\log\left(\sum_{\mathbf{h}} e^{-E(\mathbf{v},\mathbf{h})}\right). \tag{5.9}$$

Using $F(\mathbf{v})$, we can write the marginal probability $P(\mathbf{v})$ as

$$\begin{aligned}
P(\mathbf{v}) &= \sum_{\mathbf{h}} P(\mathbf{v}, \mathbf{h}) \\
&= \sum_{\mathbf{h}} \frac{e^{-E(\mathbf{v},\mathbf{h})}}{Z} \\
&= \frac{\sum_{\mathbf{h}} e^{-E(\mathbf{v},\mathbf{h})}}{Z} \\
&= \frac{e^{-F(\mathbf{v})}}{\sum_{v} e^{-F(v)}}.
\end{aligned} \tag{5.10}$$

If the visible neurons take real values, the marginal probability density function is

$$p_0(\mathbf{v}) = \frac{e^{-\frac{1}{2}(\mathbf{v}-\mathbf{a})^\mathrm{T}(\mathbf{v}-\mathbf{a})}}{Z_0} \tag{5.11}$$

when the RBM has no hidden neuron. This is a unit-variance Gaussian distribution centered at vector **a** as shown in Fig. 5.2a. Note that

$$
\begin{aligned}
p_n(\mathbf{v}) &= \frac{\sum_\mathbf{h} e^{-E_n(\mathbf{v},\mathbf{h})}}{Z_n} \\
&= \frac{\prod_{i=1}^n \sum_{h_i=0}^1 e^{b_i h_i + h_i \mathbf{W}_{i,*}\mathbf{v}}}{Z_n} \\
&= \frac{\prod_{i=1}^{n-1} \sum_{h_i=0}^1 e^{b_i h_i + h_i \mathbf{W}_{i,*}\mathbf{v}} \left(1 + e^{b_n + \mathbf{W}_{n,*}\mathbf{v}}\right)}{Z_n} \\
&= p_{n-1}(\mathbf{v}) \frac{Z_{n-1}}{Z_n} \left(1 + e^{b_n + \mathbf{W}_{n,*}\mathbf{v}}\right) \\
&= p_{n-1}(\mathbf{v}) \frac{Z_{n-1}}{Z_n} + P_{n-1}(\mathbf{v}) \frac{Z_{n-1}}{Z_n} e^{b_n + \mathbf{W}_{n,*}\mathbf{v}}, \tag{5.12}
\end{aligned}
$$

where n is the number of hidden neurons. This means that when a new hidden neuron is added while other model parameters are fixed, the original distribution is scaled and a copy of the same distribution is placed along the direction determined by $\mathbf{W}_{n,*}$. Figure 5.2b–d shows the marginal probability density distribution with 1–3 hidden neurons. It is obvious that RBMs represent the visible inputs as a Gaussian mixture model with exponential number of unit-variance Gaussian components. Compared to the conventional Gaussian mixture models (GMMs) RBMs use much more mixture components. However, the conventional GMMs use different variances for different

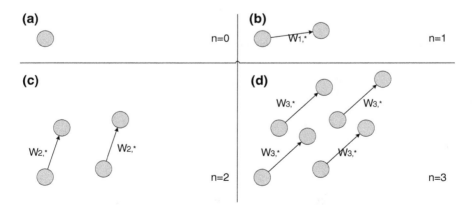

Fig. 5.2 Marginal probability density function represented by a Gaussian-Bernoulli RBM

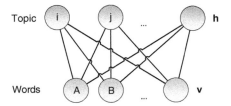

Fig. 5.3 Use RBMs to learn word co-occurrence relationship. Each visible neuron represents a *word* and takes value 1 if the word occurred in the document and takes value 0 otherwise. After the training, each hidden neuron represents a *topic*

Gaussian components to represent the distribution. Since Gaussian-Bernoulli RBMs can represent distributions of real-valued data similar to GMMs, RBMs can be used in generative models in place of GMMs. For example, RBMs have been successfully used in some recent text-to-speech (TTS) systems [14].

Given the training samples that are represented with visible neurons, the RBM can learn the correlation between different dimensions of the features. For example, if the visible neurons represent the words occur in a document as shown in Fig. 5.3, after the training, the hidden neurons will represent topics. Each hidden neuron groups words that occur together in the documents while each word links to the topics in which it may occur. The intervisible neuron correlation can thus be represented through the hidden neurons they connect to. An advanced topic model that exploits this property can be found in [11]. Since the hidden layer can be considered as a different representation of the raw feature, which is also represented by the visible layer, the RBM can also be used to learn feature representations [3].

Although the neurons are separated into visible and hidden ones in our description, the visible neurons may not be observable and the hidden neurons may be observable in some applications. For example, if the visible layer represents the users and the hidden layer represents the movies, the link between the visible neuron A and hidden neuron i may represent "user A likes movie i." In many applications, such as in the collaborative filtering [16], some pairs are observable but others are not. The RBM learned from the training set can be used to predict the pairs that are not observable and can recommend movies to the users.

5.1.2 RBM Parameter Learning

To train an RBM, we perform stochastic gradient descent (SGD) [2] to minimize the negative log likelihood (NLL)

$$J_{\text{NLL}}\left(\mathbf{W}, \mathbf{a}, \mathbf{b}; \mathbf{v}\right) = -\log P(\mathbf{v}) = F(\mathbf{v}) + \log \sum_{v} e^{-F(v)} \qquad (5.13)$$

and update the parameters as

$$\mathbf{W}_{t+1} \leftarrow \mathbf{W}_t - \varepsilon \triangle \mathbf{W}_t, \tag{5.14}$$

$$\mathbf{a}_{t+1} \leftarrow \mathbf{a}_t - \varepsilon \triangle \mathbf{a}_t, \tag{5.15}$$

$$\mathbf{b}_{t+1} \leftarrow \mathbf{b}_t - \varepsilon \triangle \mathbf{b}_t, \tag{5.16}$$

where ε is the learning rate, and

$$\triangle \mathbf{W}_t = \rho \triangle \mathbf{W}_{t-1} + (1-\rho)\frac{1}{M_b}\sum_{m=1}^{M_b} \nabla_{\mathbf{W}_t} J_{\mathrm{NLL}}\left(\mathbf{W}, \mathbf{a}, \mathbf{b}; \mathbf{v}^m\right), \tag{5.17}$$

$$\triangle \mathbf{a}_t = \rho \triangle \mathbf{a}_{t-1} + (1-\rho)\frac{1}{M_b}\sum_{m=1}^{M_b} \nabla_{\mathbf{a}_t} J_{\mathrm{NLL}}\left(\mathbf{W}, \mathbf{a}, \mathbf{b}; \mathbf{v}^m\right), \tag{5.18}$$

$$\triangle \mathbf{b}_t = \rho \triangle \mathbf{b}_{t-1} + (1-\rho)\frac{1}{M_b}\sum_{m=1}^{M_b} \nabla_{\mathbf{b}_t} J_{\mathrm{NLL}}\left(\mathbf{W}, \mathbf{a}, \mathbf{b}; \mathbf{v}^m\right), \tag{5.19}$$

where ρ is the momentum parameter, M_b is the minibatch size, and $\nabla_{\mathbf{W}_t} J_{\mathrm{NLL}}$ $(\mathbf{W}, \mathbf{a}, \mathbf{b}; \mathbf{v}^m)$, $\nabla_{\mathbf{a}_t} J_{\mathrm{NLL}}(\mathbf{W}, \mathbf{a}, \mathbf{b}; \mathbf{v}^m)$, and $\nabla_{\mathbf{b}_t} J_{\mathrm{NLL}}(\mathbf{W}, \mathbf{a}, \mathbf{b}; \mathbf{v}^m)$ are the gradients of the NLL criterion on model parameters $\mathbf{W}$, $\mathbf{a}$, and $\mathbf{b}$, respectively.

Unlike DNN, in an RBM the gradient of the log likelihood of the data is not feasible to compute exactly. The general form of the derivative of the NLL with regard to model parameters is

$$\nabla_\theta J_{\mathrm{NLL}}\left(\mathbf{W}, \mathbf{a}, \mathbf{b}; \mathbf{v}\right) = -\left[\left\langle\frac{\partial E(\mathbf{v}, \mathbf{h})}{\partial \theta}\right\rangle_{\mathrm{data}} - \left\langle\frac{\partial E(\mathbf{v}, \mathbf{h})}{\partial \theta}\right\rangle_{\mathrm{model}}\right], \tag{5.20}$$

where θ is some model parameter, and $\langle x\rangle_{\mathrm{data}}$ and $\langle x\rangle_{\mathrm{model}}$ are the expectation of x estimated from the data and from the model, respectively. In particular, for the visible-hidden weight updates we have:

$$\nabla_{w_{ji}} J_{\mathrm{NLL}}\left(\mathbf{W}, \mathbf{a}, \mathbf{b}; \mathbf{v}\right) = -\left[\langle v_i h_j\rangle_{\mathrm{data}} - \langle v_i h_j\rangle_{\mathrm{model}}\right]. \tag{5.21}$$

The first expectation, $\langle v_i h_j\rangle_{\mathrm{data}}$, is the frequency with which the visible neuron v_i and the hidden neuron h_j fire together in the training set and $\langle v_i h_j\rangle_{\mathrm{model}}$ is that same expectation under the distribution defined by the model. Unfortunately, the term $\langle.\rangle_{\mathrm{model}}$ takes exponential time to compute exactly when the hidden values are unknown, so we are forced to use approximated methods.

The most widely used efficient approximated learning algorithm for the RBM training is contrastive divergence (CD) as described in [9]. The one-step contrastive

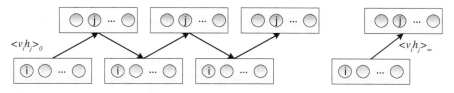

Fig. 5.4 Illustration of the contrastive divergence algorithm

divergence approximation for the gradient with regard to the visible-hidden weights is

$$\nabla_{w_{ji}} J_{\text{NLL}} \left(\mathbf{W}, \mathbf{a}, \mathbf{b}; \mathbf{v} \right) = - \left[\langle v_i h_j \rangle_{\text{data}} - \langle v_i h_j \rangle_\infty \right]$$
$$\approx - \left[\langle v_i h_j \rangle_{\text{data}} - \langle v_i h_j \rangle_1 \right], \qquad (5.22)$$

where $\langle . \rangle_\infty$ and $\langle . \rangle_1$ denote the expectation computed with samples generated by running the Gibbs sampler infinite steps and one step, respectively. The computation of $\langle v_i h_j \rangle_{\text{data}}$ and $\langle v_i h_j \rangle_1$ is often referred as the positive and negative phases, respectively.

Figure 5.4 illustrates the sampling process and the contrastive divergence algorithm. At the first step, the Gibbs sampler is initialized at a data sample. It then generates a hidden sample from the visible sample based on the posterior probability $P(\mathbf{h}|\mathbf{v})$ defined by Eq. (5.6). This hidden sample is further used to generate a visible sample based on the posterior probability $P(\mathbf{v}|\mathbf{h})$ defined by Eq. (5.7) for the Bernoulli-Bernoulli RBMs or Eq. (5.8) for the Gaussian-Bernoulli RBMs. This process continues and may run many steps. If the Gibbs sampler runs for infinite steps the true expectation $\langle v_i h_j \rangle_{\text{model}}$ can be estimated from the samples generated after the burn-in stage as

$$\langle v_i h_j \rangle_{\text{model}} \approx \frac{1}{N} \sum_{n=N_{\text{burn}}+1}^{N_{\text{burn}}+N} v_i^n h_j^n, \qquad (5.23)$$

where N_{burn} is the number of steps needed for burn-in and N is the number (huge) of samples generated after the burn-in stage. However, running Gibbs sampler many steps is not efficient. Instead, we can replace it with a very rough estimate $\langle v_i h_j \rangle_1$ by running the Gibbs sampler just one step and estimate $\langle v_i h_j \rangle_{\text{model}}$ as

$$\langle v_i h_j \rangle_{\text{model}} \approx \langle v_i h_j \rangle_1 = v_i^1 h_j^1. \qquad (5.24)$$

However, $\langle v_i h_j \rangle_1$ has high variance. To reduce the variance, we can estimate $\langle v_i h_j \rangle_{\text{model}}$ based on

$$\mathbf{h}^0 \sim P(\mathbf{h}|\mathbf{v}^0), \qquad (5.25)$$

$$\mathbf{v}^1 = \mathbb{E}(\mathbf{v}|\mathbf{h}^0) = P(\mathbf{v}|\mathbf{h}^0), \qquad (5.26)$$

$$\mathbf{h}^1 = \mathbb{E}(\mathbf{h}|\mathbf{v}^1) = P(\mathbf{h}|\mathbf{v}^1), \tag{5.27}$$

where $\sim$ means *sample from*, $\mathbf{v}_0$ is a sample from the training set, and the expectation operator is applied element-wise. Instead of sampling from $P(\mathbf{v}|\mathbf{h}^0)$ and $P(\mathbf{h}|\mathbf{v}^1)$, as in the naive solution, we use the mean-field approximation to generate the samples $\mathbf{v}^1$ and $\mathbf{h}^1$. In other words, these samples now take real values instead of binary values. The same trick can be applied to

$$\langle v_i h_j \rangle_{\text{data}} \approx \langle v_i h_j \rangle_0 = v_i^0 \mathbb{E}_j(\mathbf{h}|\mathbf{v}^0) = v_i^0 P_j(\mathbf{h}|\mathbf{v}^0). \tag{5.28}$$

If N (which typically is a small value) steps of contrastive divergence is used, the expected value is used whenever the visible vector is to be generated, and the sampling technique can be used whenever the hidden vector is needed, except the last step, in which the expected vector is used.

Similar update rules for the model parameters $\mathbf{a}$ and $\mathbf{b}$ in the Bernoulli-Bernoulli RBM can be derived by simply replacing $\frac{\partial E(\mathbf{v},\mathbf{h})}{\partial \theta}$ in Eq. (5.20) with the appropriate gradients. The complete gradient estimation in the matrix format is

$$\nabla_{\mathbf{W}} J_{\text{NLL}}(\mathbf{W}, \mathbf{a}, \mathbf{b}; \mathbf{v}) = -\left[\langle \mathbf{hv}^{\text{T}} \rangle_{\text{data}} - \langle \mathbf{hv}^{\text{T}} \rangle_{\text{model}}\right], \tag{5.29}$$

$$\nabla_{\mathbf{a}} J_{\text{NLL}}(\mathbf{W}, \mathbf{a}, \mathbf{b}; \mathbf{v}) = -\left[\langle \mathbf{v} \rangle_{\text{data}} - \langle \mathbf{v} \rangle_{\text{model}}\right], \tag{5.30}$$

$$\nabla_{\mathbf{b}} J_{\text{NLL}}(\mathbf{W}, \mathbf{a}, \mathbf{b}; \mathbf{v}) = -\left[\langle \mathbf{h} \rangle_{\text{data}} - \langle \mathbf{h} \rangle_{\text{model}}\right]. \tag{5.31}$$

The CD algorithm can also be applied to train the Gaussian-Bernoulli RBM. The only difference is that in the Gaussian-Bernoulli RBM, we use Eq. (5.8) to estimate the expected value of the posterior distribution $\mathbb{E}(\mathbf{v}|\mathbf{h})$. Algorithm 5.1 summarizes the keys steps in training RBMs using the contrastive divergence algorithm.

Similar to the DNN training, effective RBM training also requires many practical considerations. Many of the discussions in Sect. 4.3 can be applied to the RBM training. A comprehensive practical guide in training RBMs can be found in [7].

5.2 Deep Belief Network Pretraining

An RBM can be considered as a generative model with infinite number of layers, in which all the layers share the same weight matrix as shown in Fig. 5.5a, b. If we separate the bottom layer from the deep generative model described in Fig. 5.5b, the rest layers form another generative model, which also has infinite number of layers with shared weight matrices. These rest layers equal to another RBM whose visible layer and hidden layer is switched as shown in Fig. 5.5c. This model is a special generative model named deep belief network (DBN), in which the top layer is an undirected RBM, and the bottom layers form a directed generative model. We apply the same reasoning to Fig. 5.5c and can see that it equals to the DBN illustrated in Fig. 5.5d.

Algorithm 5.1 The contrastive divergence algorithm for training RBMs.

1: **procedure** TRAINRBMWITHCD($\mathbb{S} = \{\mathbf{o}^m | 0 \leq m < M\}, N$)
                                    ▷ $\mathbb{S}$ is the training set with M samples, N is the CD steps
2:     Randomly initialize $\{\mathbf{W}_0, \mathbf{a}_0, \mathbf{b}_0\}$
3:     **while** Stopping Criterion Not Met **do**
                     ▷ Stop if reached max iterations or the training criterion improvement is small
4:         Randomly select a minibatch $\mathbf{O}$ with M_b samples.
5:         $\mathbf{V}^0 \leftarrow \mathbf{O}$ ▷ Positive phase
6:         $\mathbf{H}^0 \leftarrow P(\mathbf{H}|\mathbf{V}^0)$ ▷ Applied column-wise
7:         $\nabla_{\mathbf{W}} J \leftarrow \mathbf{H}^0 \left(\mathbf{V}^0\right)^{\mathrm{T}}$
8:         $\nabla_{\mathbf{a}} J \leftarrow$ sumrow $\left(\mathbf{V}^0\right)$ ▷ Sum along rows
9:         $\nabla_{\mathbf{b}} J \leftarrow$ sumrow $\left(\mathbf{H}^0\right)$
10:       **for** $n \leftarrow 0; n < N; n \leftarrow n+1$ **do**   ▷ Negative phase
11:             $\mathbf{H}^n \leftarrow \mathbb{I}\left(\mathbf{H}^n > \text{rand}(0, 1)\right)$ ▷ Sampling, $\mathbb{I}(\bullet)$ is the indicator function
12:             $\mathbf{V}^{n+1} \leftarrow P(\mathbf{V}|\mathbf{H}^n)$
13:             $\mathbf{H}^{n+1} \leftarrow P(\mathbf{H}|\mathbf{V}^{n+1})$
14:       **end for**
15:       $\nabla_{\mathbf{W}} J \leftarrow \nabla_{\mathbf{W}} J - \mathbf{H}^N \left(\mathbf{V}^N\right)^{\mathrm{T}}$ ▷ Subtract negative statistics
16:       $\nabla_{\mathbf{a}} J \leftarrow \nabla_{\mathbf{a}} J -$ sumrow $\left(\mathbf{V}^0\right)$
17:       $\nabla_{\mathbf{b}} J \leftarrow \nabla_{\mathbf{b}} J -$ sumrow $\left(\mathbf{H}^0\right)$
18:       $\mathbf{W}_{t+1} \leftarrow \mathbf{W}_t + \frac{\varepsilon}{M_b} \triangle \mathbf{W}_t$ ▷ Update $\mathbf{W}$
19:       $\mathbf{a}_{t+1} \leftarrow \mathbf{a}_t + \frac{\varepsilon}{M_b} \triangle \mathbf{a}_t$ ▷ Update $\mathbf{a}$
20:       $\mathbf{b}_{t+1} \leftarrow \mathbf{b}_t + \frac{\varepsilon}{M_b} \triangle \mathbf{b}_t$ ▷ Update $\mathbf{b}$
21:     **end while**
22:     Return $rbm = \{\mathbf{W}, \mathbf{a}, \mathbf{b}\}$
23: **end procedure**

    The relationship between the RBM and DBN suggests a layer-wise procedure to train very deep generative models [8]. Once we have trained an RBM, we cause the RBM to re-represent the data. For each data vector, $\mathbf{v}$, we compute a vector of expected hidden neuron activations (which equal to the probabilities) $\mathbf{h}$. We use these hidden expectations as training data for a new RBM. Thus each set of RBM weights can be used to extract features from the output of the previous layer. Once we stop training RBMs, we have the initial values for all the weights of the hidden layers of a DBN with a number of hidden layers equal to the number of RBMs we trained. The DBN can be further fine-tuned using the algorithms such as the wake-sleep algorithm [10].

    In the above procedure, we have assumed that the dimensions in the RBMs are fixed. In this setup, the DBN would perform exactly as the RBM if the RBM is perfectly trained. However, this assumption is not necessary and we can stack RBMs with different dimensions. This allows for flexibility in the DBN architecture, and stacking additional layers can potentially improve the upper bound of the likelihood.

    The DBN weights can be used as the initial weights in the sigmoidal DNN. This is because the conditional probability $P(\mathbf{h}|\mathbf{v})$ in the RBM has the same form as that in the DNN if the sigmoid nonlinear activation function is used. The DNN described in Chap. 4 can be viewed as a statistical graphical model, in which each hidden

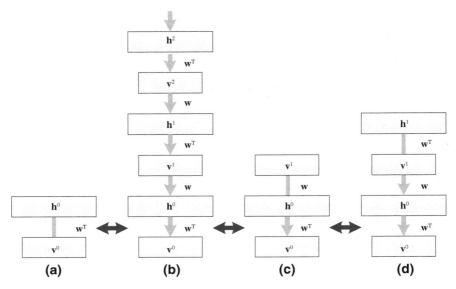

Fig. 5.5 An RBM (**a**) is equivalent to a generative model with infinite number of layers (**b**), in which all the layers share the same weight matrix. By replacing top layers with RBMs, (**b**) equals to the DBNs (**c**) and (**d**)

layer $0 < \ell < L$ models posterior probabilities of conditionally independent hidden binary neurons $\mathbf{h}^\ell$ given input vectors $\mathbf{v}^{\ell-1}$ as Bernoulli distribution

$$P\left(\mathbf{h}^\ell | \mathbf{v}^{\ell-1}\right) = \sigma\left(\mathbf{z}^\ell\right) = \sigma\left(\mathbf{W}^\ell \mathbf{v}^{\ell-1} + \mathbf{b}^\ell\right), \tag{5.32}$$

and the output layer approximates the label $\mathbf{y}$ conditioned on the input as a multinomial probability distribution as

$$P\left(\mathbf{y} | \mathbf{v}^{L-1}\right) = \text{softmax}\left(\mathbf{z}^L\right) = \text{softmax}\left(\mathbf{W}^L \mathbf{v}^{L-1} + \mathbf{b}^L\right). \tag{5.33}$$

Given the observed feature $\mathbf{o}$ and the label $\mathbf{y}$, the precise modeling of $P(\mathbf{y}|\mathbf{o})$ requires integration over all possible values of $\mathbf{h}$ across all layers which is infeasible. An effective practical trick is to replace the marginalization with the mean-field approximation [17]. In other words, we define

$$\mathbf{v}^\ell = \mathbb{E}(\mathbf{h}^\ell | \mathbf{v}^{\ell-1}) = P\left(\mathbf{h}^\ell | \mathbf{v}^{\ell-1}\right) = \sigma\left(\mathbf{W}^\ell \mathbf{v}^{\ell-1} + \mathbf{b}^\ell\right), \tag{5.34}$$

and we get the conventional nonstochastic description of the DNN discussed in Chap. 4.

Based on this view of the sigmoidal DNN, we can see that the DBN weights can be used as the initial weights of the DNN. The only difference between DBN and

DNN is that in the DNN we have labels. As such, in the DNN, when the pretraining finishes, we add a randomly initialized softmax output layer and use backpropagation to fine-tune all the weights in the network discriminatively.

Initializing DNN weights with generative pretraining may potentially improve the performance of the DNN on the testing set. This is due to three reasons. First, the DNN is highly nonlinear and non-convex. The initialization point may greatly affect the final model especially if the batch mode training algorithm is used. Second, the generative criterion used in the pretraining stage is different from the discriminative criterion used in the backpropagation phase. Starting the BP training from the generatively pretrained model thus implicitly regularizes the model. Third, since only the supervised fine-tuning phase requires labeled data, we can potentially leverage a large quantity of unlabeled data during pretraining. Experiments have shown that generative pretraining often helps and never hurts the training of DNN, except that pretraining takes additional time. The generative pretraining is particularly helpful when the training set is small.

The DBN pretraining is not important when only one hidden layer is used and it typically works best with two hidden layers [18, 19]. When the number of hidden layers increases, the effectiveness often decreases. This is because DBN-pretraining employs two approximations. First, the mean-field approximation is used as the generation target when training the next layer. Second, the approximated contrastive divergence algorithm is used to learn the model parameters. Both these approximations introduce modeling errors for each additional layer. As the number of layers increases, the integrated errors increase and the effectiveness of DBN-pretraining decreases. It is obvious that although we can still use the DBN-pretrained model as the initial model for DNNs with rectified linear units, the effectiveness is greatly discounted since there is no direct link between two.

5.3 Pretraining with Denoising Autoencoder

In the layer-wise generative DBN-pretraining RBM is used as the building block. However, RBM is not the only technique that can be used to generatively pretrain the model. An equally effective approach is denoising autoencoder as illustrated in Fig. 5.6. In the autoencoder, the goal is to find an N_h-dimensional hidden

Fig. 5.6 Denoising autoencoder: the goal is to train a hidden representation which can be used to reconstruct the original input from a stochastically corrupted version of it

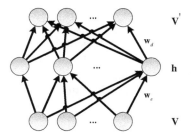

representation $\mathbf{h} = \mathbf{f}(\mathbf{v}) \in \mathbb{R}^{N_h \times 1}$, from which the original N_v-dimensional signal $\mathbf{v}$ can be reconstructed as $\tilde{\mathbf{v}} = g(\mathbf{h})$ with minimum mean square error (MSE)

$$J_{MSE}(\mathbf{W}, \mathbf{b}; \mathbb{S}) = \frac{1}{M} \sum_{m=1}^{M} \frac{1}{2} \left\| \tilde{\mathbf{v}}^m - \mathbf{v}^m \right\|^2 \qquad (5.35)$$

over an unlabeled training set $\mathbb{S} = \{(\mathbf{v}^m) \,|\, 1 \leq m \leq M\}$. In theory, the deterministic encoding function $f(\mathbf{v})$ and the deterministic decoding function $g(\mathbf{h})$ can be any function. In practice, some form of functional is selected to reduce the complexity of the optimization problem.

In its simplest form, one linear hidden layer can be used to represent the input signal. Under this condition, the hidden neurons learn to project the input in the span of the first N_h principal components of the data. If the hidden layer is nonlinear, which is typically the case for our purposes, the autoencoder behaves differently from the principal component analysis (PCA) and has the potential to capture multimodal properties of the input distribution.

Since our interest here is to use autoencoder to initialize the weights in the sigmoidal DNN, we choose

$$\mathbf{h} = f(\mathbf{v}) = \sigma(\mathbf{W}_e \mathbf{v} + \mathbf{b}), \qquad (5.36)$$

where $\mathbf{W}_e \in \mathbb{R}^{N_h \times N_v}$ is the encoding matrix, and $\mathbf{b} \in \mathbb{R}^{N_h \times 1}$ is the hidden layer bias vector. If the input feature $\mathbf{v} \in \{0, 1\}^{N_v \times 1}$ takes binary values, we can choose

$$\tilde{\mathbf{v}} = g(\mathbf{h}) = \sigma(\mathbf{W}_d \mathbf{h} + \mathbf{a}), \qquad (5.37)$$

where $\mathbf{a} \in \mathbb{R}^{N_v \times 1}$ is the reconstruction layer bias vector. If the input feature $\mathbf{v} \in \mathbb{R}^{N_v \times 1}$ takes real values, we can choose

$$\tilde{\mathbf{v}} = g(\mathbf{h}) = \mathbf{W}_d \mathbf{h} + \mathbf{a}. \qquad (5.38)$$

Note that unlike in RBMs, in autoencoders the weight matrices $\mathbf{W}_e$ and $\mathbf{W}_d$ may be different although they are typically tied as $\mathbf{W}_e = \mathbf{W}$ and $\mathbf{W}_d = \mathbf{W}^T$. No matter whether the input feature is binary or real-valued and no matter what encoding and decoding functions are used, the backpropagation algorithm described in Chap. 4 can be used to learn the parameters in the autoencoder.

In the autoencoder, the hope is that the distributed hidden representation $\mathbf{h}$ can capture the main factors of variation in the data. Since the autoencoder is trained to minimize the reconstruction error on the training set, it gives low reconstruction error to test samples drawn from the same distribution as the training samples, but relatively higher reconstruction error to other samples.

There is a potential problem with the autoencoder when the dimension of the hidden representation is higher than that of the input feature. If there is no other constraint than the minimization of the reconstruction error, the autoencoder may just learn the identity function and does not extract any statistical regularities present in the training set.

This problem can be resolved with many different approaches. For example, we can add sparsity constraint to the hidden representation by forcing a large percentage of neurons in the hidden representation to be zero. Alternatively, we can add random noise to the representation learning process. This is exploited in the denoising autoencoder [1], which forces the hidden layer to discover more robust features [21] and prevents it from simply learning the identity function, by reconstructing the input from a corrupted version of it.

The input can be corrupted in many ways, the simplest mechanism is randomly selecting entries (as many as half) of the input and setting them to zero. A denoising autoencoder does two things: preserve the information in the input, and undo the effect of the stochastic corruption process. The latter can only be done by capturing the statistical dependencies between the inputs. Note that in the contrastive divergence training procedure of the RBM, the sampling step essentially does the stochastic corruption of the input.

Similar to using RBMs, we cause the denoising autoencoder to pretrain a DNN [1]. We can first train a denoising autoencoder and use the encoding weight matrix as weight in the first hidden layer. We can then use the hidden representation as the input to a second denoising autoencoder which, when is trained, can be used as the second hidden layer weight matrix in the DNN. This process can continue until the desired number of hidden layers is reached.

The denoising autoencoder-based pretraining has similar properties to the DBN (RBM) pretraining procedure. It is a generative procedure and does not require labeled data. As such, it can potentially bring the DNN weights to a relatively good initial point and implicitly regularize the DNN training procedure with the generative pretraining criterion.

5.4 Discriminative Pretraining

Both the DBN and the denoising autoencoder-based pretraining are generative pre-training techniques. Alternatively, the DNN parameters can be initialized completely discriminatively using the discriminative pretraining, or DPT. An obvious approach is layer-wise BP (LBP) as shown in Fig. 5.7. With layer-wise BP, we first train a one-hidden-layer DNN (Fig. 5.7a) to full convergence using labels discriminatively, then insert another hidden layer (dotted box in Fig. 5.7b) between layer v_1 and the output layer with randomly initialized weights (indicated with green arrows in Fig. 5.7b), again discriminatively train the network to full convergence, and so on until the desired number of hidden layers is all trained. This is similar to greedy layer-wise training [1], but differs in that greedy layer-wise training only updates newly added hidden layers while in the layer-wise BP all layers are jointly updated each time a new hidden layer is added. For this reason, under most conditions layer-wise BP

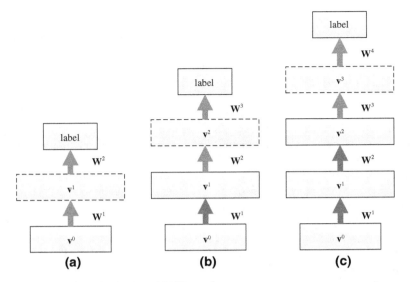

Fig. 5.7 Discriminative pretraining of DNNs

outperforms greedy layer-wise training since in the latter case the lower-layer weights are learned without knowing the upper layers. The layer-wise BP, however, has the drawback that some hidden neurons may operate in the saturation range after it is trained to convergence and thus are difficult to further update when new hidden layers are added. This limitation can be alleviated by not training the model to convergence every time when a new hidden layer is added. A typical heuristic is that we run the DPT by going through the data $\frac{1}{L}$ times of the total number of the data passes needed to converge, where L is the total number of layers in the final model. In the DPT, the goal is to bring the weights close to a good local optimum. It does not have the regularization effect which is available in the generative DBN pretraining. As such, DPT is best used when large amount of training data are available.

5.5 Hybrid Pretraining

Both the generative and the discriminative pretraining have their drawbacks. The generative pretraining is not tied to the task specific objective function. It helps to reduce overfitting but is not guaranteed to help the discriminative fine-tuning. The discriminative pretraining directly minimizes the objective function (e.g., cross-entropy). However, if the training is not scheduled well, weights learned in lower layers are potentially tuned too much to the final objective without considering additional layers that will be added later. To alleviate these problems, we can adopt a hybrid pretraining approach, in which a weighted sum of the generative and discriminative criteria is optimized [15]. A typical hybrid pretraining criterion is

$$J_{\mathrm{HYB}}\left(\mathbf{W}, \mathbf{b}; \mathbb{S}\right) = J_{\mathrm{DISC}}\left(\mathbf{W}, \mathbf{b}; \mathbb{S}\right) + \alpha J_{\mathrm{GEN}}\left(\mathbf{W}, \mathbf{b}; \mathbb{S}\right), \tag{5.39}$$

where α is an interpolation weight between the discriminative criterion J_{DISC} $(\mathbf{W}, \mathbf{b}; \mathbb{S})$ and the generative criterion $J_{\mathrm{GEN}}\left(\mathbf{W}, \mathbf{b}; \mathbb{S}\right)$. The discriminative criterion can be cross-entropy for classification tasks or mean square error for regression tasks. The generative criterion can be negative log-likelihood for RBMs or reconstruction error for autoencoder-based pretraining. Intuitively, the generative component acts as a data-dependent regularizer for the discriminative component [13]. It is obvious that this hybrid criterion can be used not only in the pretraining phase but also in the fine-tuning phase in which case it is referred to as HDRBM [13].

It has been shown that the generative pretraining often helps for training deep architectures [4–6, 19]. When the model becomes deeper, however, the discriminative pretraining performs equally well or even better than the generative pretraining [18]. The hybrid pretraining can outperform both the generative and discriminative pretraining [15]. It has been observed that when the training set size is large enough, pretraining becomes less important [18, 22]. However, even under this condition, pretraining may still help to make the training procedure robust to different random seeds.

5.6 Dropout Pretraining

In Sect. 4.3.4, we have introduced dropout [12] as a technique to improve the generalization ability of DNNs. We mentioned that dropout can be viewed as a way to reduce the capacity of the DNN by randomly dropping neurons. Alternatively, as pointed out by Hinton et al. [12], dropout can be considered as a bagging technique that averages over a large amount of models with tied parameters. In other words, dropout can generate smoother objective surface than the DNNs without using dropout. Since a smoother objective surface has less ill-conditioned local optimum than a sharper surface it is less likely to be trapped into a very bad local optimum. This suggests that we may pretrain a DNN using dropout to quickly find a relatively good initial point and then fine-tune the DNN without using dropout.

This is exactly what suggested by Zhang et al. [23] who showed that we can pretrain a DNN by going through the training data 10–20 passes using a dropout rate of 0.3–0.5 and then continue training the DNN by setting dropout rate to 0. DNNs initialized in this way outperformed DNNs pretrained using RBM-pretraining by 3 % relative error reduction. Our internal experiments indicate that we can achieve similar improvements on other tasks. Note that dropout pretraining also requires labeled data and achieves similar performance as the discriminative pretraining we discussed in Sect. 5.4 but is much more easier to implement and control than DPT.

References

1. Bengio, Y., Lamblin, P., Popovici, D., Larochelle, H.: Greedy layer-wise training of deep networks. In: Proceedings of the Neural Information Processing Systems (NIPS), pp. 153–160 (2006)
2. Bottou, L.: Online learning and stochastic approximations. On-line Learn. Neural Netw. **17**, 9 (1998)
3. Coates, A., Ng, A.Y., Lee, H.: An analysis of single-layer networks in unsupervised feature learning. In: Proceedings of the International Conference on Artificial Intelligence and Statistics (AISTATS), pp. 215–223 (2011)
4. Dahl, G.E., Yu, D., Deng, L., Acero, A.: Context-dependent pre-trained deep neural networks for large-vocabulary speech recognition. IEEE Trans. Audio, Speech Lang. Process. **20**(1), 30–42 (2012)
5. Erhan, D., Bengio, Y., Courville, A., Manzagol, P.A., Vincent, P., Bengio, S.: Why does unsupervised pre-training help deep learning? J. Mach. Learn. Res. (JMLR) **11**, 625–660 (2010)
6. Erhan, D., Manzagol, P.A., Bengio, Y., Bengio, S., Vincent, P.: The difficulty of training deep architectures and the effect of unsupervised pre-training. In: Proceedings of the International Conference on Artificial Intelligence and Statistics (AISTATS), pp. 153–160 (2009)
7. Hinton, G.: A practical guide to training restricted Boltzmann machines. Technical Report UTML TR 2010-003, University of Toronto (2010)
8. Hinton, G., Osindero, S., Teh, Y.W.: A fast learning algorithm for deep belief nets. Neural Comput. **18**, 1527–1554 (2006)
9. Hinton, G.E.: Training products of experts by minimizing contrastive divergence. Neural Comput. **14**(8), 1771–1800 (2002)
10. Hinton, G.E., Dayan, P., Frey, B.J., Neal, R.M.: The wake-sleep algorithm for unsupervised neural networks. SCIENCE-NEW YORK THEN WASHINGTON- pp. 1158–1158 (1995)
11. Hinton, G.E., Salakhutdinov, R.: Replicated softmax: an undirected topic model. In: Proceedings of the Neural Information Processing Systems (NIPS), pp. 1607–1614 (2009)
12. Hinton, G.E., Srivastava, N., Krizhevsky, A., Sutskever, I., Salakhutdinov, R.R.: Improving neural networks by preventing co-adaptation of feature detectors. arXiv preprint arXiv:1207.0580 (2012)
13. Larochelle, H., Bengio, Y.: Classification using discriminative restricted Boltzmann machines. In: Proceedings of the International Conference on Machine Learning (ICML), pp. 536–543 (2008)
14. Ling, Z.H., Deng, L., Yu, D.: Modeling spectral envelopes using restricted Boltzmann machines and deep belief networks for statistical parametric speech synthesis. IEEE Trans. Audio, Speech Lang. Process. **21**(10), 2129–2139 (2013)
15. Sainath, T., Kingsbury, B., Ramabhadran, B.: Improving training time of deep belief networks through hybrid pre-training and larger batch sizes. In: Proceedings of the Neural Information Processing Systems (NIPS) Workshop on Log-linear Models (2012)
16. Salakhutdinov, R., Mnih, A., Hinton, G.: Restricted boltzmann machines for collaborative filtering. In: Proceedings of the International Conference on Machine Learning (ICML), pp. 791–798 (2007)
17. Saul, L.K., Jaakkola, T., Jordan, M.I.: Mean field theory for sigmoid belief networks. J. Artif. Intell. Res. (JAIR) **4**, 61–76 (1996)
18. Seide, F., Li, G., Chen, X., Yu, D.: Feature engineering in context-dependent deep neural networks for conversational speech transcription. In: Proceedings of the IEEE Workshop on Automfatic Speech Recognition and Understanding (ASRU), pp. 24–29 (2011)
19. Seide, F., Li, G., Yu, D.: Conversational speech transcription using context-dependent deep neural networks. In: Proceedings of the Annual Conference of International Speech Communication Association (INTERSPEECH), pp. 437–440 (2011)
20. Smolensky, P.: Information processing in dynamical systems: foundations of harmony theory. Department of Computer Science, University of Colorado, Boulder (1986)

21. Vincent, P., Larochelle, H., Bengio, Y., Manzagol, P.A.: Extracting and composing robust features with denoising autoencoders. In: Proceedings of the International Conference on Machine Learning (ICML), pp. 1096–1103 (2008)
22. Yu, D., Deng, L., Dahl, G.: Roles of pre-training and fine-tuning in context-dependent DBN-HMMs for real-world speech recognition. In: Proceedings of the Neural Information Processing Systems (NIPS) Workshop on Deep Learning and Unsupervised Feature Learning (2010)
23. Zhang, S., Bao, Y., Zhou, P., Jiang, H., Li-Rong, D.: Improving deep neural networks for LVCSR using dropout and shrinking structure. In: Proceedings of the International Conference on Acoustics, Speech and Signal Processing (ICASSP), pp. 6899–6903 (2014)

Part III
Deep Neural Network-Hidden Markov Model Hybrid Systems for Automatic Speech Recognition

Chapter 6
Deep Neural Network-Hidden Markov Model Hybrid Systems

Abstract In this chapter, we describe one of the several possible ways of exploiting deep neural networks (DNNs) in automatic speech recognition systems—the deep neural network-hidden Markov model (DNN-HMM) hybrid system. The DNN-HMM hybrid system takes advantage of DNN's strong representation learning power and HMM's sequential modeling ability, and outperforms conventional Gaussian mixture model (GMM)-HMM systems significantly on many large vocabulary continuous speech recognition tasks. We describe the architecture and the training procedure of the DNN-HMM hybrid system and point out the key components of such systems by comparing a range of system setups.

6.1 DNN-HMM Hybrid Systems

6.1.1 Architecture

The DNNs we described in Chap. 4 cannot be directly used to model speech signals since speech signals are time series signals while DNNs require fixed-size inputs. To exploit the strong classification ability of DNNs in speech recognition, we need to find a way to handle the variable length problem in speech signals.

The combination of artificial neural networks (ANNs) and HMMs as an alternative paradigm for ASR started between the end of the 1980s and the beginning of the 1990s. A variety of different architectures and training algorithms were proposed at that time (see the comprehensive survey in [28]). This line of research was resurrected recently after the strong representation learning power of DNNs became well known.

One of the approaches that have been proven to work well is to combine DNNs with HMMs in a framework called DNN-HMM hybrid system as illustrated in Fig. 6.1. In this framework, the dynamics of the speech signal is modeled with HMMs and the observation probabilities are estimated through DNNs. Each output neuron of the DNN is trained to estimate the posterior probability of continuous density HMMs' state given the acoustic observations. In addition to their inherently discriminative nature, DNN-HMMs have two additional advantages: training can be performed using the embedded Viterbi algorithm and decoding is generally quite efficient.

© Springer-Verlag London 2015
D. Yu and L. Deng, *Automatic Speech Recognition*,
Signals and Communication Technology, DOI 10.1007/978-1-4471-5779-3_6

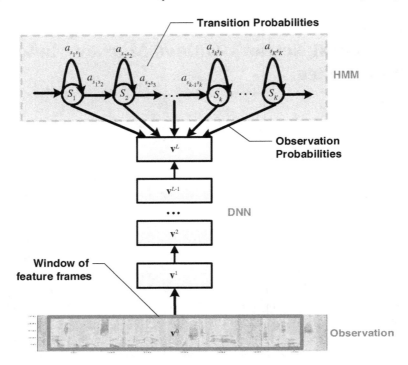

Fig. 6.1 Architecture of the DNN-HMM hybrid system. The HMM models the sequential property of the speech signal, and the DNN models the scaled observation likelihood of all the senones (tied triphone states) [13]. The same DNN is replicated over different points in time

These kinds of hybrid models were proposed and seen as a promising technique for LVCSR in the early to mid-1990s and were referred as ANN-HMM hybrid models in the literature [5, 17, 18]. Most early work on the hybrid approach used context-independent phone states as labels for ANN training and considered small vocabulary tasks. ANN-HMMs were later extended to model context-dependent phones [4] and were applied to mid-vocabulary and some large vocabulary ASR tasks [24], which also employed recurrent neural network architectures.

However, in the earlier work on context-dependent ANN-HMM hybrid architectures [4], the posterior probability of the context-dependent phone was modeled as either

$$p(s_i, c_j|\mathbf{x}_t) = p(s_i|\mathbf{x}_t)p(c_i|s_j, \mathbf{x}_t) \tag{6.1}$$

or

$$p(s_i, c_j|\mathbf{x}_t) = p(c_i|\mathbf{x}_t)p(s_i|c_j, \mathbf{x}_t), \tag{6.2}$$

where $\mathbf{x}_t$ is the acoustic observation at time t, c_j is one of the clustered context classes $\{c_1, \ldots, c_J\}$, and s_i is either a context-independent phone or a state in a context-independent phone. ANNs were used to estimate both $p(s_i|\mathbf{x}_t)$ and $p(c_i|s_j, \mathbf{x}_t)$ (alternatively $p(c_i|\mathbf{x}_t)$ and $p(s_i|c_j, \mathbf{x}_t)$). Although these types of context-dependent ANN-HMM models outperformed GMM-HMMs for some tasks, the improvements were small.

These earlier hybrid models have some important limitations. For example, ANNs with more than two hidden layers were rarely exploited due to the computational power limitations, and the context-dependent model described above does not take advantage of the numerous effective techniques developed for GMM-HMMs.

The recent advancement [6, 7, 12, 15, 25, 29] indicates that significant recognition accuracy improvement can be achieved if we replace the traditional shallow neural networks with deeper (optionally pretrained) ones, and use senones (tied triphone states) [13] instead of monophone states as the output units of the neural network. This improved ANN-HMM hybrid model is referred as CD-DNN-HMM [7]. Modeling senones directly also carries two other advantages: First, we can implement a CD-DNN-HMM system with only minimal modifications to an existing CD-GMM-HMM system. Second, any advancement of modeling units (e.g., cross-word triphone models) in the CD-GMM-HMM systems is accessible to the CD-DNN-HMM system since the improvement can be directly reflected in the output units of the DNNs.

In the CD-DNN-HMM, a single DNN is trained to estimate the conditional state posterior probability $p(q_t = s|\mathbf{x}_t)$ for all states $s \in [1, S]$. This is different from the GMMs in which a different GMM is used to model each different state. In addition, the input to the DNN is typically not a single frame but a window of $2\varpi + 1$ (typically 9–13) frames of input features $\mathbf{x}_t = \left[\mathbf{o}_{\max(0,t-\varpi)} \cdots \mathbf{o}_t \cdots \mathbf{o}_{\min(T,t+\varpi)}\right]$ to exploit information in the neighboring frames.

6.1.2 Decoding with CD-DNN-HMM

Since the HMM requires the likelihood $p(\mathbf{x}_t|q_t)$ instead of the posterior probability during the decoding process, we need to convert the posterior probability to the likelihood as

$$p(\mathbf{x}_t|q_t = s) = p(q_t = s|\mathbf{x}_t)p(\mathbf{x}_t)/p(s), \tag{6.3}$$

where $p(s) = \frac{T_s}{T}$ is the prior probability of each state (senone) estimated from the training set, T_s is the number of frames labeled as state s, and T is the total number of frames. $p(\mathbf{x}_t)$ is independent of the word sequence and thus can be ignored, which leads to the scaled likelihood $\bar{p}(\mathbf{x}_t|q_t) = p(q_t = s|\mathbf{x}_t)/p(s)$ [18]. Although dividing by the prior probability $p(s)$ may not give improved recognition accuracy under some conditions, it can be important in alleviating the label bias problem, especially when the training utterances contain long silence segments.

As a summary, in the CD-DNN-HMM the decoded word sequence $\hat{w}$ is determined as

$$\hat{w} = \arg \max_w p(w|\mathbf{x}) = \arg \max_w p(\mathbf{x}|w) p(w)/p(\mathbf{x})$$
$$= \arg \max_w p(\mathbf{x}|w) p(w), \qquad (6.4)$$

where $p(w)$ is the language model (LM) probability, and

$$p(\mathbf{x}|w) = \sum_q p(\mathbf{x}|q, w) p(q|w) \qquad (6.5)$$

$$\approx \max \pi(q_0) \prod_{t=1}^{T} a_{q_{t-1}q_t} \prod_{t=0}^{T} p(q_t|\mathbf{x}_t)/p(q_t) \qquad (6.6)$$

is the acoustic model (AM) probability, where $p(q_t|\mathbf{x}_t)$ is computed from the DNN, $p(q_t)$ is the state prior probability, and $\pi(q_0)$ and $a_{q_{t-1}q_t}$ are the initial state probability and state transition probability, respectively, determined by the HMM. Similar to that in the GMM-HMM case, an LM weight λ is often used to balance between the AM and LM scores. The final decoding path is thus determined by

$$\hat{w} = \arg \max_w \left[\log p(\mathbf{x}|w) + \lambda \log p(w)\right]. \qquad (6.7)$$

6.1.3 Training Procedure for CD-DNN-HMMs

CD-DNN-HMMs can be trained using the embedded Viterbi algorithm, the main steps of which are summarized in Algorithm 6.1.

The CD-DNN-HMM has three components, a DNN *dnn*, an HMM *hmm*, and a state prior probability distribution *prior*. Since the CD-DNN-HMM shares the phoneme tying structure and the HMM with the GMM-HMM system, the first step in the CD-DNN-HMM training is to train a GMM-HMM system using the training data. Since the DNN training labels are generated from the GMM-HMM system and the label quality can affect the performance of the DNN system, it is important to train a good GMM-HMM system as the seed model.

Once the GMM-HMM model *hmm0* is available, we can parse the model and build a mapping from the state name to the *senoneID*. The mapping *stateToSenoneIDMap* may not be trivial to build. This is because the logical triphone HMMs that are effectively equivalent are clustered and represented by a physical triphone. In other words, several logical triphones are mapped to the same physical triphone. Each physical triphone has several (typically 3) states which are tied and represented by senones.

With the GMM-HMM model *hmm0* we can generate a forced alignment at the state level using the Viterbi algorithm on the training set. By using the *stateToSenoneIDMap* we can convert the state names in the alignment to the *senoneIDs*. We can then generate the *featureSenoneIDPairs* from the alignment and use them to train the DNN. The same *featureSenoneIDPairs* can also be used to estimate the senone prior probabilities *prior*.

With the GMM-HMM model *hmm0* we can also generate an HMM *hmm*, which contains the same state transition probabilities as that in *hmm0*, for use in the DNN system. A simple approach is to replace each GMM (which models one senone) in *hmm0* with a (pseudo) single one-dimensional Gaussian. The variance (or precision) of the Gaussian is irrelevant and can be set to any positive value (e.g., always set to 1). The value of each senone's mean is set to the corresponding *senoneID*. Using this trick, evaluating each senone is equivalent to a table lookup of the features (log-likelihood) produced by the DNN with the index indicated by the *senoneID*.

In this procedure, it is assumed that a CD-GMM-HMM exists and is used to generate senone alignment. The decision tree used to cluster tri-phone states is also built using the GMM-HMM. However, this is not necessary. If we would completely remove GMM-HMM from the picture, we can build a monophone DNN-HMM by evenly segmenting the utterances (called flat-start) and using that information as the training label. We can then realign the utterances using the monophone DNN-HMM. After that, we can build a single Gaussian model for each monophone state and build the decision tree as usual. In fact, this GMM-free CD-DNN-HMM training procedure has recently been reported in [26] with success.

Algorithm 6.1 Main steps involved in training CD-DNN-HMMs

1: **procedure** TRAINCD- DNN- HMM($\mathbb{S}$)   ▷ $\mathbb{S}$ is the training set
2:     $hmm0 \leftarrow$ TrainCD-GMM-HMM($\mathbb{S}$);   ▷ $hmm0$ is used in the GMM system
3:     $stateAlignment \leftarrow$ ForcedAlignmentWithGMMHMM($\mathbb{S}$, $hmm0$);
4:     $stateToSenoneIDMap \leftarrow$ GenerateStateTosenoneIDMap($hmm0$);
5:     $featureSenoneIDPairs \leftarrow$ GenerateDNNTrainingSet($stateToSenoneIDMap$,
         $stateAlignment$);
6:     $ptdnn \leftarrow$ PretrainDNN($\mathbb{S}$);   ▷ Optional
7:     $hmm \leftarrow$ ConvertGMMHMMToDNNHMM($hmm0$, $stateToSenoneIDMap$);
                                                        ▷ hmm is used in the DNN system
8:     $prior \leftarrow$ EstimatePriorProbability($featureSenoneIDPairs$)
9:     $dnn \leftarrow$ Backpropagate($ptdnn$, $featureSenoneIDPairs$);
10:     Return $dnnhmm = \{dnn, hmm, prior\}$
11: **end procedure**

For each utterance with T frames, the embedded Viterbi training algorithm minimizes the average cross entropy, which is equivalent to the negative log-likelihood

$$J_{\text{NLL}}(\mathbf{W}, \mathbf{b}; \mathbf{x}, \mathbf{q}) = -\sum_{t=1}^{T} \log p(q_t | \mathbf{x}_t; \mathbf{W}, \mathbf{b}).$$ (6.8)

If the new model $\left(\mathbf{W}', \mathbf{b}'\right)$ improves the training criterion over the old model $(\mathbf{W}, \mathbf{b})$, we have

$$-\sum_{t=1}^{T} \log p\left(q_t | \mathbf{x}_t; \mathbf{W}', \mathbf{b}'\right) < -\sum_{t=1}^{T} \log p\left(q_t | \mathbf{x}_t; \mathbf{W}, \mathbf{b}\right).$$

The score of the aligned utterance

$$
\begin{aligned}
\log p(\mathbf{x}|w; \mathbf{W}', \mathbf{b}') &= \log \pi(q_0) + \sum_{t=1}^{T} \log\left(a_{q_{t-1}q_t}\right) + \sum_{t=1}^{T}\left[\log p\left(q_t|\mathbf{x}_t; \mathbf{W}', \mathbf{b}'\right) - \log p(q_t)\right] \\
&> \log \pi(q_0) + \sum_{t=1}^{T} \log\left(a_{q_{t-1}q_t}\right) + \sum_{t=1}^{T}\left[\log p\left(q_t|\mathbf{x}_t; \mathbf{W}, \mathbf{b}\right) - \log p(q_t)\right] \\
&= \log p(\mathbf{x}|w; \mathbf{W}, \mathbf{b}).
\end{aligned}
\tag{6.9}
$$

In other words, the new model not only improves the frame-level cross entropy but also the likelihood score of the utterance given the correct word sequence. This proves the correctness of the embedded Viterbi training algorithm. A different justification of the embedded Viterbi training algorithm can be found in [10]. It is to be noted though that the score associated with each competitive word is not guaranteed to decrease although the sum of scores of all competitive words decreases. In addition, the above argument is true on average but may not be true for each individual utterance. If the average cross entropy improvement is very small, it is possible that the recognition accuracy may actually decrease, especially if the small improvement is from better modeling of the silence segments. A more principled yet more complex approach to train the CD-DNN-HMM system is to use the sequence discriminative training criterion which we will discuss in Chap. 8.

6.1.4 Effects of Contextual Window

As we have mentioned in Sect. 6.1.1, it is important to use a window of (typically 9–13) frames of input features in the CD-DNN-HMM to achieve great results. It is obvious that by including a long window of frames the DNN model can exploit information in the neighboring frames. Contrary to common sense, however, by including neighboring frames the DNN also models correlations between frames of features and thus partially alleviates the violation to the observation independence assumption made in HMMs.

Note that the score of each word sequence is evaluated as

$$\log p(\mathbf{x}|w) = \log \pi(q_0) + \sum_{t=1}^{T} \log \left(a_{q_{t-1}q_t}\right)$$
$$+ \sum_{n=1}^{N} \left[\log p\left(\mathbf{o}_{t_n}, \ldots, \mathbf{o}_{t_{n+1}-1}|s_n\right)\right], \tag{6.10}$$

where T is the length of the feature frames, $N \le T$ is the length of the state sequence, s_n is the nth state in the state sequence, q_t is the state at time t, and t_n is the start time of the nth state. Here we have assumed that the state duration can be modeled using a Markov chain.[1] Note that the observation score $\log p\left(\mathbf{o}_{t_n}, \ldots, \mathbf{o}_{t_{n+1}-1}|s_n\right)$ measures the log-likelihood of observing a segment of features given the state s_n and is what have been used in segmental models [19]. In the HMM, each feature frame is assumed to be independent and thus

$$\log p\left(\mathbf{o}_{t_n}, \ldots, \mathbf{o}_{t_{n+1}-1}|s_n\right) \simeq \sum_{t=t_n}^{t_{n+1}-1} \left[\log p\left(\mathbf{o}_t|s_n\right)\right]. \tag{6.11}$$

This assumption is known to be untrue in reality since the adjacent frames do correlate with each other given the same state.[2] To model the correlations between the frames, the segment score should be evaluated as

$$\log p\left(\mathbf{o}_{t_n}, \ldots, \mathbf{o}_{t_{n+1}-1}|s_n\right) = \sum_{t=t_n}^{t_{n+1}-1} \left[\log p\left(\mathbf{o}_t|s_n, \mathbf{o}_{t_n}, \ldots, \mathbf{o}_{t-1}\right)\right]. \tag{6.12}$$

As we know, if two frames are too far (e.g., more than M frames) away, they can be considered uncorrelated. For this reason, the above score can be approximated as

$$\log p\left(\mathbf{o}_{t_n}, \ldots, \mathbf{o}_{t_{n+1}-1}|s_n\right) \simeq \sum_{t=t_n}^{t_{n+1}-1} \left[\log p\left(\mathbf{o}_t|s_n, \mathbf{o}_{t-M}, \ldots, \mathbf{o}_{t-1}\right)\right]$$
$$= \sum_{t=t_n}^{t_{n+1}-1} \left[\log p\left(s_n|\mathbf{x}_t^M\right) - \log p\left(s_n|\mathbf{x}_{t-1}^{M-1}\right)\right] + c$$
$$\simeq \sum_{t=t_n}^{t_{n+1}-1} \left[\log p\left(s_n|\mathbf{x}_t^M\right) - \log p\left(s_n\right)\right] + c, \tag{6.13}$$

---

[1] For the desired segmental model, this duration model is very rough.

[2] The independence assumption made in the HMM is one of the reasons why language model weighting is needed. Assuming one doubles the features by extracting a feature for each 5 ms instead of 10 ms, the acoustic model score will be doubled and so the language model weight will also need to be doubled.

where c is a constant irrelevant to s_n, $\mathbf{x}_t^M = \{\mathbf{o}_{t-M}, \ldots, \mathbf{o}_{t-1}, \mathbf{o}_t\}$ is the feature vector concatenated from M frames, and the state prior is assumed to be independent of the observation. We can see that by including the neighbor frames in the DNN model (i.e., estimate $p\left(s_n|\mathbf{x}_t^M\right)$) we can more accurately estimate the segment score while still enjoying the benefit provided by the independence assumption made in the HMM.

6.2 Key Components in the CD-DNN-HMM and Their Analysis

CD-DNN-HMMs outperform GMM-HMMs in many large vocabulary continuous speech recognition (LVCSR) tasks. It is important to understand which components or procedures have contributed to the performance improvement. In this section, we examine how different decisions affect recognition accuracy. In particular, we empirically compare the performance difference between using a mono-phone alignment and a tri-phone alignment, using monophone state labels and tri-phone senone labels, using shallow and deep networks, and tuning the HMM transition probabilities and otherwise. We show, by experimental results summarized from a series of studies reported in [6, 7, 25, 27, 29], that the three key components contributing most to the performance improvement are: (1) using deep neural networks with sufficient depth, (2) using a long window of frames, and (3) model senones directly. In all the experiments, the CD-DNN-HMM system replaces the Gaussian-mixture component with likelihoods derived from the MLP (DNN) posteriors, while leaving everything else the same.

6.2.1 Datasets and Baselines for Comparisons and Analysis

6.2.1.1 Bing Mobile Voice Search Dataset

The Bing mobile voice search application allows users to do USA-wide business and web search from their mobile phones via voice. The business search dataset used in the experiments was collected under real usage scenarios in 2008, at which time the application was restricted to do location and business lookup [30]. All audio files collected were sampled at 8 kHz and encoded with the GSM codec. This is a challenging task since the dataset contains all kinds of variations: noise, music, side-speech, accents, sloppy pronunciation, hesitation, repetition, interruption, and different audio channels.

The dataset was split into a training set, a development set, and a test set. To simulate the real data collection and training procedure, and to avoid having overlap between training, development, and test sets, the dataset was split based on the time stamp of the queries. All queries in the training set were collected before those in the development set, which were in turn collected before those in the test set. The public lexicon from Carnegie Mellon University was used. The normalized nationwide

Table 6.1 Bing mobile voice search dataset

	Hours	Number of utterances
Training set	24	32,057
Development set	6.5	8,777
Test set	9.5	12,758

Table 6.2 The CD-GMM-HMM baseline sentence error rate (SER) on the voice search dataset (Summarized from Dahl et al. [7])

Criterion	Dev SER (%)	Test SER (%)
ML	37.1	39.6
MMI	34.9	37.2
MPE	34.5	36.2

language model (LM) used in the evaluation contains 65 K word unigrams, 3.2 million word bi-grams, and 1.5 million word tri-grams, and was trained using the data feed and collected query logs; the perplexity is 117.

Table 6.1 summarizes the number of utterances and total duration of audio files (in hours) in the training, development, and test sets. All 24 h of training data included in the training set are manually transcribed.

Performance on this task was evaluated using sentence error rate (SER) instead of word error rate (WER). The average sentence length is 2.1 tokens, so sentences are typically quite short. Also, the users care most about whether they can find the business or location they seek for with fewest attempts. They typically will repeat what they have said if one of the words is mis-recognized. Additionally, there is significant inconsistency in spelling that makes using sentence accuracy more convenient. For example, "Mc-Donalds" sometimes is spelled as "McDonalds," "Walmart" sometimes is spelled as "Wal-mart", and "7-eleven" sometimes is spelled as "7 eleven" or "seven-eleven". The sentence out-of-vocabulary (OOV) rate using the 65 K vocabulary LM is 6 % on both the development and test sets. In other words, the best possible SER we can achieve is 6 % using this setup.

The clustered cross-word triphone GMM-HMMs were trained with the maximum likelihood (ML), maximum mutual information (MMI) [3, 14, 20], and minimum phone error (MPE) [20, 23] criteria. The 39-dim features used in the experiments include the 13-dim static Mel-frequency cepstral coefficient (MFCC) (with C0 replaced with energy) and its first and second derivatives. The features were pre-processed with the cepstral mean normalization (CMN) algorithm.

The baseline systems were optimized by tuning the tying structures, number of senones, and Gaussian splitting strategies on the development set. All systems have 53 K logical and 2 K physical tri-phones with 761 shared states (senones), each of which is a GMM with 24 mixture components. The GMM-HMM baseline results are summarized in Table 6.2.

For all CD-DNN-HMM experiments cited for the VS dataset, 11 frames (5-1-5) of MFCCs were used as the input features of the DNNs. During DNN pretraining a learning rate of $1.5e^{-4}$ per sample was used for all layers. For fine-tuning, a learning rate of $3e^{-3}$ per sample was used for the first 6 epochs and a learning rate of $8e^{-5}$

per sample was used for the last 6 epochs. In all the experiments, a minibatch size of 256 and a momentum of 0.9 was used. The hyperparameters were selected by hand, based on preliminary single hidden layer experiments so it may be possible to obtain even better performance with the deeper models using a more exhaustive hyperparameter search strategy.

6.2.1.2 Switchboard Dataset

The Switchboard (SWB) dataset [8, 9] is a corpus of conversational telephone speech. It has three setups whose training set sizes are 30 h (a random subset of Switchboard-I training set), 309 h (full Switchboard-I training set), and 2,000 h (+ Fisher training sets), respectively. For all configurations the 1,831-segment SWB part of the NIST 2,000 Hub5 eval set and the FSH half of the 6.3 h Spring 2003 NIST rich transcription set (RT03S) are used as the evaluation sets. The system uses 13-dimensional PLP features with rolling-window mean-variance normalization and up to third-order derivatives, which is reduced to 39 dimensions by heteroscedastic linear discriminant analysis (HLDA) [16]. The speaker-independent 3-state cross-word triphones share 1,504 (40 mixture), 9,304 (40 mixture), and 18,004 (72 mixture) CART-tied states, respectively, on the 30, 309, and 2,000 h setups, optimized for the GMM-HMM systems. The trigram language model was trained on the 2,000 h Fisher transcripts and interpolated with a written-text trigram. Test-set perplexity with the 58 k lexicon is 84.

The DNN system was trained using stochastic gradient descent in mini-batches. The mini-batch sizes were 1,024 frames except for the first mini-batches for which 256 samples were used. For DBN-pretraining the mini-batch size was 256.

For pretraining, the learning rate is $1.5e^{-4}$ per sample. For the first 24 h of training data the learning rate is $3e^{-3}$ per sample, and was reduced to $8e^{-5}$ per sample after 3 epochs. The momentum of 0.9 was used. These learning rates are the same as that used in the VS dataset.

6.2.2 Modeling Monophone States or Senones

As mentioned at the beginning of this section, there are three key components in the CD-DNN-HMMs. Direct modeling of the context-dependent phone states (i.e., senones) is one of them. Modeling senones directly allows to take advantage of information encoded in the fine-grained labels and to alleviate overfitting. It also reduces the confusable state transitions in the HMM and thus decreases the ambiguity in decoding, even though increasing the DNN output nodes would reduce the frame classification accuracy. The benefit of modeling senones over monophone states is clearly demonstrated in Table 6.3 which indicates a 15 % relative SER reduction on the VS development set when using a DNN with three hidden layers each of which has 2 K neurons (the 3×2 K setup), and Table 6.4, which indicates an over 50 %

Table 6.3 Sentence error rate (SER) on the voice search development set when the context-independent monophone state labels and context-dependent triphone senone labels are used

Model	Monophone state (%)	Senone (761) (%)
CD-GMM-HMM (MPE)	–	34.5
DNN-HMM (3 × 2 K)	35.8	30.4

(Summarized from Dahl et al. [7])

Table 6.4 Word error rate (WER) on Hub5'00-SWB using 309 h training set and ML alignment label

Model	Monophone state (%)	Senone (9304) (%)
CD-GMM-HMM (BMMI)	–	23.6
DNN-HMM (7 × 2 K)	34.9	17.1

When the context-independent monophone state labels and context-dependent triphone senone labels are used. (Summarized from Seide et al. [25])

relative word error rate (WER) reduction on the 309 h SWB task when using a DNN with seven hidden layers each of which has 2 K neurons (the 7×2 K setup). The difference in the relative error reduction is partially attributed to the fact that in the SWB more senones are used. Using senone labels has been the single largest source of improvement of all the design decisions we analyzed.

6.2.3 Deeper Is Better

Using DNNs instead of shallow MLPs is another key component in CD-DNN-HMMs. Table 6.5 demonstrates how the sentence error rate reduces as more layers are added in the CD-DNN-HMM. When only one hidden layer is used the SER is 31.9 %. When three hidden layers were used, the error rate was reduced to 30.4 %. The error rate was further reduced to 29.8 % with four hidden layers and 29.7 % with five hidden layers. Overall, the five hidden layer model provides us with a 2.2 % SER reduction over the single hidden layer system when the same alignment is used.

In order to demonstrate the efficiency of parameterization enjoyed by deeper neural networks, the result using a single hidden layer neural network with 16 K hidden neurons was also reported in Table 6.5. Since the output layer has 761 neurons, this shallow model requires a little more space to store than the five hidden layer model that uses 2 K hidden neurons. The development set SER of 31.4 % obtained using this wide and shallow model is slightly better than 31.9 % achieved with the single hidden layer model with 2 K hidden neurons, but well below even 30.5 % obtained with the two hidden layer model (let alone 29.7 % obtained with the five hidden layer model).

Table 6.6 summarizes the WER results on SWB Hub5'00-SWB test set using the 309 h training data whose senone alignment label was generated from the

Table 6.5 Deeper is better. SER on the voice search development set achieved using DNNs with different number of hidden layers

LxN	DBN-PT (%)	1xN	DBN-PT (%)
$1 \times 2\,k$	31.9		
$2 \times 2\,k$	30.5		
$3 \times 2\,k$	30.4		
$4 \times 2\,k$	29.8		
$5 \times 2\,k$	29.7	$1 \times 16\,K$	31.4

ML alignment and DBN-pretraining were used in all the experiments. (Summarized from Dahl et al. [7])

Table 6.6 Deeper is better. WER on Hub5'00-SWB achieved using DNNs with different number of hidden layers

LxN	DBN-PT (%)	1xN	DBN-PT (%)
$1 \times 2\,k$	24.2		
$2 \times 2\,k$	20.4		
$3 \times 2\,k$	18.4		
$4 \times 2\,k$	17.8		
$5 \times 2\,k$	17.2	$1 \times 3{,}772$	22.5
$7 \times 2\,k$	17.1	$1 \times 4{,}634$	22.6
		$1 \times 16\,K$	22.1

309 h SWB training set, ML alignment and DBN-pretraining were used in all the experiments. (Summarized from Seide et al. [25])

maximum likelihood (ML) trained GMM system. From this table we can make several observations. First, deeper networks do perform better than shallow ones. In other words, deeper models have stronger discriminative ability than the shallow models. The WER keeps reducing as we increase the number of hidden layers if each layer contains 2 K hidden neurons. More interestingly, if you compare the $5 \times 2\,K$ setup with the $1 \times 3{,}772$ setup, or the $7 \times 2\,K$ setup with the $1 \times 4{,}634$ setup that have the same number of model parameters, the deep model also significantly outperforms the shallow model. Even if the hidden layer size on the single-hidden layer MLP is further increased to 16 K we can only achieve a WER of 22.1 %, which is significantly worse than the 17.1 % WER obtained using a $7 \times 2\,K$ DNN under the same condition. Note that as we further grow the number of hidden layers the performance improvement decreases and it saturates after using nine hidden layers. In practice, a tradeoff needs to be made between the additional gains in WER and the increased cost in training and decoding as we increase the number of hidden layers.

Table 6.7   Compare the effectiveness of using neighboring frames

Model	1 frame (%)	11 frames (%)
CD-DNN-HMM 1×4,634	26.0	22.4
CD-DNN-HMM 7×2k	23.2	17.1

WER on Hub5'00-SWB using 309 h training set and ML alignment label. GMM-HMM baseline with BMMI training is 23.6 %. (Summarized from Seide et al. [25])

6.2.4 Exploit Neighboring Frames

Table 6.7 compares WER on the 309 h SWB task with and without using neighboring frames. It can be clearly seen that regardless of whether shallow or deep networks are used, using information in the neighboring frames significantly boosts accuracy. However, deep networks get more gains from using neighboring frames as indicated by a 24 % relative WER reduction compared to the 14 % relative WER reduction obtained using shallow models although both models have the same number of parameters. Also observable is that if only a single frame is used, the DNN system only slightly outperforms the BMMI [21] trained GMM system (23.2 versus 23.6 %). Note however, the DNN system's performance can be further improved if sequence discriminative training similar to BMMI is conducted [27]. To use the neighboring frames in the GMM system, complicated techniques such as fMPE [22], HLDA [16], region-dependent transformation [31], or tandem structure [11, 32] will need to be used. This is because to use diagonal covariance matrices in GMMs each dimension should be uncorrelated. DNNs, however, are discriminative models that can freely use any features even if they are correlated.

6.2.5 Pretraining

Before 2011, people believed that pretraining was critical to train deep neural networks. Later, however, researchers found that pretraining, although can provide additional gains sometimes, is not critical. This can be observed from Table 6.8. The table suggests that the DBN-pretraining, which does not rely on labels, does provide significant gains over the system trained without any pretraining when the hidden layer number is smaller than five. However, as the number of layers increases, the gain is decreased and eventually diminishes. This is against the original motivation of using pretraining. Researchers have conjectured that as the number of hidden layers increases, we should see more gains from using DBN-pretraining instead of less as indicated in the table. This behavior is partly explained by the fact that the stochastic gradient descent algorithm used to train the DNNs has the ability to jump out of the local optima. In addition, when large amount of training data are used overfitting, which is partially addressed by pretraining, is less of a concern.

Table 6.8 Compare the effectiveness of pretraining procedures

LxN	NOPT (%)	DBN-PT (%)	DPT (%)
1×2 k	24.3	24.2	24.1
2×2 k	22.2	20.4	20.4
3×2 k	20.0	18.4	18.6
4×2 k	18.7	17.8	17.8
5×2 k	18.2	17.2	17.1
7×2 k	17.4	17.1	16.8

WER on Hub5'00-SWB using 309 h training set and ML alignment label. NOPT: no pretraining; DBN-PT: with DBN-pretraining, DPT: with discriminative pretraining. (Summarized from Seide et al. [25])

On the other hand, the benefit of generative pretraining is likely to decrease when the number of layers increases. This is because DBN-pretraining employs two approximations. First, the mean-field approximation is used as the generation target when training the next layer. Second, the approximated contrastive divergence algorithm is used to learn the model parameters. Both these approximations introduce modeling errors for each additional layer. As the number of layers increases, the integrated errors increase and the effectiveness of DBN-pretraining decreases. The discriminative pretraining (DPT) is an alternative pretraining technique. According to Table 6.8, it performs at least as good as the DBN-pretraining procedure, especially when the DNN contains more than five hidden layers. However, even with DPT, the performance gains against the system with pure backpropagation (BP) is still quite small compared to the gains we get from modeling senones and from using deep networks. Although the WER reduction is less than what people have hoped by using pretraining techniques, these techniques can secure a more robust training procedure. By applying these techniques, we can avoid very bad initializations and employ some kind of implicit regularization in the training process so that good performance can be obtained even with a small training set.

6.2.6 Better Alignment Helps

In the embedded Viterbi training procedure, the training label of the samples are determined using forced alignment. Intuitively, if the labels are generated from a more accurate model the trained DNN should perform better. This is empirically confirmed by Table 6.9. As we can see, when alignments generated with MPE-trained CD-GMM-HMMs were used, 29.3 and 31.2 % SER on the development and test sets, respectively, can be obtained. These results are 0.4 % better than those achieved using the CD-GMM-HMM ML alignment trained CD-DNN-HMM. Since CD-DNN-HMM performs better than CD-GMM-HMM we can further improve the label quality by using the best CD-DNN-HMM to generate alignments. Table 6.9 shows that the

Table 6.9 Compare the effectiveness of the alignment quality and transition probability tuning on voice search dataset

Alignment	GMM transition		DNN tuned transition	
	Dev SER (%)	Test SER (%)	Dev SER (%)	Test SER (%)
from CD-GMM-HMM ML	29.7	31.6	–	–
from CD-GMM-HMM MPE	29.3	31.2	29.0	31.0
from CD-DNN-HMM	28.3	30.4	28.2	30.4

5×2K model is used. SER on the development and test sets. (Summarized from Dahl et al. [7])

SER on the development and test sets can be reduced to 28.3 and 30.4 %, respectively, with the CD-DNN-HMM alignment.

Similar observations can be made on SWB. When the 7×2k DNN is trained using the labels generated from the CD-GMM-HMM system we get 17.1 % WER on the Hub5'00 test set. If the DNN is trained using the improved label generated from the CD-DNN-HMM the WER can be reduced to 16.4 %.

6.2.7 Tuning Transition Probability

Table 6.9 also demonstrates that tuning the transition probabilities in the CD-DNN-HMMs helps slightly. Tuning the transition probabilities comes with another benefit. When the transition probabilities are directly borrowed from the CD-GMM-HMMs, the best decoding performance usually was obtained when the AM weight was set to 2. However, after tuning the transition probabilities, tuning the AM weights is no longer needed on the voice search task.

6.3 Kullback-Leibler Divergence-Based HMM

In the hybrid DNN/HMM systems, the observation probability is true probability that satisfies the appropriate constraints. However, we may remove this constraint and replace the state log-likelihood with other scores. In the Kullback-Leibler divergence-based HMM (KL-HMM) [1, 2] the state score is computed as

$$S_{\mathrm{KL}}(s, \mathbf{z}_t) = \mathrm{KL}(\mathbf{y}_s \parallel \mathbf{z}_t) = \sum_{d=1}^{D} y_s^d \ln \frac{y_s^d}{z_t^d}, \tag{6.14}$$

where s is a state (e.g., senone), $z_t^d = P(a_d|\mathbf{x}_t)$ is the posterior probability of some class a_d given the observation $\mathbf{x}_t$, D is the number of classes, and $\mathbf{y}_s$ is a probability distribution that represents the state s. Theoretically, a_d can be any class. In practice though, context-independent phones or states are typically chosen as a_d. For example, $\mathbf{z}_t$ can be the output of a DNN whose output neurons represent monophones.

Different from the hybrid DNN/HMM systems, in KL-HMM $\mathbf{y}_s$ is an additional model parameter that needs to be estimated for each state. In [1, 2] $\mathbf{y}_s$ is optimized to minimize the average frame score defined in Eq. 6.14 while keeping $\mathbf{z}_t$ (and thus the DNN or MLP) fixed.

Alternatively, the reverse KL (RKL) distance

$$S_{\text{RKL}}(s, \mathbf{z}_t) = \text{KL}(\mathbf{z}_t \parallel \mathbf{y}_s) = \sum_{d=1}^{D} z_t^d \ln \frac{z_t^d}{y_s^d}, \qquad (6.15)$$

or the symmetric KL (SKL) distance

$$S_{\text{SKL}}(s, \mathbf{z}_t) = \text{KL}(\mathbf{y}_s \parallel \mathbf{z}_t) + \text{KL}(\mathbf{z}_t \parallel \mathbf{y}_s) \qquad (6.16)$$

can be used as the state score.

Note that, KL-HMM can be considered as a special DNN/HMM in which a_d serves as a neuron in a D-dimensional bottleneck layer in the DNN and the softmax layer is replaced with the KL distance. For this reason, when comparing the hybrid DNN/HMM systems with the KL-HMM systems an additional layer should be added to the DNN/HMM hybrid system for fair comparison.[3]

Compared to the simpler hybrid DNN/HMM systems, KL-HMM has other two drawbacks: the parameters $\mathbf{y}_s$ are estimated separately from the DNN model instead of jointly optimized as in the hybrid systems and the sequence-discriminative training (which we will discuss in Chap. 8) in KL-HMM is not as straightforward as that in the hybrid systems. For these reasons, this book focuses on the hybrid DNN/HMM system instead of the KL-HMM although it is an interesting model as well.

References

1. Aradilla, G., Bourlard, H., Magimai-Doss, M.: Using KL-based acoustic models in a large vocabulary recognition task. In: Proceedings of Annual Conference of International Speech Communication Association (INTERSPEECH), pp. 928–931 (2008)
2. Aradilla, G., Vepa, J., Bourlard, H.: An acoustic model based on kullback-leibler divergence for posterior features. In: Proceedings of International Conference on Acoustics, Speech and Signal Processing (ICASSP), vol. 4, pp. IV–657 (2007)
3. Bahl, L., Brown, P., De Souza, P., Mercer, R.: Maximum mutual information estimation of hidden markov model parameters for speech recognition. In: Proceedings of International Conference on Acoustics, Speech and Signal Processing (ICASSP), vol. 11, pp. 49–52 (1986)
4. Bourlard, H., Morgan, N., Wooters, C., Renals, S.: CDNN: a context dependent neural network for continuous speech recognition. In: Proceedings of International Conference on Acoustics, Speech and Signal Processing (ICASSP), vol. 2, pp. 349–352 (1992)

---

[3] Unfair comparison was conducted in several papers that compare the hybrid DNN/HMM system and the KL-HMM system. The conclusions in these papers are thus questionable.

5. Bourlard, H., Wellekens, C.J.: Links between Markov models and multilayer perceptrons. IEEE Trans. Pattern Anal. Mach. Intell. (PAMI) **12**(12), 1167–1178 (1990)
6. Dahl, G.E., Yu, D., Deng, L., Acero, A.: Large vocabulary continuous speech recognition with context-dependent DBN-HMMs. In: Proceedings of International Conference on Acoustics, Speech and Signal Processing (ICASSP), pp. 4688–4691 (2011)
7. Dahl, G.E., Yu, D., Deng, L., Acero, A.: Context-dependent pre-trained deep neural networks for large-vocabulary speech recognition. IEEE Trans. Audio, Speech Lang. Process. **20**(1), 30–42 (2012)
8. Godfrey, J.J., Holliman, E.: Switchboard-1 Release 2. Linguistic Data Consortium, Philadelphia (1997)
9. Godfrey, J.J., Holliman, E.C., McDaniel, J.: Switchboard: telephone speech corpus for research and development. In: Proceedings of International Conference on Acoustics, Speech and Signal Processing (ICASSP), vol. 1, pp. 517–520 (1992)
10. Hennebert, J., Ris, C., Bourlard, H., Renals, S., Morgan, N.: Estimation of global posteriors and forward-backward training of hybrid hmm/ann systems (1997)
11. Hermansky, H., Ellis, D.P., Sharma, S.: Tandem connectionist feature extraction for conventional HMM systems. In: Proceedings of International Conference on Acoustics, Speech and Signal Processing (ICASSP), vol. 3, pp. 1635–1638 (2000)
12. Hinton, G., Deng, L., Yu, D., Dahl, G.E., Mohamed, A.R., Jaitly, N., Senior, A., Vanhoucke, V., Nguyen, P., Sainath, T.N., et al.: Deep neural networks for acoustic modeling in speech recognition: the shared views of four research groups. IEEE Signal Process. Mag. **29**(6), 82–97 (2012)
13. Hwang, M., Huang, X.: Shared-distribution hidden Markov models for speech recognition. IEEE Trans. Speech Audio Process. **1**(4), 414–420 (1993)
14. Kapadia, S., Valtchev, V., Young, S.: MMI training for continuous phoneme recognition on the TIMIT database. In: Proceedings of International Conference on Acoustics, Speech and Signal Processing (ICASSP), vol. 2, pp. 491–494 (1993)
15. Kingsbury, B., Sainath, T.N., Soltau, H.: Scalable minimum bayes risk training of deep neural network acoustic models using distributed hessian-free optimization. In: Proceedings of Annual Conference of International Speech Communication Association (INTERSPEECH) (2012)
16. Kumar, N., Andreou, A.G.: Heteroscedastic discriminant analysis and reduced rank HMMs for improved speech recognition. Speech Commun. **26**(4), 283–297 (1998)
17. Morgan, N., Bourlard, H.: Continuous speech recognition using multilayer perceptrons with hidden Markov models. In: Proceedings of International Conference on Acoustics, Speech and Signal Processing (ICASSP), pp. 413–416 (1990)
18. Morgan, N., Bourlard, H.A.: Neural networks for statistical recognition of continuous speech. Proc. IEEE **83**(5), 742–772 (1995)
19. Ostendorf, M., Digalakis, V.V., Kimball, O.A.: From HMM's to segment models: a unified view of stochastic modeling for speech recognition. IEEE Trans. Speech Audio Process. **4**(5), 360–378 (1996)
20. Povey, D.: Discriminative Training for Large Vocabulary Speech Recognition. Ph.D. thesis, Cambridge University Engineering Department, Cambridge (2003)
21. Povey, D., Kanevsky, D., Kingsbury, B., Ramabhadran, B., Saon, G., Visweswariah, K.: Boosted MMI for model and feature-space discriminative training. In: Proceedings of International Conference on Acoustics, Speech and Signal Processing (ICASSP), pp. 4057–4060 (2008)
22. Povey, D., Kingsbury, B., Mangu, L., Saon, G., Soltau, H., Zweig, G.: FMPE: discriminatively trained features for speech recognition. In: Proceedings of International Conference on Acoustics, Speech and Signal Processing (ICASSP), vol. 1, pp. 961–964 (2005)
23. Povey, D., Woodland, P.C.: Minimum phone error and I-smoothing for improved discriminative training. In: Proceedings of International Conference on Acoustics, Speech and Signal Processing (ICASSP), vol. 1, pp. I-105 (2002)
24. Robinson, A.J., Cook, G., Ellis, D.P., Fosler-Lussier, E., Renals, S., Williams, D.: Connectionist speech recognition of broadcast news. Speech Commun. **37**(1), 27–45 (2002)

25. Seide, F., Li, G., Yu, D.: Conversational speech transcription using context-dependent deep neural networks. In: Proceedings of Annual Conference of International Speech Communication Association (INTERSPEECH), pp. 437–440 (2011)
26. Senior, A., Heigold, G., Bacchiani, M., Liao, H.: GMM-free DNN training. In: Proceedings of International Conference on Acoustics, Speech and Signal Processing (ICASSP) (2014)
27. Su, H., Li, G., Yu, D., Seide, F.: Error back propagation for sequence training of context-dependent deep networks for conversational speech transcription. In: Proceedings of International Conference on Acoustics, Speech and Signal Processing (ICASSP) (2013)
28. Trentin, E., Gori, M.: A survey of hybrid ANN/HMM models for automatic speech recognition. Neurocomputing $37(1)$, 91–126 (2001)
29. Yu, D., Deng, L., Dahl, G.: Roles of pre-training and fine-tuning in context-dependent DBN-HMMs for real-world speech recognition. In: Proceedings of Neural Information Processing Systems (NIPS) Workshop on Deep Learning and Unsupervised Feature Learning (2010)
30. Yu, D., Ju, Y.C., Wang, Y.Y., Zweig, G., Acero, A.: Automated directory assistance system-from theory to practice. In: Proceedings of Annual Conference of International Speech Communication Association (INTERSPEECH), pp. 2709–2712 (2007)
31. Zhang, B., Matsoukas, S., Schwartz, R.: Discriminatively trained region dependent feature transforms for speech recognition. In: Proceedings of International Conference on Acoustics, Speech and Signal Processing (ICASSP), vol. 1,pp. I–I (2006)
32. Zhu, Q., Chen, B., Morgan, N., Stolcke, A.: Tandem connectionist feature extraction for conversational speech recognition. In: Machine Learning for Multimodal Interaction, vol. 3361, pp. 223–231. Springer, Berlin (2005)

Chapter 7
Training and Decoding Speedup

Abstract Deep neural networks (DNNs) have many hidden layers each of which has many neurons. This greatly increases the total number of parameters in the model and slows down both the training and decoding. In this chapter, we discuss algorithms and engineering techniques that speedup the training and decoding. Specifically, we describe the parallel training algorithms such as pipelined backpropagation algorithm, asynchronous stochastic gradient descend algorithm, and augmented Lagrange multiplier algorithm. We also introduce model size reduction techniques based on low-rank approximation which can speedup both training and decoding, and techniques such as quantization, lazy evaluation, and frame skipping that significantly speedup the decoding.

7.1 Training Speedup

Speech recognition systems are typically trained with thousand or even tens of thousand hours of speech data. Since we extract one frame of feature every 10 ms, every 24 h of data translate to

$$24\,\text{h} \times 60\,\text{min/h} \times 60\,\text{s/min} \times 100\,\text{frames/s} = 8.64\,\text{million frames,}$$

and 1,000 h of data are equivalent to 360 million frames (or samples). This is huge especially considering at the same time the size of the deep neural networks (DNNs). Techniques that can speedup the training is thus very important.

The training speed may be boosted by using high-performance computing devices such as general purpose graphical processing units (GPGPUs) and/or using better algorithms that can reduce model parameters, converge faster, or take advantage of multiple processing units. From our experience, we have found that GPGPUs perform significantly faster than multicore CPUs and are preferred platforms for training DNNs.

© Springer-Verlag London 2015

D. Yu and L. Deng, *Automatic Speech Recognition*,
Signals and Communication Technology, DOI 10.1007/978-1-4471-5779-3_7

7.1.1 Pipelined Backpropagation Using Multiple GPUs

It is well known that minibatch based stochastic gradient descend (SGD) training algorithm can easily scale to large datasets on a single computing device. However, the need to use minibatches of only few hundred samples makes parallelization difficult. This is because the model is updated after each minibatch and requires prohibitive bandwidth if the naive parallelization across data is used. For example, a typical 2kx7 (7 hidden layers each with 2k neurons) CD-DNN-HMM has around 50–100 million floating parameters or 200–400 million bytes. Each minibatch would require the distribution of 400 MB worth of gradients and gathering another 400 MB of model parameters, per server. If each minibatch takes about 500 ms to compute which is easy to achieve on GPGPUs, we get close to the PCIe-2 limit (about 6 GB/s).

The minibatch size, however, is typically determined by two factors. Smaller minibatch size means more frequent update of the model but also means less effective usage of the GPU's computing power. Larger minibatch can be more efficiently computed but the whole training process requires more passes over the training set. An optimal minibatch size can be obtained by balancing these two factors. Figure 7.1 shows runtime (right y-axis) and early frame accuracies (left y-axis) for different minibatch sizes (x-axis) after seeing 12 h of data. In these experiments, if the minibatch size is larger than 256 the minibatch size was set to 256 for the first 2.4 h of data and then increased to the actual size. It can be seen that the optimal minibatch size is in the 256–1,024 range.

If NVidia S2090 (hosted on T620), the best GPU listed in Fig. 7.1, is used, getting a good performing DNN trained with the cross-entropy criterion and 300 h of training data costs about 15 days. If 2,000 h of data are used, the total time is increased to 45 days. Note that the total time is increased by about 3 times instead of 2,000/300≈ 6.7 times. This is because although each pass of the data costs 6.7 times, it takes

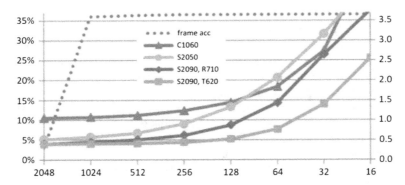

Fig. 7.1 Relative runtime for different minibatch sizes and GPU/CPU model types, and corresponding frame accuracy measured after seeing 12 h of data. *Left y-axis* frame accuracy; *right y-axis* runtime relative to C1060 with minibatch size of 2,048 (Figure from Chen et al. [5], permitted to use by ISCA)

less passes when trained with SGD. The training time can be reduced to about 20 days if the newer GPU model such as K20X is used. Even with this new GPU, however, training with 20 K hours of data still takes over two months. Parallel training algorithm thus is very important to support training with large datasets.

Let's denote K the number of GPUs (for example, 4), T the size of a mini-batch (like 1,024), N the dimensions of all hidden layers , e.g., 2,048, and J the output dimension (number of senones), e.g., 9,304. If we use the simple classic map-reduce [7] approach which achieves parallelization by splitting the training data, it would require accumulation/redistribution of gradients/models of the dimension of the entire model to/from a master server to the other $K - 1$ GPUs for each minibatch. On the shared bus between GPGPUs, bandwidth per minibatch is of $\mathcal{O}(N \cdot (T + 2(L \cdot N + J)(K - 1)))$. A tree-structured communication architecture could reduce it to $\mathcal{O}(N \cdot (T + 2(L \cdot N + J)\lceil \log_2 K \rceil))$, where $\lceil x \rceil$ is the minimum integer that is larger than or equal to x.

Alternatively, we can partition each layer's model parameters into stripes and distribute stripes in different computing nodes. In this node parallelization approach, each GPU holds one out of K vertical stripes of each layer's parameters and gradients. Model update happens only locally within each GPU. In forward computation, each layer's input $v^{\ell-1}$ gets distributed to all GPUs, each of which computes a slice of the output vector v^ℓ. The slices are then distributed to all other GPUs for computing the next layer. In backpropagation, error vectors are parallelized as slices, but the resulting matrix products from each slice are partial sums that need to be further summed up. As a result, in both forward computation and backpropagation, each vector is transferred $K - 1$ times. The bandwidth is of $\mathcal{O}(N \cdot (K - 1) \cdot T \cdot (2L + 1))$.

The pipelined backpropagation [5, 18] avoids the multiple copying of data vectors of the striping method, by distributing the layers across GPUs to form a pipeline. Data, instead of model, flows from GPU to GPU. All GPUs work simultaneously on the data they have. As an example, in Fig. 7.2, the two-hidden-layer DNN is split and

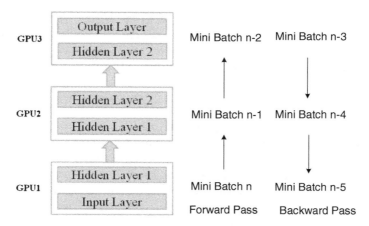

Fig. 7.2 Illustration of a pipelined parallelization paradigm

stored in three GPUs. When the first batch of training data comes in, it's processed in GPU1. The activations (outputs) of hidden layer 1 is then passed to GPU2 for processing. At the same time, a new batch comes in and is processed in GPU1. After three batches, all GPUs are occupied. This suggests a speedup of three if all layers are balanced. The backpropagation pass is processed in the similar way. After six batches, all GPUs process both a forward batch and a backward batch. Since GPUs use the single instruction multiple data (SIMD) architecture, we can update the model first and then do the forward pass at each layer. This guarantees that the most recently updated weights are used when the forward computation is conducted to reduce the delayed-update problem we will mention below.

In the pipeline architecture, each vector travels twice per GPU, once for forward computation and once for backpropagation. The bandwidth is $\mathscr{O}(N \cdot T \cdot (2K - 1))$, which is cheaper than data parallelization and striping. If the number of layers L is larger than the number of GPUs K, you may group several layers in the same GPU. Lastly, asynchronous data transfer and appropriate order of execution allows most data transfers to happen in parallel to computation, which can reduce the effective communication time to close to zero.

Note that the efficiency comes with a cost. This is because there is a mismatch between the weights used to do forward computation and that used to do backpropagation. For example, the weights used for forward computation of batch n on GPUs 1, 2, and 3 are updated after batch $n - 5$, $n - 3$, and $n - 1$, respectively. However, when computing the gradients, these weights have already been further updated on batches $n - 1$ and $n - 2$, respectively on GPUs 2 and 1, although on GPU3 they are the same. This means in lower layers, due to the delay in the pipeline process, the gradients calculated are not accurate. Based on this analysis, we can consider the delayed update as a special complicated momentum technique in which the update (smoothed gradient) is a function of previous models and gradients. For this reason, when the pipeline is too long performance degradation can be observed if the same minibatch size is used [5]. To alleviate the side effect of the delayed update, we need to cut the minibatch size.

The key to achieve great speedup is to balance the computation on each GPU. If the number of layers is a plural of the number of GPUs and all layers have the same dimension, balancing is trivial. However, in CD-DNN-HMMs the softmax layer typically dominates the number of parameters. This is because the number of senones is often around 10 K and the hidden layer size is typically around 2 K. To balance the computation, we need to use stripe for the softmax layer and pipeline for the rest.

Table 7.1, quoted from Chen et al. [5], shows training runtime using up to 4 GPUs (NVidia Tesla S2090) in a single server (Dell PowerEdge T620), measured for 429 input feature dimension, $L = 7$ hidden layers, $N = 2,048$ hidden dimensions, and $J = 9,304$ senones. From this table, we can observe that speedups of 1.7–1.9 can be achieved on dual GPUs (e.g., reducing runtime from 61 to 33 min for minibatch size of 512), at no accuracy loss despite its delayed-update approximation. To achieve this speedup, GPU1 contains five weight matrices and GPU2 has only two due to the unbalanced softmax layer. The computation time ratio on these two GPUs

Table 7.1 Training runtime in minutes per 24 h of data for different parallelization configurations using pipelined backpropagation

Method	# GPUs	Minibatch size		
		256	512	1024
Baseline (single GPU)	1	68	61	**59**
Pipeline (0..5; 6..7)	2	36	33	**31**
Pipeline (0..2; 3..4; 5..6;7)	4	32	**29**	[27]
Pipeline + striped top layer (0..3; 4..6; 7L; 7R)	4	20	**18**	[[18]]

[[·]] denotes divergence, and [·] denotes a greater than 0.1 % word error rate (WER) loss on the test set (Quoted from Chen et al. [5])

are $(429 + 5 \times 2{,}048) \times 2{,}048 : (2{,}048 + 9{,}304) \times 2{,}048 = 0.94{:}1$ and thus are very balanced. Going to 4 GPUs using pipeline alone barely helps. The overall speedup remains below 2.2 (e.g.,~61 vs.~29 min). This is because the softmax layer is 4.5 times larger $(9{,}304 \times 2{,}048$ parameters) than the hidden layers $(2{,}048^2)$, and is thus the limiting bottleneck. The computation time ratio on the four GPUs is $(429 + 2 \times 2{,}048) \times 2{,}048 : (2 \times 2{,}048) \times 2{,}048 : (2 \times 2{,}048) \times 2048{:}9304 \times 2048 = 1.1{:}1{:}1{:}2.27$. In other words, GPU4 takes twice as time as other GPUs. However, if pipelined BP is combined with the striping method, which is applied only to the softmax layer, significant speedup can be achieved. In this configuration, four GPUs are assigned with layers (0..3; 4..6; 7L; 7R) where L and R denote left and right stripe respectively. In other words, two GPUs jointly form the top stage of the pipeline, while the lower 7 layers are pipelined on the other two GPUs. Under this condition, similar calculation indicates that the computation cost ratio on the four GPUs is $(429 + 3 \times 2{,}048) \times 2{,}048 : (3 \times 2{,}048) \times 2{,}048 : 4{,}652 \times 2{,}048 : 4{,}652 \times 2{,}048 = 1.07{:}1{:}0.76{:}0.76$. At no word error rate (WER) loss, the fastest pipelined system (18 min to process 24 h of data with minibatch size 512) runs 3.3 times faster on four GPUs than the fastest single-GPU baseline (59 min to process 24 h of data with minibatch size 1024), a 3.3 times speedup.

The drawback of the pipelined backpropagation is obvious. The overall speedup heavily depends on whether you can find a way to balance the computation across GPUs. In addition, due to the delayed-update effect, It is not easy to extend the same speedup to more GPUs.

7.1.2 Asynchronous SGD

The model training can be parallelized using another technique referred as asynchronous SGD (ASGD) [12, 17, 26]. It was first proposed to run across CPU servers as shown in Fig. 7.3. In this architecture, the DNN model is stored across several (3 in the figure) computers referred as parameter server pool. The parameter server pool is the master. The master sends the model parameters to the slaves, each

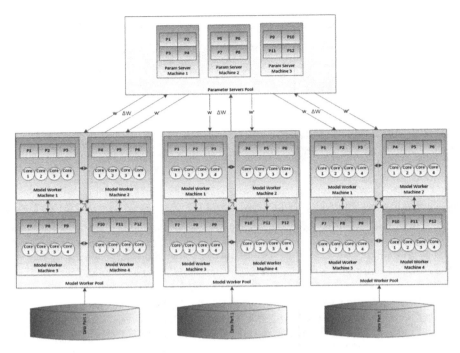

Fig. 7.3 Illustration of asynchronous SGD. Shown in the figure is a master server pool and three slave clusters (Picture courtesy of Erdinc Basci)

of which is consisted of several (4 in the figure) computers. Each slave works on a subset of the training data. It calculates and sends the gradients of each minibatch to the master. The master updates the parameters and sends the new values back to the slave.

Since each computer in the slave contains part of the model, the activations need to be copied across computers. To reduce the communication cost, the models need to have sparse connection between the components stored on different computers. Having several computers on the master helps to reduce the communication cost between the master and slaves since each computer pair only needs to transfer a subset of the parameters. The key to the success of the ASGD, however, is to use asynchronous lock-free update. In other words, the parameters on the server is not locked during updating. When the master gets gradients from several slaves, it updates the model independently in different threads. When the master sends the new parameter to the slave, part of the parameters may be updated using gradients sent from several slaves. At the first glance, this may cause convergence problem. In practice, however, the parameters converge fine and the model training time is significantly reduced since each slave does not need to wait for other slaves to finish [12, 17]. The model gradually evolves as it is exposed to random records from the training dataset. A proof on the convergence of ASGD is given in [17].

There are several practical issues to consider in the ASGD. First, some slaves may take longer than others to finish an epoch. As a result, the gradient calculated on these slaves might be based on a very old model. The simplest approach to handle this problem is to send a time stamp in all communications. The master just abandons the outdated gradient and sends the slave the most updated model if the time stamp difference between that sent from the slave and that on the master exceeds a threshold. If a slave is consistently slower, the data assigned to that slave needs to be redistributed to other slaves. This can be easily done by fetching minibatches from the same pool. Second, it is obvious that the delayed-update problem that happens in the pipelined BP also happens in ASGD. For this reason, we need to either reduce the number of slaves or reduce the learning rate to compensate for the problem. Either of these solutions, however, will slow down the training. Third, the delayed-update problem manifests itself most when the gradients are large, which typically happens at the early stage of the model training. This problem can be alleviated by a technique called warm start, which starts the ASGD training from a model trained with one pass of SGD.

Although ASGD works on CPU clusters [12], the communication cost is very high and can become the bottleneck. For example, running ASGD on 1,000 distributed CPU cores perform approximately as fast as 8 GPUs on a single machine. The main reason to use ASGD on CPUs is to take advantage of the existing CPU clusters and to train models that cannot fit to GPU memory.

Alternatively, the ASGD algorithm can be applied to GPUs on a single hosting machine [26]. Since in speech recognition the DNN model can be fit into both CPU and GPU memory, we can use the hosting machine (CPU) as the master and each GPU as a slave. Note that with GPU-based ASGD the overall speed is significantly improved. This is because each minibatch takes much less time on GPUs and the communication between the GPUs and the hosting machine (through PCIe bus) is significantly faster than that between CPU machines. Even on GPUs the communication may still become the bottleneck if the minibatch is too small. This problem can be addressed by reducing the frequency of data transmission between the master and the slaves. Instead of updating the model after every minibatch, the GPU slave can accumulate updated gradients and send them to the master every three to four minibatches. Since this essentially increased the minibatch size, the learning rate may need to be reduced to compensate for it.

Table 7.2, which is extracted from [26], compares the character error rate (CER) and training time on 10 h of data between SGD and ASGD on a Chinese task. The 42-dimensional feature is formed from 13-dimensional PLP and a 1-dim pitch with the first- and second-order derivatives appended. The DNN is trained on a 130-h dataset and is tuned by another one hour data as development set. Concatenations of 11 frames are used as input to the DNN, which has 5 hidden layers each with 2,048 neurons. The output layer has 10,217 senones. The systems are evaluated on two individual test sets, namely clean7k and noise360, which were collected through mobile microphone under clean and noise environments, respectively. The NVidia GeForce GTX 690 was used for training. This table indicates that ASGD achieves a 3.2 times speedup on four GPUs compared to the SGD running on a single GPU.

Table 7.2 Compare character error rate (CER) and training times in minutes per 10 h of data on a Chinese speech recognition task (Summarized from [26])

	CER		Time (min)
	Clean7K (%)	Noise360 (%)	
GMM BMMI	11.30	36.56	–
DNN SGD	9.27	26.99	195.1
DNN ASGD (4 GPU)	9.05	25.98	61.1

7.1.3 Augmented Lagrangian Methods and Alternating Directions Method of Multipliers

It is typically not practical to have more than four GPUs on a single hosting machine. Even if it does, the speedup achieved with ASGD and pipelined BP is less desirable on more than four GPUs due to the delayed update. To support training with even more data we still need to take advantage of multiple GPU/CPU machines. Augmented Lagrangian methods (ALMs) [2, 9, 19] aim to address this request.

In the DNN training, we optimize the model parameters θ to minimize the empirical risk $J(\theta; \mathbb{S})$ on the training set $\mathbb{S}$. Note that

$$J(\theta; \mathbb{S}) = \sum_{k=1}^{K} J(\theta; \mathbb{S}_k), \tag{7.1}$$

where $\mathbb{S}_k$ is the k-th subset of the training data, $\mathbb{S}_k \cap \mathbb{S}_i = \emptyset, \forall k \neq i$, and $\bigcup_{k=1}^{K} \mathbb{S}_k = \mathbb{S}$. This suggests we may optimize the model parameters on different subsets using different processors (either on the same or different computers) independently if we can guarantee that models trained on different subsets are the same, which can be enforced by equality constraints. The distributed training problem can thus be converted into the constrained optimization problem

$$\min_{\theta, \theta_k} J(\theta, \theta_k; \mathbb{S}) = \sum_{k=1}^{K} \min_{\theta, \theta_k} J(\theta_k; \mathbb{S}_k), \ s.t. \ \theta_k = \theta, \tag{7.2}$$

where θ_k are the local model parameters and θ are the global consensus model parameters.

This constrained problem can be converted into an unconstrained optimization problem

$$\min_{\theta, \theta_k, \lambda_k} J(\theta, \theta_k, \lambda_k; \mathbb{S}) = \sum_{k=1}^{K} \min_{\theta, \theta_k, \lambda_k} \left[J(\theta_k; \mathbb{S}_k) + \lambda_k^T (\theta_k - \theta) \right] \tag{7.3}$$

using the Lagrange multiplier method, where λ_k are Lagrange multipliers or dual parameters. This constrained problem can be solved using the dual ascent method if the training criterion is strictly convex and is finite. However, the training criteria such as cross entropy and mean square error used in the DNNs are not convex. As the result, the convergence property of the dual ascent formulation is very poor.

To bring robustness to the dual ascent method and to improve convergence property, we can add a penalty term to the unconstrained optimization problem so that it becomes

$$\min_{\theta,\theta_k,\lambda_k} J_{\text{ALM}}(\theta,\theta_k,\lambda_k;\mathbb{S}) = \sum_{k=1}^{K} \min_{\theta,\theta_k,\lambda_k} \left[J(\theta_k;\mathbb{S}_k) + \lambda_k^T(\theta_k - \theta) + \frac{\rho}{2}\|\theta_k - \theta\|_2^2 \right],$$

(7.4)

where $\rho > 0$ is called the penalty parameter. This new formulation is called augmented Lagrangian multiplier. Note that Eq. 7.4 can be viewed as the (nonaugmented) Lagrangian associated with the problem

$$\min_{\theta_k} J(\theta_k;\mathbb{S}) + \frac{\rho}{2}\|\theta_k - \theta\|_2^2, \quad s.t. \, \theta_k = \theta,$$

(7.5)

which has the solution as that of the original problem Eq. 7.2. This problem can be solved using the alternating directions method of multipliers (ADMM) [3] as

$$\theta_k^{t+1} = \min_{\theta_k} \left[J(\theta_k;\mathbb{S}_k) + \left(\lambda_k^t\right)^T(\theta_k - \theta^t) + \frac{\rho}{2}\|\theta_k - \theta^t\|_2^2 \right],$$

(7.6)

$$\theta^{t+1} = \frac{1}{K}\sum_{k=1}^{K}\left(\theta_k^{t+1} + \frac{1}{\rho}\lambda_k^t\right),$$

(7.7)

$$\lambda_k^{t+1} = \lambda_k^t + \rho\left(\theta_k^{t+1} - \theta^{t+1}\right).$$

(7.8)

Equation 7.6 solves the primal optimization problem using the SGD algorithm distributively. Equation 7.7 collects updated local parameters from each processor and estimates the new global model parameters on the parameter server. Equation 7.8 retrieves the global model parameters from the parameter server and updates the dual parameters.

This algorithm is a batch algorithm. However, it can be applied to minibatches larger than that typically used in SGD to speedup the training process without introducing too much communication overhead. Since DNN training is a nonconvex problem, however, ALM/ADMM often converges to a point that performs slightly worse than that achieved with the normal SGD algorithm. To close the gap, we need to initialize the model with SGD (often called warm start), then use ALM/ADMM, and then finalize the training with SGD, L-BFGS, or Hessian-free algorithm. This algorithm can also be combined with ASGD or pipelined BP to further speed up the training process. The ALM/ADMM algorithm can be used to train not only DNNs but also other models.

7.1.4 Reduce Model Size

The training speed can be improved not only by using better training algorithms but also by using smaller models. The naive way of reducing the model size is to use less hidden layers and less neurons at each layer, which, unfortunately often reduces the recognition accuracy [6, 21]. An effective model size reduction technique is based on the low-rank factorization [20, 24].

There are two indications that the weights in the DNNs are approximately low rank. First, in CD-DNN-HMMs, the DNN's output layer typically has a large number of neurons (i.e., 5,000–10,000), equal to or more than the number of senones (tied triphone states) of an optimal GMM/HMM system, to achieve good recognition performance. As the result, the last layer contributes about 50 % of the parameters and training computation in the system. However, during the decoding typically only few output targets are active. It is thus reasonable to believe that those activated outputs are correlated (i.e., belong to the same set of confusable context-dependent HMM states). This suggests that the weight matrix of the output layer has low rank. Second, it has been shown that only the largest 30–40 % of the weights in any layer of the DNNs are important. The DNN performance does not decrease if the rest of the weights are set to zero [25]. This indicates that each weight matrix can be approximated with low-rank factorization without sacrificing the recognition accuracy.

With low-rank factorization, each weight matrix can be factorized into two smaller matrices, thereby significantly reducing the number of parameters in the DNN. Using low-rank factorization not only reduces the model size but also can constrain the parameter space and helps to make the optimization more efficient and reduce the number of training iterations.

Let us denote an $m \times n$ low-rank matrix by $\mathbf{W}$. If $\mathbf{W}$ has rank r, then there exists a factorization $\mathbf{W} = \mathbf{W}_2\mathbf{W}_1$ where $\mathbf{W}_2$ is a rank r matrix of size $m \times r$ and $\mathbf{W}_1$ is a rank r matrix of size $r \times n$. If we replace matrix $\mathbf{W}$ by the multiplication of matrices $\mathbf{W}_2$ and $\mathbf{W}_1$, we may reduce the model size and speedup the training if $m \times r + r \times n < m \times n$, or alternatively $r < \frac{m \times n}{m+n}$. If we want to reduce the parameter by a fraction p, we need $r < \frac{p \times m \times n}{m+n}$. When $\mathbf{W}$ is replaced by $\mathbf{W}_1$ and $\mathbf{W}_2$, it is equivalent to introduce a linear layer $\mathbf{W}_1$ followed by a nonlinear layer $\mathbf{W}_2$ as shown in Fig. 7.4. This is because

$$\mathbf{y} = f(\mathbf{W}\mathbf{v}) = f(\mathbf{W}_2\mathbf{W}_1\mathbf{v}) = f(\mathbf{W}_2\mathbf{h}), \tag{7.9}$$

where $\mathbf{h} = \mathbf{W}_1\mathbf{v}$ is a linear transformation. It is shown in [20] that there is typically no performance degradation if only the softmax layer is factorized to matrices with ranks between 128 and 512 for different tasks.

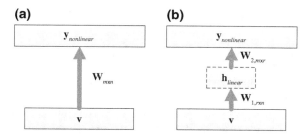

Fig. 7.4 Illustration of low-rank factorization. Replacing weight matrix **W** with the multiplication of two smaller weight matrices **W**₂ and **W**₁ is equivalent to replacing the nonlinear layer **W** with a linear layer **W**₁ followed by a nonlinear layer **W**₂

7.1.5 Other Approaches

Other approaches have also been proposed. For example, one can train 4 DNNs each with 1/4 of the data and then train a top layer network to combine the predictions from the 4 separate DNNs. Further reduction can be achieved by clustering the senones and then training a classifier to classify the clusters and a DNN for each cluster of the senones [27]. Since the output layer size is smaller (only contains senones that belong to the cluster) and the training set for the individual DNN is also smaller (only frames related to the output senones are used), the training speed can be significantly reduced. It was reported that a 5 times speedup with only 1–2 % relative word error reduction can be achieved on four GPUs [27]. This approach, however, requires additional computation power at the decoding time since the clustered DNNs are separated and typically have the same number of hidden layers and hidden layer neurons as the singe DNN system. In other words, this approach trades decoding time for training time. Another popular approach is to use the Hessian-free training algorithm [10, 15, 16], which takes much larger batch size and can be easily parallelized across machines. However, this algorithm is much slower than the SGD on a single GPU and requires a lot of practical tricks to make it work.

7.2 Decoding Speedup

The large number of parameters that need to be evaluated at every frame also has the disadvantage of making real-time inference more computationally challenging. However, careful engineering and clever techniques can significantly improve the decoding speed. In this section, we discuss the quantization and parallel computation techniques proposed in [23], the sparse DNN proposed in [25], the low-rank factorization technique proposed in [24], and the multiframe DNN proposed in [22]. Each of these techniques can improve the decoding speed. The best practice is to combine these techniques.

7.2.1 Parallel Computation

The obvious solution to speeding up the decoding time is to parallelize the DNN computation. This is trivial on GPUs. However, in many tasks, it is more economically efficient to use commodity CPU hardware. Fortunately, modern CPUs often support instruction sets that can support single instruction multiple data (SIMD) low-level parallelization. These instructions perform the same multiple operations in parallel on contiguous data. On Intel and AMD CPUs of the x86 family, they typically operate on 16 bytes (e.g., 2 doubles, 4 floats, 8 shorts, or 16 bytes) worth of data at a time. By taking advantage of these new instruction sets, we can greatly improve the decoding speed.

Table 7.3, extracted from [23], summarizes the techniques applicable to the CPU decoder and the real-time factors (RTFs, defined as the processing time divided by the playback time) achievable with a DNN configured as 440:2000X5:7969. This is a typical DNN with 11 frames of features as input. Each frame consists of 40 log-energies extracted from filterbanks on a Mel frequency scale. Sigmoid nonlinearity is used in all the 5 hidden layers each of which has 2000 neurons. The output layer has 7,969 senones. The results were obtained on an Intel Xeon DP Quad Core E5640 machine with Ubuntu OS. CPU scaling was disabled and each run was performed a minimum of 5 times and averaged.

From the table, it is clear that the naive implementation would require 3.89 real time to just compute the posterior probabilities from the DNN. Using the floating-point SSE2 instruction set, which operates on 4 floats at a time, the decoding time can be reduced significantly to 1.36 real time. However, this is still very expensive and slower than real time. In contrast, we may alternatively linearly quantize the 4-byte floating-point values of the hidden activations (constrained within (0, 1) if sigmoid activation function is used) to unsigned char (1 byte) and the weight values to signed char (1 byte). The biases can be encoded as 4-byte integer, and the input remains the floating point. This quantization technique can reduce the time to 1.52 real time

Table 7.3 Real-time factor (RTF) on a typical DNN (440:2000X5:7969) used in speech recognition with different engineering optimization techniques (Summarized from Vanhoucke et al. [23])

Technique	Real-time factor	Note
Floating-point baseline	3.89	Baseline
Floating-point SSE2	1.36	4-way parallelization (16 bytes)
8-bit quantization	1.52	Activation: unsigned char; Weight: signed char
Integer SSSE3	0.51	16-way parallelization
Integer SSE4	0.47	Faster 16-32 conversion
Batching	0.36	Batches over tens of milliseconds
Lazy evaluation	0.26	Assume 30 % active senones
Batched lazy evaluation	0.21	Combine batching and lazy evaluation

even without using SIMD instructions. Quantization also reduced the model size to 1/3–1/4.

When the integer SSSE3 instruction set is applied to the 8-bit quantized values, which allows for 16-way parallel computation, additional 2/3 of time is reduced and the overall computation time is reduced to 0.51 real time. Using the integer SSE4 instruction set, which introduces one small optimization with a single instruction for 16–32-bit conversion, a slight gain can be observed and the time is reduced to 0.47 real time.

In speech recognition, even in the online recognition mode, it is common to incorporate a lookahead of a few hundred milliseconds, especially in the beginning of an utterance, to help improve runtime estimates of speech and noise statistics. This means process frames in small batches over tens of milliseconds will not affect latency too much. To take full advantage of batching, the batches have to be propagated through the neural network layers in bulk, so that every linear computation becomes a matrix–matrix multiply which can take advantage of CPU caching of both weights and activations. Using batching can further reduce the computation time to 0.36 real time.

One last trick to further improve the decoding speed is to compute the senone posteriors only if needed. It is well known that during decoding, at every frame, only a fraction (25 to 35 %) of the state scores ever need to be computed. In the GMM–HMM system, this can be easily exploited since every state has its own, small, set of Gaussians. In the DNN, however, all the hidden layers are shared and need to be computed even if only one state is active, except the last layer, in which only the neurons corresponding to the necessary state posteriors need to be computed. This means we can lazily evaluate the output layer. Evaluating the output layer in a lazy manner, however, adds inefficiency to the matrix computation and hence introduces a small fixed cost of about 22 % relatively. Overall, using lazy evaluation (without batching) can reduce the time to 0.26 real time since the output layer dominates (typically account for around 50 %) the computation.

With lazy evaluation, however, one can no longer compute batches of output scores for all outputs across multiple frames although we can continue to batch the computation of all hidden layers. Furthermore, since a state is very likely to be needed at frame $t + 1$ if it is needed by the decoder at frame t, it is still possible to compute a batch of these posteriors for consecutive frames at the same time while the weights are in cache. Combining lazy evaluation with batch further reduced the DNN computation time to 0.21 real time.

Overall, the engineering optimization techniques achieved near 20 times speedup (reduced from 3.89 real time to 0.21 real time) compared to the naive implementation. Note that this 0.21 real time is only the DNN posterior probability computation time. The decoder also needs to search over all possible state sequences which typically adds another 0.2–0.3 real time on average and 06–0.7 real time for extreme cases depending on the language model perplexity and beam used in the search. Overall, the complete decoding time is within real time without loss of accuracy.

7.2.2 Sparse Network

On some devices, such as smart phones, SIMD instructions may not be available. Under those conditions, we still can use 8-bit quantization to improve the decoding speed. However, we cannot use many other parallel computation techniques we discussed in the Sect. 7.2.1.

Fortunately, by inspecting the fully connected DNNs after the training we can notice that a large portion of all connections have very small weights. For example, in a typical DNN used in speech recognition, the magnitude of 70 % of weights is below 0.1 [25]. This means we may reduce model size and increase decoding speed by removing connections with small weight magnitude. Note that we do not observe similar patterns on bias parameters. This is expected since nonzero bias terms indicate the shift of hyperplanes from the origin. However, given that the number of bias parameters is very small compared to that of weight parameters, keeping bias parameters intact does not affect the final model size and decoding speed in a noticeable way.

There are many ways to generate sparse models. For example, the task of enforcing sparseness can be formulated as a multiobjective optimization problem since we want to minimize the cross entropy and the number of nonzero weights at the same time. This two-objective optimization problem can be converted into a single objective optimization problem with L1 regularization. Unfortunately, this formulation does not work well with the stochastic gradient descent (SGD) training algorithm often employed in the DNN training [25]. This is because the subgradient update does not lead to precise sparse solutions. To enforce a sparse solution, one often truncates the solutions after each T steps by forcing parameters with magnitude smaller than a threshold θ to zero [11]. This truncation step, however, is somewhat arbitrary and T is difficult to select correctly. In general, it is not desirable to take a small T (e.g., 1), especially when the minibatch size is small, since in that case each SGD update step only slightly modifies weights. When a parameter is close to zero, it remains so after several SGD updates and will be rounded back to zero if T is not sufficiently large. Consequently, truncation can be done only after (a reasonably large) T steps in the hopes that nonzero coefficients have sufficient time to go above θ. On the other hand, a large T means that every time the parameters are truncated, the training criterion will be reduced and will require a similar number of steps to get the loss compensated.

Another well-known work [8, 13] pruned the weights after training converges based on the second-order derivatives. Unfortunately, these algorithms are difficult to scale up to large training set we typically use in speech recognition and their advantages vanish if additional training iterations are carried out upon the pruned weights.

A third approach, which performs well and generates good model is to formulate the problem as an optimization problem with a convex constraint

$$\|\mathbf{W}\|_0 \leq q, \tag{7.10}$$

where q is a threshold value for the maximal number of nonzero weights allowed.

This constrained optimization problem is hard to solve. However, an approximate solution can be found following two observations: First, after sweeping through the full training set several times the weights become relatively stable—they tend to remain either large or small magnitudes. Second, in a stabilized model, the importance of the connection is approximated well by the magnitudes of the weights.[1] This leads to the very simple yet efficient and effective algorithm.

We first train a fully connected DNN by sweeping through the full training set several times. We then keep only the connections whose weight magnitudes are in top q. Continue training the DNN and keep the same sparse connections unchanged. This can be achieved either by masking the pruned connections or round weights with magnitude below $\min\{0.02, \theta/2\}$ to zero, where θ is the minimal weight magnitude that survived the pruning and 0.02 is determined by examining the patterns of weights in the fully connected network. The masking approach is cleaner but requires storage of a huge masking matrix. The rounding alternative is cheaper but trickier since it is important to round only weights smaller than $\min\{0.02, \theta/2\}$, instead of θ, to zero. This is because the weights may shrink and be suddenly removed if not doing so. In addition, after the pruning it is very important to continue training to remedy the accuracy degradation caused by sudden removal of the small weights.

Tables 7.4 and 7.5, provided in [25], summarize the experimental results on the voice search (VS) and Switchobard (SWB) datasets described in Sect. 6.2.1. By exploiting the sparseness property in the model, we can obtain 0.2–0.3 % error reduction and simultaneously reduce the connections to only 30 % on both the VS and SWB datasets. Alternatively, we can reduce the number of weights to 12 % and

Table 7.4 Model size, computation time, and sentence error rate (SER) with and without sparseness constraints on the VS dataset

Acoustic model	# nonzero params	% nonzero params	Hub5'00 FSH (%)	RT03S SWB (%)
GMM MPE	1.5M	–	34.5	36.2
DNN, CE	19.2M	Fully connected	28.0	30.4
	12.8M	67 %	27.9	30.3
	8.8M	46 %	27.7	30.1
	6.0M	31 %	27.7	30.1
	4.0M	21 %	27.8	30.2
	2.3M	12 %	27.9	30.4
	1.0M	5 %	29.7	31.7

The fully connected DNN contains 5 hidden layers each with 2,048 neurons. The OOV rate for both the dev and test sets is about 6 % (Summarized from Yu et al. [25])

---

[1] More precisely, it can be approximated by the magnitudes of the product of the weights and the input values. However, the magnitude of the input values are relatively uniform within each layer since on the input layer, features are normalized to zero-mean and unit-variance, and hidden layer values are probabilities.

Table 7.5 Model size, computation time, and word error rate (WER) with and without sparseness constraints on the SWB dataset

Acoustic model	# nonzero params	% nonzero params	Hub5'00 FSH	RT03S SWB (%)
GMM, BMMI	29.4M	–	23.6%	27.4
DNN, CE	45.1M	Fully connected	16.4%	18.6
	31.1M	69%	16.2v	18.5
	23.6M	52%	16.1%	18.5
	15.2M	34%	16.1%	18.4
	11.0M	24%	16.2%	18.5
	8.6M	19%	16.4%	18.7
	6.6M	5%	16.5%	18.7

The fully connected DNN contains 7 hidden layers each with 2,048 neurons (Summarized from Yu et al. [25])

19%, respectively, on the VS and SWB datasets, without sacrificing recognition accuracy. In that case, the CD-DNN-HMM is only 1.5 and 0.3 times as large as the CD-GMM-HMM on the VS and SWB datasets, respectively, and takes only 18% and 29% of the model size compared to the fully connected models. This translates to reducing the DNN computation to only 14% and 23% of that needed by the fully connected models on the VS and SWB datasets respectively if SIMD instructions are not available.

The sparse weights learned generally have random patterns. This prevents it from being very efficient both in storage and in computation even if high degree of sparseness can be achieved, especially when SIMD parallelization is used.

7.2.3 Low-Rank Approximation

The low-rank matrix factorization technique can reduce not only the training time but also the decoding time. In Sect. 7.1.4, we mentioned that we can replace the softmax layer with two smaller low-rank matrices even before training started. However, this approach has several limitations. First, it is not easy to know beforehand the rank to keep and thus it requires building models with different rank r. Second, if we use the low-rank technique in the lower layers, the performance of the final model may become very bad [20]. In other words, we cannot reduce the size and decoding time for the lower layers. However, the lower layers need to be calculated even if there is only one activated senone. It is thus very important to reduce the size of the lower layers.

If we only care about the decoding time, we can determine the rank r by using singular value decomposition (SVD) [24]. Once we have trained a fully connected model, we can convert each $m \times n$ ($m \geq n$) weight matrix $\mathbf{W}$ with SVD

$$\mathbf{W}_{m \times n} = \mathbf{U}_{m \times n} \mathbf{\Sigma}_{n \times n} \mathbf{V}_{n \times n}^T, \tag{7.11}$$

where $\mathbf{\Sigma}$ is a diagonal matrix consisted of nonnegative singular values in the decreasing order, $\mathbf{U}$ and $\mathbf{V}^T$ are unitary, the columns of which form a set of orthonormal vectors, which can be regarded as basis vectors. The m columns of $\mathbf{U}$ and the n columns of $\mathbf{V}$ are called the left-singular vectors and right-singular vectors of $\mathbf{W}$, respectively. Since we have discussed that a large percentage of the weights are close to zero, many singular values should be close to zero. Experiments have shown that for a typical weight matrix in the DNN, 40 % of the top singular values account for 80 % of total singular value. If we only keep top k singular values, the weight matrix $\mathbf{W}$ can be approximated with two smaller matrices

$$\mathbf{W}_{m \times n} \simeq \mathbf{U}_{m \times k} \mathbf{\Sigma}_{k \times k} \mathbf{V}_{k \times n}^T = \mathbf{W}_{2,m \times k} \mathbf{W}_{1,r \times n}, \tag{7.12}$$

where $\mathbf{W}_{2,m \times k} = \mathbf{U}_{m \times k}$ and $\mathbf{W}_{1,r \times n} = \mathbf{\Sigma}_{k \times k} \mathbf{V}_{k \times n}^T$. Note that after dropping small singular values, the approximation error increases. It thus is important to continue training the model after the low-rank factorization similar to the sparse network approaches. Experiments have shown that no or little performance loss was observed if we only keep 30 % of the model size [24]. This result also agrees with the observation made in the sparse network [25]. However, the low-rank factorization approach is preferred between two since it can take advantage of SIMD architecture easily and achieve better speedup overall across different computing devices.

7.2.4 Teach Small DNN with Large DNN

The low-rank approximation technique can only reduce the model size and decoding time by 2/3. To further reduce the size so that the model can be run on the small devices without sacrificing accuracy, some other techniques need to be exploited. The most effective approach is to teach the small DNN with output from a large DNN so that the small DNN can generate the same output as that of the large DNN. This technique was first proposed by Buciluǎ et al. [4] to compress models. It was later proposed by Ba and Caruana [1] to emulate the output of DNNs with shallow MLPs. By combining DNNs trained with different random seeds, they can get a very complicated model that performs much better than the single DNN. They then pass the whole training set through this complicated model to generate the outputs and train the shallow model with the minimum square error criterion to approach the outputs from the large model. By doing this, the single shallow model can perform as good as the single DNN.

Li et al. [14] further extended this approach in two ways. First, they trained the small model by minimizing the DL divergence between the output distribution of the small and large models. Second, they not only pass the labeled data but also unlabeled data through the large model to generate training data to train the small model. They found out that using the additional unlabeled data is very helpful to reduce the performance gap between the large and small models.

7.2.5 Multiframe DNN

In CD-DNN-HMMs, we estimate the senone posterior probability for each 10 ms frame covering a window of 9–13 frames of input. However, speech is a rather stationary process when analyzed at a 10 ms frame rate. It is natural to believe that the predictions between adjacent frames are very similar. A simple and computationally efficient approach to take advantage of time correlations between feature frames is to simply copy the predictions from previous frame and thus cut the computation by 2. This simple approach, discussed in [22] and referred as frame-asynchronous DNN, performs surprisingly well.

An improved approach, referred as multiframe DNN (MFDNN), is proposed in [22]. Instead of copying the state predictions from the previous frame as that in the frame-asynchronous DNN, the MFDNN predicts both the frame label at time t and that at adjacent frames with the same input window at frame t. This is done by replacing the single softmax layer in the conventional DNN with several softmax layers each for a different frame label. Since all the softmax layers share the same hidden layers, MFDNN can cut the hidden layer computation time.

For example, in [22], the MFDNN jointly predicts labels for frames t to $t - K$, where K is the number of lookback frames. This is a special case of multitask learning. Note that here the MFDNN jointly predicts past $(t - 1, \ldots, t - K)$ instead of future $(t + 1, \ldots, t + K)$ frames. This is because in [22] a much longer context window in the past (20 frames) than that in the future (5 frames) was used. As the result, the input of the DNN provides more balanced context for predicting past frame labels than future frame labels. In most implementations of DNNs [6, 21], the input window has balanced context of the past and future frames. For these DNNs, the jointly predicted frame labels can from both past and future frames. Note that if K is large the overall latency of the system will be increased. Since latency will affect users' experience, K is typically set to a value less than 4 so that the additional latency introduced by the MFDNN is less than 30 ms.

Training such MFDNNs can be performed by backpropagating the errors from all softmax layers jointly through the network. When doing so the gradient magnitudes will increase since the error signal is multiplied by the number of jointly predicted frames. To maintain the convergence property the learning rate might have to be reduced.

The MFDNN performs better than the frame-asynchronous DNN and performs as good as the baseline system. It was reported [22] that a system which predicts jointly 2 frames at a time achieved a 10 % improvement in the query processing rate at no cost in accuracy or median latency, compared to an equivalent baseline system.

A system which predicts jointly 4 frames achieved a further 10 % improvement in the query processing rate at a cost of a 0.4 % absolute increase in word error rate.

References

1. Ba, L.J., Caruana, R.: Do deep nets really need to be deep? arXiv preprint http://arxiv.org/abs/1312.6184arXiv:1312.6184 (2013)
2. Bertsekas, D.P.: Constrained optimization and lagrange multiplier methods. Computer Science and Applied Mathematics, vol. 1982, p. 1. Academic Press, Boston (1982)
3. Boyd, S., Parikh, N., Chu, E., Peleato, B., Eckstein, J.: Distributed optimization and statistical learning via the alternating direction method of multipliers. Found. Trends® Mach. Learn. **3**(1), 1–122 (2011)
4. Bucilua, C., Caruana, R., Niculescu-Mizil, A.: Model compression. In: Proceedings of the International Conference on Knowledge Discovery and Data Mining (SIGKDD), pp. 535–541. ACM (2006)
5. Chen, X., Eversole, A., Li, G., Yu, D., Seide, F.: Pipelined back-propagation for context-dependent deep neural networks. In: Proceedings of the Annual Conference of International Speech Communication Association (INTERSPEECH) (2012)
6. Dahl, G.E., Yu, D., Deng, L., Acero, A.: Context-dependent pre-trained deep neural networks for large-vocabulary speech recognition. IEEE Trans. Audio, Speech Lang. Process. **20**(1), 30–42 (2012)
7. Dean, J., Ghemawat, S.: Mapreduce: simplified data processing on large clusters. Commun. ACM **51**(1), 107–113 (2008)
8. Hassibi, B., Stork, D.G., et al.: Second order derivatives for network pruning: optimal brain surgeon. Proc. Neural Inf. Process. Syst. (NIPS) **5**, 164–164 (1993)
9. Hestenes, M.R.: Multiplier and gradient methods. J. Optim. Theory Appl. **4**(5), 303–320 (1969)
10. Kingsbury, B. and Sainath, T.N., Soltau, H.: Scalable minimum bayes risk training of deep neural network acoustic models using distributed hessian-free optimization (INTERSPEECH) (2012)
11. Langford, J., Li, L., Zhang, T.: Sparse online learning via truncated gradient. J. Mach. Learn. Res. (JMLR) **10**, 777–801 (2009)
12. Le, Q.V., Ranzato, M., Monga, R., Devin, M., Chen, K., Corrado, G.S., Dean, J., Ng, A.Y.: Building high-level features using large scale unsupervised learning. arXiv preprint arXiv:1112.6209 (2011)
13. LeCun, Y., Denker, J.S., Solla, S.A., Howard, R.E., Jackel, L.D.: Optimal brain damage. Proc. Neural Inf. Process. Syst. (NIPS) **2**, 598–605 (1989)
14. Li, J., Zhao, R., Huang, J.T., Gong, Y.: Learning small-size DNN with output-distribution-based criteria. In: Proceedings of the Annual Conference of International Speech Communication Association (INTERSPEECH) (2014)
15. Martens, J.: Deep learning via Hessian-free optimization. In: Proceedings of the International Conference on Machine Learning (ICML), pp. 735–742 (2010)
16. Martens, J., Sutskever, I.: Learning recurrent neural networks with Hessian-free optimization. In: Proceedings of the International Conference on Machine Learning (ICML), pp. 1033–1040 (2011)
17. Niu, F., Recht, B., Ré, C., Wright, S.J.: Hogwild!: A lock-free approach to parallelizing stochastic gradient descent. arXiv preprint arXiv:1106.5730 (2011)
18. Petrowski, A., Dreyfus, G., Girault, C.: Performance analysis of a pipelined backpropagation parallel algorithm. IEEE Trans. Neural Netw. **4**(6), 970–981 (1993)
19. Powell, M.J.: A method for non-linear constraints in minimization problems. UKAEA (1967)

20. Sainath, T.N., Kingsbury, B., Sindhwani, V., Arisoy, E., Ramabhadran, B.: Low-rank matrix factorization for deep neural network training with high-dimensional output targets. In: Proceedings of the International Conference on Acoustics, Speech and Signal Processing (ICASSP), pp. 6655–6659 (2013)
21. Seide, F., Li, G., Yu, D.: Conversational speech transcription using context-dependent deep neural networks. In: Proceedings of the Annual Conference of International Speech Communication Association (INTERSPEECH), pp. 437–440 (2011)
22. Vanhoucke, V., Devin, M., Heigold, G.: Multiframe deep neural networks for acoustic modeling. In: Acoustics, Speech and Signal Processing (ICASSP), 2013 IEEE International Conference on, pp. 7582–7585. IEEE (2013)
23. Vanhoucke, V., Senior, A., Mao, M.Z.: Improving the speed of neural networks on CPUs. In: Proceedings of the NIPS Workshop on Deep Learning and Unsupervised Feature Learning (2011)
24. Xue, J., Li, J., Gong, Y.: Restructuring of deep neural network acoustic models with singular value decomposition. In: Proceedings of the Annual Conference of International Speech Communication Association (INTERSPEECH) (2013)
25. Yu, D., Seide, F., G.Li, Deng, L.: Exploiting sparseness in deep neural networks for large vocabulary speech recognition. In: Proceedings of the International Conference on Acoustics, Speech and Signal Processing (ICASSP), pp. 4409–4412 (2012)
26. Zhang, S., Zhang, C., You, Z., Zheng, R., Xu, B.: Asynchronous stochastic gradient descent for DNN training. In: Proceedings of the International Conference on Acoustics, Speech and Signal Processing (ICASSP), pp. 6660–6663 (2013)
27. Zhou, P., Liu, C., Liu, Q., Dai, L., Jiang, H.: A cluster-based multiple deep neural networks method for large vocabulary continuous speech recognition. In: Proceedings of the International Conference on Acoustics, Speech and Signal Processing (ICASSP), pp. 6650–6654 (2013)

Chapter 8
Deep Neural Network Sequence-Discriminative Training

Abstract The cross-entropy criterion discussed in the previous chapters treats each frame independently. However, speech recognition is a sequence classification problem. In this chapter, we introduce the sequence-discriminative training techniques that match better to the problem. We describe the popular maximum mutual information (MMI), boosted MMI (BMMI), minimum phone error (MPE), and minimum Bayes risk (MBR) training criteria, and discuss the practical techniques, including lattice generation, lattice compensation, frame dropping, frame smoothing, and learning rate adjustment, to make DNN sequence-discriminative training effective.

8.1 Sequence-Discriminative Training Criteria

In the previous chapters, deep neural networks (DNNs) for speech recognition are trained to classify individual frames based on a cross-entropy (CE) criterion, which minimizes the expected frame error. However, speech recognition is a sequence classification problem. Sequence-discriminative training [8, 9, 11, 15, 16] seeks to better match the maximum a posteriori (MAP) decision rule of large vocabulary continuous speech recognition (LVCSR) by considering sequence (inter-frame) constraints from hidden Markov models (HMMs), dictionary, and the language model (LM). Intuitively better recognition accuracy can be achieved if the CD-DNN-HMM speech recognizer is trained using sequence-discriminative criteria such as maximum mutual information (MMI) [1, 7], boosted MMI (BMMI) [13], minimum phone error (MPE) [14], or minimum Bayes risk (MBR) [2] that have been proven to obtain state-of-the-art results in the GMM-HMM framework. Experimental results have shown that sequence-discriminative training can obtain from 3 to 17 % relative error rate reduction against the CE-trained models depends on the implementation and the dataset.

8.1.1 Maximum Mutual Information

The MMI criterion [1, 7] used in automatic speech recognition (ASR) systems aims at maximizing the mutual information between the distributions of the observation

© Springer-Verlag London 2015   137
D. Yu and L. Deng, *Automatic Speech Recognition*,
Signals and Communication Technology, DOI 10.1007/978-1-4471-5779-3_8

sequence and the word sequence, which is highly correlated with minimizing the expected sentence error. Let us denote $\mathbf{o}^m = \mathbf{o}_1^m, \ldots, \mathbf{o}_t^m, \ldots, \mathbf{o}_{T_m}^m$ and $\mathbf{w}^m = \mathbf{w}_1^m, \ldots, \mathbf{w}_t^m, \ldots, \mathbf{w}_{N_m}^m$ the observation sequence and the correct word transcription of the mth utterance, respectively, where T_m is the total number of frames in utterance m, and N_m is the total number of words in the transcription of the same utterance. The MMI criterion over the training set $\mathbb{S} = \{(\mathbf{o}^m, \mathbf{w}^m) | 0 \leq m < M\}$ is:

$$
\begin{aligned}
J_{MMI}(\theta; \mathbb{S}) &= \sum_{m=1}^{M} J_{MMI}(\theta; \mathbf{o}^m, \mathbf{w}^m) \\
&= \sum_{m=1}^{M} \log P(\mathbf{w}^m | \mathbf{o}^m; \theta) \\
&= \sum_{m=1}^{M} \log \frac{p(\mathbf{o}^m | \mathbf{s}^m; \theta)^\kappa P(\mathbf{w}^m)}{\sum_{\mathbf{w}} p(\mathbf{o}^m | \mathbf{s}^w; \theta)^\kappa P(\mathbf{w})},
\end{aligned}
\tag{8.1}
$$

where θ is the model parameter including DNN weight matrices and biases, $\mathbf{s}^m = s_1^m, \ldots, s_t^m, \ldots, s_{T_m}^m$ is the sequence of states corresponding to $\mathbf{w}^m$, and κ is the acoustic scaling factor. Theoretically the sum in the denominator should be taken over all possible word sequences. In practice, however, the sum is constrained by the decoded speech lattice for utterance m to reduce the computational cost. Note that the gradient of the criterion (8.1) with regard to the model parameters θ can be computed as

$$
\begin{aligned}
\nabla_\theta J_{MMI}(\theta; \mathbf{o}^m, \mathbf{w}^m) &= \sum_m \sum_t \nabla_{\mathbf{z}_{mt}^L} J_{MMI}(\theta; \mathbf{o}^m, \mathbf{w}^m) \frac{\partial \mathbf{z}_{mt}^L}{\partial \theta} \\
&= \sum_m \sum_t \ddot{\mathbf{e}}_{mt}^L \frac{\partial \mathbf{z}_{mt}^L}{\partial \theta},
\end{aligned}
\tag{8.2}
$$

where the error signal $\ddot{\mathbf{e}}_{mt}^L$ is defined as $\nabla_{\mathbf{z}_{mt}^L} J_{MMI}(\theta; \mathbf{o}^m, \mathbf{w}^m)$ and $\mathbf{z}_{mt}^L$ is the softmax layer's excitation (the value before softmax is applied) for utterance m at frame t. Since $\frac{\partial \mathbf{z}_{mt}^L}{\partial \theta}$ is irrelevant to the training criterion, the only difference the new training criterion introduces compared to the frame-level cross-entropy training criterion (4.11) is the way the error signal is calculated. In the MMI training, the error signal becomes

$$
\begin{aligned}
\ddot{e}_{mt}^L(i) &= \nabla_{\mathbf{z}_{mt}^L(i)} J_{MMI}(\theta; \mathbf{o}^m, \mathbf{w}^m) \\
&= \sum_r \frac{\partial J_{MMI}(\theta; \mathbf{o}^m, \mathbf{y}^m)}{\partial \log p(\mathbf{o}_t^m | r)} \frac{\partial \log p(\mathbf{o}_t^m | r)}{\partial \mathbf{z}_{mt}^L(i)} \\
&= \sum_r \kappa \left(\delta(r = s_t^m) - \frac{\sum_{\mathbf{w}: s_t = r} p(\mathbf{o}^m | \mathbf{s})^\kappa P(\mathbf{w})}{\sum_{\mathbf{w}} p(\mathbf{o}^m | \mathbf{s}^w)^\kappa P(\mathbf{w})} \right)
\end{aligned}
$$

$$\times \frac{\partial \log P\left(r|\mathbf{o}_t^m\right) - \log P\left(r\right) + \log p\left(\mathbf{o}_t^m\right)}{\partial \mathbf{z}_{mt}^L\left(i\right)}$$

$$= \sum_r \kappa \left(\delta\left(r = s_t^m\right) - \ddot{\gamma}_{mt}^{\mathrm{DEN}}\left(r\right)\right) \frac{\partial \log \mathbf{v}_{mt}^L\left(r\right)}{\partial \mathbf{z}_{mt}^L\left(i\right)}$$

$$= \kappa \left(\delta\left(i = s_t^m\right) - \ddot{\gamma}_{mt}^{\mathrm{DEN}}\left(i\right)\right), \tag{8.3}$$

where $\ddot{\mathbf{e}}_{mt}^L\left(i\right)$ is the ith element of the error signal, $\mathbf{v}_{mt}^L\left(r\right) = P\left(r|\mathbf{o}_t^m\right) =$ softmax$_r\left(\mathbf{z}_{mt}^L\right)$ is the rth output of the DNN,

$$\ddot{\gamma}_{mt}^{\mathrm{DEN}}\left(r\right) = \frac{\sum_{\mathbf{w}:s_{t=r}} p\left(\mathbf{o}^m|\mathbf{s}\right)^\kappa P\left(\mathbf{w}\right)}{\sum_{\mathbf{w}} p\left(\mathbf{o}^m|\mathbf{s}^w\right)^\kappa P\left(\mathbf{w}\right)} \tag{8.4}$$

is the posterior probability of being in state r at time t, computed over the denominator lattices for utterance m, $P\left(r\right)$ is the prior probability of state r, $p\left(\mathbf{o}_t^m\right)$ is the prior probability of observing $\mathbf{o}_t^m$, and $\delta\left(\bullet\right)$ is the Kronecker delta. Both $P\left(r\right)$ and $p\left(\mathbf{o}_t^m\right)$ are independent of $\mathbf{z}_{mt}^L$. Here we assumed that the nominator reference state labels are obtained through a forced alignment of the acoustics with the word transcript. If we consider all possible state sequences that lead to the reference transcription $\mathbf{w}^m$, we can use the forward-backward algorithm over the word reference to obtain the numerator occupancies $\ddot{\gamma}_{mt}^{\mathrm{NUM}}\left(i\right)$ to replace $\delta\left(i = s_t^m\right)$.

If your DNN training algorithm is defined to minimize an objective function, you can, instead of maximizing the mutual information, minimize $J_{NMMI}\left(\theta; \mathbb{S}\right) = -J_{MMI}\left(\theta; \mathbb{S}\right)$, in which case the error signals are negated. Note that criterion similar to MMI has been explored in the early ANN/HMM hybrid work [6].

8.1.2 Boosted MMI

The boosted MMI (BMMI) [13] criterion

$$J_{BMMI}\left(\theta; \mathbb{S}\right) = \sum_{m=1}^M J_{BMMI}\left(\theta; \mathbf{o}^m, \mathbf{w}^m\right)$$

$$= \sum_{m=1}^M \log \frac{P\left(\mathbf{w}^m|\mathbf{o}^m\right)}{\sum_{\mathbf{w}} P\left(\mathbf{w}|\mathbf{o}^m\right) e^{-bA\left(\mathbf{w},\mathbf{w}^m\right)}} \tag{8.5}$$

$$= \sum_{m=1}^M \log \frac{p\left(\mathbf{o}^m|\mathbf{s}^m\right)^\kappa P\left(\mathbf{w}^m\right)}{\sum_{\mathbf{w}} p\left(\mathbf{o}^m|\mathbf{s}^w\right)^\kappa P\left(\mathbf{w}\right) e^{-bA\left(\mathbf{w},\mathbf{w}^m\right)}}$$

is a variant of the MMI objective (8.1) to boost the likelihood of paths that contain more errors, where b, whose typical value is 0.5, is the boosting factor, and $A\left(\mathbf{w}, \mathbf{w}^m\right)$

is a raw accuracy measure between word sequences $\mathbf{w}$ and $\mathbf{w}^m$ and can be computed at word level, phoneme level, or state level. For example, if it is measured at the phone level, it equals to the number of correct phones minus the number of insertions. This raw accuracy must be approximated efficiently. It has been shown that the BMMI criterion may be interpreted as incorporating a margin term in the MMI objective [13]. Since the only difference between the MMI and BMMI objective functions is the boosting term $e^{-bA(\mathbf{w},\mathbf{w}^m)}$ at the denominator, the error signal $\ddot{\mathbf{e}}_{mt}^L(i)$ can be similarly derived as

$$
\begin{aligned}
\ddot{\mathbf{e}}_{mt}^L(i) &= \nabla_{\mathbf{z}_{mt}^L(i)} J_{BMMI}\left(\theta; \mathbf{o}^m, \mathbf{w}^m\right) \\
&= \kappa \left(\delta\left(i = s_t^m\right) - \ddot{\gamma}_{mt}^{\text{DEN}}(i)\right),
\end{aligned} \tag{8.6}
$$

where, different from the MMI criterion, the denominator posterior probability is computed as

$$
\ddot{\gamma}_{mt}^{\text{DEN}}(i) = \frac{\sum_{\mathbf{w}:s_t=i} p\left(\mathbf{o}^m|\mathbf{s}\right)^\kappa P(\mathbf{w}) e^{-bA(\mathbf{w},\mathbf{w}^m)}}{\sum_{\mathbf{w}} p\left(\mathbf{o}^m|\mathbf{s}^w\right)^\kappa P(\mathbf{w}) e^{-bA(\mathbf{w},\mathbf{w}^m)}}. \tag{8.7}
$$

The extra computation involved in BMMI as opposed to MMI is very small if $A(\mathbf{w}, \mathbf{w}^m)$ can be efficiently estimated. The only change occurs in the forward-backward algorithm on the denominator lattice. For each arc in the denominator lattice, we subtract from the acoustic log-likelihood $bA(\mathbf{s}, \mathbf{s}^m)$ that is corresponding to the arc. This behave is similar to modifying the language model contribution on each arc.

8.1.3 MPE/sMBR

The MBR [2, 8] family of objective functions aims at minimizing the expected error corresponding to different granularity of labels. For example, the MPE criterion aims to minimize the expected phone error, while state MBR (sMBR) aims to minimize the expected state error when HMM topology and language models are taken into consideration. In general, the MBR objective can be written as

$$
\begin{aligned}
J_{MBR}(\theta; \mathbb{S}) &= \sum_{m=1}^M J_{MBR}\left(\theta; \mathbf{o}^m, \mathbf{w}^m\right) \\
&= \sum_{m=1}^M \sum_{\mathbf{w}} P\left(\mathbf{w}|\mathbf{o}^m\right) A\left(\mathbf{w}, \mathbf{w}^m\right) \\
&= \sum_{m=1}^M \frac{\sum_{\mathbf{w}} p\left(\mathbf{o}^m|\mathbf{s}^w\right)^\kappa P(\mathbf{w}) A(\mathbf{w}, \mathbf{w}^m)}{\sum_{\mathbf{w}'} p\left(\mathbf{o}^m|\mathbf{s}^{w'}\right)^\kappa P(\mathbf{w}')},
\end{aligned} \tag{8.8}
$$

where $A\left(\mathbf{w}, \mathbf{w}^m\right)$ is the raw accuracy between word sequences $\mathbf{w}$ and $\mathbf{w}^m$. For example, for MPE, it is the number of correct phone labels, while for sMBR it is the number of correct state labels. Similar to that in the MMI/BMMI case, the error signal is

$$
\begin{aligned}
\ddot{\mathbf{e}}_{mt}^L(i) &= \nabla_{\mathbf{z}_{mt}^L(i)} J_{MBR}\left(\theta; \mathbf{o}^m, \mathbf{w}^m\right) \\
&= \sum_r \frac{\partial J_{MBR}\left(\theta; \mathbf{o}^m, \mathbf{w}^m\right)}{\partial \log p\left(\mathbf{o}_t^m | r\right)} \frac{\partial \log p\left(\mathbf{o}_t^m | r\right)}{\partial \mathbf{z}_{mt}^L(i)} \\
&= \sum_r \kappa \, \ddot{\gamma}_{mt}^{\text{DEN}}(r)\left(\bar{A}^m\left(r = s_t^m\right) - \bar{A}^m\right) \frac{\partial \log \mathbf{v}_{mt}^L(r)}{\partial \mathbf{z}_{mt}^L(i)} \\
&= \kappa \, \ddot{\gamma}_{mt}^{\text{DEN}}(i)\left(\bar{A}^m\left(i = s_t^m\right) - \bar{A}^m\right),
\end{aligned}
\tag{8.9}
$$

where $\bar{A}^m$ is the average accuracy of all paths in the lattice, $\bar{A}^m\left(r = s_t^m\right)$ is the average accuracy of all paths in the lattice for utterance m that passes through state r at time t, and $\ddot{\gamma}_{mt}^{\text{DEN}}(r)$ is the MBR *occupancy* statistics. For sMBR,

$$
\ddot{\gamma}_{mt}^{\text{DEN}}(r) = \sum_{\mathbf{s}} \delta\left(r = s_t\right) P\left(\mathbf{s} | \mathbf{o}^m\right)
\tag{8.10}
$$

$$
A\left(\mathbf{w}, \mathbf{w}^m\right) = A\left(\mathbf{s}^w, \mathbf{s}^m\right) = \sum_t \delta\left(s_t^w = s_t^m\right)
\tag{8.11}
$$

$$
\bar{A}^m\left(r = s_t^m\right) = \mathbb{E}\left\{A\left(\mathbf{s}, \mathbf{s}^m\right) | s_t = r\right\} = \frac{\sum_{\mathbf{s}} \delta\left(r = s_t\right) P\left(\mathbf{s} | \mathbf{o}^m\right) A\left(\mathbf{s}, \mathbf{s}^m\right)}{\sum_{\mathbf{s}} \delta\left(r = s_t\right) P\left(\mathbf{s} | \mathbf{o}^m\right)}
\tag{8.12}
$$

and

$$
\bar{A}^m = \mathbb{E}\left\{A\left(\mathbf{s}, \mathbf{s}^m\right)\right\} = \frac{\sum_{\mathbf{s}} P\left(\mathbf{s} | \mathbf{o}^m\right) A\left(\mathbf{s}, \mathbf{s}^m\right)}{\sum_{\mathbf{s}} P\left(\mathbf{s} | \mathbf{o}^m\right)}.
\tag{8.13}
$$

8.1.4 A Uniformed Formulation

There can be other sequence-discriminative training criteria $J_{SEQ}\left(\theta; \mathbf{o}, \mathbf{w}\right)$. If the criteria are formulated as objective functions to be maximized (e.g., MMI/BMMI) instead of loss functions to be minimized we can always derive a loss function to be minimized by multiplying the original objective function by -1. Such loss functions can always be formulated as a ratio of values computed from two lattices: the numerator lattice that represents the reference transcription and the denominator lattice that represents competing hypotheses. The expected occupancies $\gamma_{mt}^{\text{NUM}}(i)$

and $\gamma_{mt}^{\text{DEN}}\,(i)$ for each state i required by the extended Baum-Welch (EBW) algorithm are computed with forward-backward passes over the numerator and denominator lattices, respectively.

Note that the gradient of the loss with respect to state log-likelihood is

$$\frac{\partial J_{SEQ}\,(\theta;\,\mathbf{o}^m,\,\mathbf{w}^m)}{\partial \log p\,(\mathbf{o}_t^m|r)} = \kappa\,\left(\gamma_{mt}^{\text{DEN}}\,(r) - \gamma_{mt}^{\text{NUM}}\,(r)\right). \tag{8.14}$$

Since $\log p\,(\mathbf{o}_t^m|r) = \log P\,(r|\mathbf{o}_t^m) - \log P\,(r) + \log p\,(\mathbf{o}_t^m)$, by the chain rule,

$$\frac{\partial J_{SEQ}\,(\theta;\,\mathbf{o}^m,\,\mathbf{w}^m)}{\partial P\,(r|\mathbf{o}_t^m)} = \kappa\,\frac{\left(\gamma_{mt}^{\text{DEN}}\,(r) - \gamma_{mt}^{\text{NUM}}\,(r)\right)}{P\,(r|\mathbf{o}_t^m)}. \tag{8.15}$$

Given that $P\,(r|\mathbf{o}_t^m) = \text{softmax}_r\,(\mathbf{z}_{mt}^L)$ we get

$$\mathbf{e}_{mt}^L\,(i) = \frac{\partial J_{SEQ}\,(\theta;\,\mathbf{o}^m,\,\mathbf{w}^m)}{\partial \mathbf{z}_{mt}^L\,(i)} = \kappa\,\left(\gamma_{mt}^{\text{DEN}}\,(i) - \gamma_{mt}^{\text{NUM}}\,(i)\right). \tag{8.16}$$

This formula can be applied to all the above sequence-discriminative training criteria as well as new ones [8, 15, 16]. The only difference is the way the occupancy statistics $\gamma_{mt}^{\text{NUM}}\,(i)$ and $\gamma_{mt}^{\text{DEN}}\,(i)$ are computed.

8.2 Practical Considerations

The above discussion seems to indicate that the sequence-discriminative training can be trivial. The only difference between the sequence-discriminative training and the frame-level cross-entropy training is the more complicated error-signal computation, which now involves numerator and denominator lattices. In practice, however, many practical techniques would help and sometimes are critical to obtain good recognition accuracy.

8.2.1 Lattice Generation

Similar to training GMM-HMM systems, the first step in the sequence-discriminative training of DNNs is to generate the numerator and denominator lattices. As we have pointed out above, the numerator lattices are often reduced to forced alignment of the transcription. It was shown that it is important to generate the lattices (especially the denominator lattices) by decoding the training data with a unigram LM in the GMM-HMM [12]. This still holds in the CD-DNN-HMM. In addition, people have found that it is desirable to use the best model available to generate the lattice and to serve as

Table 8.1 The effect (WER on Hub5'00 dataset) of the quality of the seed model and the lattice to the performance of the sequence-discriminative training on SWB 300 h task

Model to generate lattice	Seed model	
	CE1	CE2
GMM	15.8 % (−2 %)	–
DNN CE1 (WER 16.2 %)	14.1 % (−13 %)	13.7 % (−15 %)
DNN CE2 (WER 15.6 %)	–	13.5 % (−17 %)

Relative WER reduction over the CE1 model is in parentheses. *CE1* trained with the alignment generated from GMM. *CE2* refinement of CE1 model with alignment generated from CE1. (Summarized from Su et al. [15])

the seed model for the sequence-discriminative training [15]. Since CD-DNN-HMMs often outperform the CD-GMM-HMMs, we should at least use the CD-DNN-HMM trained with the CE criterion as both the seed model and the model for generating the alignments and lattices of each training utterance. Since lattice generation is an expensive process, the lattice is thus typically generated once and reused across training epochs. Further improvement can be obtained if new alignment and lattices are generated after each epoch. It is important that all the lattices are regenerated using the same model when doing so.

Table 8.1, based on results extracted from [15], clearly indicates the effect of the lattice quality and the seed model to the final recognition accuracy. From the table, we can make several observations. First, compared to the CE1 model trained with the CE criterion and the alignment generated from the GMM model, the model trained with the sequence-discriminative training only obtains 2 % relative error reduction if the lattices used are generated from the GMM model. However, we can obtain 13 % relative error reduction if the lattice is generated from the CE1 model even though the same CE1 model is used as the seed model in both conditions. Second, if the same lattice generated from the CE1 model is used to generate statistics in the sequence-discriminative training, additional 2 % relative error reduction can be obtained if we use CE2 instead of CE1 model as the seed model. The best result of 17 % relative word error rate (WER) reduction can be obtained using the CE2 model as both the seed model and the model for generating the lattice.

8.2.2 Lattice Compensation

Since the error signal is the weighted difference between the statistics calculated from the denominator and numerator lattices, the quality of the lattice is thus very important. However, even if the beam width used in the lattice generation process is large it is still impossible to cover all possible competing hypotheses. Actually, if indeed all competing hypotheses are included, using lattice to constrain the computation of the denominator lattice can no longer speedup the training process.

The problem often happens when the reference hypothesis is missing from or misaligned with the denominator lattice, under which condition the gradient can be unfairly higher since $\gamma_{mt}^{\text{DEN}}(i)$ is 0. This behavior can be frequently seen for silence frames since they are likely to be missing from the denominator lattices but occur very often in the numerator lattices. One of the behaviors introduced by poor lattice quality is the runaway silence frames. The number of silence frames in the decoding results increases as the training epoch increases. This results in increased deletion errors in the decoding result.

There are several ways to fix this problem. The first approach is to remove the frames where the reference hypothesis is not in the denominator lattice when computing the gradients. For silence frames, for example, this can be achieved by counting silence frames as *incorrect* in $A(\mathbf{s}, \mathbf{s}^m)$ in sMBR which effectively sets the error signal of these frames to be zero. A more general approach, referred as frame rejection [16], is to just remove these frames directly. Another approach, which is believed to perform better, is to augment the lattices with the reference hypothesis [15]. For example, we can add artificial silence arcs to the lattice, one for each start/end node pair connected by a word arc, with an appropriate entering probability and without introducing redundant silence paths.

Figure 8.1, shown in [16], is the result on the SWB 110h dataset with and without using the frame rejection technique. From the figure, we can observe that, without frame rejection, the MMI training starts overfitting after epoch 3. However, when frame rejection is used the training is stable even after epoch 8 and better test accuracy can be achieved.

Figure 8.2, based on the results from [15], compares the WER on Hub5'00 using 300h training set with and without using lattice compensation for silences. In this figure, a relatively large learning rate was used. Without using the silence treatment, severe overfitting can be observed even at the first epoch. However, when the silence frames are specially handled we can see great WER reduction at the first epoch and relatively stable results after that.

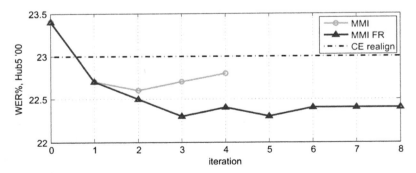

Fig. 8.1 Effect of frame rejection (FR) measured as WER on Hub5'00 when the SWB 110h training set is used. (Figure from Vesely et al. [16], permitted to use by ISCA.)

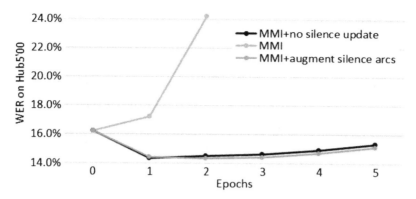

Fig. 8.2 WER on Hub5'00 with and without using silence treatment. (Based on the results from [15])

8.2.3 Frame Smoothing

Even if the lattices are correctly compensated, we can still observe the quick over-fitting phenomena during the training, which can be easily identified by the diverge of the sequence training criterion and the frame accuracy computed from the DNN score alone: The training criterion continues to improve while the frame accuracy computed from the DNN score alone tanks significantly. While people have hypothesized that the overfitting problem is caused by the sparse lattice (e.g., even the fattest lattices that can be practically generated reference only about 3 % of senones [15]), we believe this is not the only reason. The overfitting problem may also attribute to the fact that sequence is in a higher dimensional space than the frames. As such, the posterior distribution estimated from the training set is more likely to be different from that in the testing. This problem can be alleviated by making the sequence-discriminative training criterion closer to the frame-discriminative training criterion, for example, by using weak LM. The problem can be further alleviated using a technique referred as frame smoothing (F-smoothing) [15], which instead of minimizing the sequence-discriminative training criterion alone, minimizing a weighted sum of the sequence and frame criteria

$$J_{FS-SEQ}(\theta; \mathbb{S}) = (1 - H) J_{CE}(\theta; \mathbb{S}) + H J_{SEQ}(\theta; \mathbb{S}), \qquad (8.17)$$

where H is a smoothing factor often set empirically. It has been shown that a frame/sequence ratio of 1:4 (or $H = 4/5$) to 1:10 (or $H = 10/11$) is often effective. F-smoothing not only reduces the possibility of overfitting but also makes the training process less sensitive to the learning rate. F-smoothing is inspired by I-smoothing [12] and similar regularization approaches for adaptation [17]. Note that normal regularization techniques such as L1 and L2 regularization do not help.

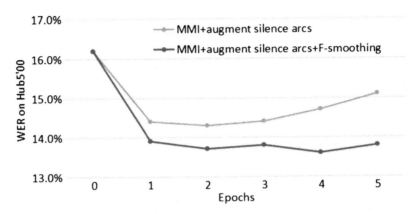

Fig. 8.3 WER on SWB Hub5'00 with and without using F-smoothing in the sequence-discriminative training of DNNs. (Based on results from [15])

Figure 8.3, based on [15], demonstrates the results on SWB Hub5'00 with and without using F-smoothing. With F-smoothing, it is much less likely to overfit to the training set. Overall, F-smoothing achieves 0.6 % absolute or 4 % relative WER reduction.

8.2.4 Learning Rate Adjustment

The learning rate used in the sequence-discriminative training should be smaller than that used in the frame cross-entropy training for two reasons. First, the sequence-discriminative training is often started from the CE-trained model which has already been well trained and thus requires smaller updates. Second, the sequence-discriminative training is more prone to overfitting. Using smaller learning rate can control the convergence more effectively. In practice, people have found that using a learning rate that is similar to that used in the final stage of CE training to be effective. For example, Vesely et al. [16] reported that an effective learning rate of $1e^{-4}$ per utterance worked well for both (B)MMI and sMBR, while Su et al. [15] showed that a learning rate of 1/128,000 per frame (or 0.002 per 256 frames) worked well when F-smoothing is used. The requirement to choose a good learning rate may be eliminated if algorithms such as Hessian-free [9] are used.

8.2.5 Training Criterion Selection

There are different observations with regard to the training criterion. Most results seem to suggest that the training criterion is not critical. For example, Table 8.2, which is extracted from [16], indicates that across MMI, BMMI, MPE, and sMBR, the WER on the SWB Hub5'00 and Hub5'01 datasets are very close although sMBR slightly

Table 8.2 The effect of different sequence-discriminative training criteria measured as WER on the Hub5'00 and Hub5'01 datasets when the 300 h training set is used. (Summarized from [16])

	Hub5'00 SWB (%)	Hub5'01 SWB (%)
GMM BMMI	18.6	18.9
DNN CE	14.2	14.5
DNN MMI	12.9	13.3
DNN BMMI	12.9	13.2
DNN MPE	12.9	13.2
DNN sMBR	12.6	13.0

outperforms other criteria. Since MMI is best understood and easiest to implement, it is thus suggested to use MMI if you need to implement one from scratch.

8.2.6 Other Considerations

Sequence-discriminative training is more computationally demanding. As such it is much slower. For example, a simple CPU-side implementation may increase runtime 12-fold compared to the frame CE training. Fortunately, with careful engineering, it is possible to achieve significant speedups through parallelized execution on a GPGPU. To speedup the acoustic-score computation, each arc can be processed at a separate CUDA thread. The lattice-level forward-backward processing, though, requires special treatment since the computation must be decomposed into sequential, dependency-free CUDA launches. In an example provided by Su et al. [15], there are 106 dependency-free node regions (=launches) for a 7.5-second lattice with 211,846 arcs and 6974 nodes at an average of 1,999 arcs (=threads per launch). Moreover, lattice forward/backward and error-signal accumulation require atomic summation of log-probabilities. This can be emulated through CUDA's atomic *compare-and-swap* instruction. To reduce target-operand collisions, it has been found to be critical to shuffle operations into a random-like order.

Further speed improvement can be obtained by using a parallel read-ahead thread to preload data and by generating lattices using a cluster of CPU computers. In the runtime experiments conducted by Su et al. [15], it was shown that the overall runtime increases only by about 70% compared to the CE training (when lattice generation is not considered) even on fat lattices of nearly 500 arcs per frame.

8.3 Noise Contrastive Estimation

In the above discussion, we assumed that the conventional minibatch-based SGD algorithm is used for sequence-discriminative training. Although careful engineering can speed up the training (e.g., as in [15]), the possible speed improvement is limited by the nature of the algorithm. In this section, we introduce noise contrastive estimation (NCE), an advanced training algorithm that can potentially further improve the training speed.

NCE was first proposed by Gutmann and Hyvarinen [3, 4] as a more reliable algorithm to estimate unnormalized statistical models. It was later successfully applied to training neural network language models (LMs) [10].

8.3.1 Casting Probability Density Estimation Problem as a Classifier Design Problem

Assume p_d is an unknown probability density function (pdf) and $\mathbf{X} = (\mathbf{x}_1, \ldots, \mathbf{x}_{T_d})$ of a random vector $\mathbf{x} \in \mathbb{R}^N$ is sampled from p_b, where T_d is the sample size. To estimate p_b we assume it belongs to a parameterized family of functions $p_m(.; \vartheta)_\vartheta$, where ϑ is a vector of parameters. In other words, $p_d(.) = p_m(.; \vartheta^*)$ for some parameter ϑ^*. The parametric density estimation problem can thus be converted into the problem of finding ϑ^* from the observed sample $\mathbf{X}$.

Typically, we require, for any ϑ,

$$\int p_m(\mathbf{u}; \vartheta)d\mathbf{u} = 1, \tag{8.18}$$

$$p_m(\mathbf{u}; \vartheta) \geq 0 \quad \forall \mathbf{u} \tag{8.19}$$

so that $p_m(.; \vartheta)$ is a valid pdf. In this case, we say the model is normalized and the maximum likelihood principle can then be used to estimate ϑ. However, in many cases, we only require that for certain ϑ (e.g., the true parameter ϑ^*) the normalization constraint is satisfied. In this case, we say that the model is unnormalized. Since we assume p_b belongs to $p_m(.; \vartheta)_\vartheta$, we know that the unnormalized model integrates to one at least for parameter ϑ^*.

Following [4], we denote by $p_m^0(.; \alpha)$ the unnormalized model parameterized by α. The unnormalized model $p_m^0(.; \alpha)$ can be converted into a normalized one as $p_m^0(.; \alpha)/Z(\alpha)$, where

$$Z(\alpha) = \int p_m^0(\mathbf{u}; \alpha)d\mathbf{u} \tag{8.20}$$

is the partition function and often is expensive to compute when $\mathbf{u}$ is of high dimension. Since for each α there is a corresponding $Z(\alpha)$, we may define a normalizing parameter $c = -\ln Z(\alpha)$ and represent the likelihood of the normalized model as

$$\ln p_m(.; \vartheta) = \ln p_m^0(.; \alpha) + c \tag{8.21}$$

with parameter $\vartheta = (\alpha, c)$. Note that at ϑ^* we have $\vartheta^* = (\alpha^*, 0)$.

The basic idea of NCE is to convert the density estimation problem to a two-class classification problem by describing properties of the observed sample $\mathbf{X}$ relative to the properties of some reference i.i.d. sample $\mathbf{Y} = (\mathbf{y}_1, \ldots, \mathbf{y}_{T_n})$ of a random variable $\mathbf{y} \in \mathbb{R}^N$ sampled from pdf p_n, which we can control, where T_n is the sample size. In [4], Gutmann and Hyvarinen proposed to use logistic regression, a popular two-class classifier, to provide the relative description on the form of the ratio p_d/p_n.

Let us construct a unified dataset $\mathbf{U} = \mathbf{X} \cup \mathbf{Y} = (\mathbf{u}_1, \ldots, \mathbf{u}_{T_d+T_n})$ and assign to each data point $\mathbf{u}_t$ a binary class label

$$C_t = \begin{cases} 1 & \text{if } \mathbf{u}_t \in \mathbf{X} \\ 0 & \text{if } \mathbf{u}_t \in \mathbf{Y} \end{cases} \tag{8.22}$$

Note that the prior probabilities are

$$P(C = 1) = \frac{T_d}{T_d + T_n}, \tag{8.23}$$

$$P(C = 0) = \frac{T_n}{T_d + T_n} \tag{8.24}$$

and the class-conditional probability densities are

$$p(\mathbf{u}|C = 1) = p_m(\mathbf{u}; \vartheta), \tag{8.25}$$

$$p(\mathbf{u}|C = 0) = p_n(\mathbf{u}) \tag{8.26}$$

The posterior probabilities for the classes are therefore

$$h(\mathbf{u}; \vartheta) \triangleq P(C = 1|\mathbf{u}; \vartheta) = \frac{p_m(\mathbf{u}; \vartheta)}{p_m(\mathbf{u}; \vartheta) + v p_n(\mathbf{u})}, \tag{8.27}$$

$$P(C = 0|\mathbf{u}; \vartheta) = 1 - h(\mathbf{u}; \vartheta) = \frac{v p_n(\mathbf{u})}{p_m(\mathbf{u}; \vartheta) + v p_n(\mathbf{u})}, \tag{8.28}$$

where

$$v \triangleq \frac{P(C = 0)}{P(C = 1)} = T_n/T_d. \tag{8.29}$$

If we further define $G(.; \vartheta)$ as the log-ratio between $p_m(.; \vartheta)$ and p_n,

$$G(\mathbf{u}; \vartheta) \triangleq \ln p_m(\mathbf{u}; \vartheta) - \ln p_n(\mathbf{u}), \tag{8.30}$$

$h(\mathbf{u}; \vartheta)$ can be written as

$$h(\mathbf{u}; \vartheta) = \frac{1}{1 + v \frac{p_n(\mathbf{u})}{p_m(\mathbf{u}; \vartheta)}} = \sigma_v(G(\mathbf{u}; \vartheta)), \tag{8.31}$$

where

$$\sigma_\nu(u) = \frac{1}{1 + \nu \exp(-u)} \tag{8.32}$$

is the logistic function parameterized by ν. If we assume the class labels C_t are independent and follow a Bernoulli distribution the conditional log-likelihood (or negative cross entropy) is

$$\ell(\vartheta) = \sum_{t=1}^{T_d+T_n} C_t \ln P(C_t = 1|\mathbf{u_t}; \vartheta) + (1 - C_t) P(C_t = 0|\mathbf{u_t}; \vartheta)$$

$$= \sum_{t=1}^{T_d} \ln[h(\mathbf{x}_t; \vartheta)] + \sum_{t=1}^{T_n} \ln\left[1 - h(\mathbf{y}_t; \vartheta)\right] \tag{8.33}$$

By optimizing $\ell(\vartheta)$ with respect to ϑ we get an estimate of p_d. In other words, the density estimation, which is an unsupervised learning problem, is now converted into a two-class supervised classifier design problem as pointed out firstly by Hastie et al. in [5].

8.3.2 Extension to Unnormalized Models

The above argument was further extended to the unnormalized model by Gutmann and Hyvarinen in [4]. They defined the criterion

$$J_T(\vartheta) = \frac{1}{T_d} \left\{ \sum_{t=1}^{T_d} \ln[h(\mathbf{x}_t; \vartheta)] + \sum_{t=1}^{T_n} \ln\left[1 - h(\mathbf{y}_t; \vartheta)\right] \right\}$$

$$= \frac{1}{T_d} \sum_{t=1}^{T_d} \ln[h(\mathbf{x}_t; \vartheta)] + \nu \frac{1}{T_n} \sum_{t=1}^{T_n} \ln\left[1 - h(\mathbf{y}_t; \vartheta)\right] \tag{8.34}$$

to find the best ϑ to estimate p_d. It is obvious that improving $J_T(\vartheta)$ means the two-class classifier can more accurately distinguish between the observed data and the reference data.

As T_d increases and by fixing ν, $T_n = \nu T_d$ also increases, $J_T(\vartheta)$ converges in probability to

$$J(\vartheta) = \mathbb{E}\{\ln[h(\mathbf{x}_t; \vartheta)]\} + \nu \mathbb{E}\left\{\ln\left[1 - h(\mathbf{y}_t; \vartheta)\right]\right\}. \tag{8.35}$$

It is proven [4], by defining $f_m(.) = \ln p_m(.; \vartheta)$ and rewriting the criterion as a function of f_m, that

- $J(\vartheta)$ attains a maximum when $p_m(.; \vartheta) = p_d$. This is the only extrema if the noise density p_n is chosen such that it is nonzero whenever p_d is nonzero. More importantly, maximization is performed without any normalization constraint for $p_m(.; \vartheta)$.
- ϑ_T, the value of ϑ which (globally) maximizes $J_T(\vartheta)$, converges to ϑ^* if the following three conditions are all satisfied: a) p_n is nonzero whenever p_d is nonzero; b) J_T uniformly convergent in probability to J; and c) for large sample sizes the objective function J_T becomes peaked enough around the true value ϑ^*.
- $\sqrt{T_d}(\hat{\vartheta}_T - \vartheta^*)$ is asymptotically normal with mean zero and a bounded covariance matrix Σ.
- For $v \to \infty$, Σ is independent of the choice of p_n.

Based on these properties, it is suggested that we should choose noise for which an analytical expression for $\ln p_n$ is available, can be sampled easily, and in some aspect similar to the data. It is also suggested that the noise sample size should be as large as computationally possible. Example noise distributions are Gaussian and uniform distributions.

8.3.3 Apply NCE in DNN Training

In the acoustic model training using the cross-entropy criterion, we estimate the distribution $P(s|\mathbf{o}; \vartheta)$ of the senone s given the observation $\mathbf{o}$. For each observation $\mathbf{o}$ with label s we generate v noise labels $y_1, \ldots, y_v$ and optimize

$$J_T(\mathbf{o}, \vartheta) = \ln[h(s|\mathbf{o}; \vartheta)] + \sum_{t=1}^{v} \ln[1 - h(y_t|\mathbf{o}; \vartheta)]. \tag{8.36}$$

Since

$$\begin{aligned}
\frac{\partial}{\partial \vartheta} \ln[h(s|\mathbf{o}; \vartheta)] &= \frac{h(s|\mathbf{o}; \vartheta)[1 - h(s|\mathbf{o}; \vartheta)]}{h(s|\mathbf{o}; \vartheta)} \frac{\partial}{\partial \vartheta} \ln P_m(s|\mathbf{o}; \vartheta) \\
&= [1 - h(s|\mathbf{o}; \vartheta)] \frac{\partial}{\partial \vartheta} \ln P_m(s|\mathbf{o}; \vartheta) \\
&= \frac{v P_n(s|\mathbf{o})}{P_m(s|\mathbf{o}; \vartheta) + v P_n(s|\mathbf{o})} \frac{\partial}{\partial \vartheta} \ln P_m(s|\mathbf{o}; \vartheta) \tag{8.37}
\end{aligned}$$

and

$$\frac{\partial}{\partial \vartheta} \ln[1 - h(y_t|\mathbf{o}; \vartheta)] = -\frac{h(y_t|\mathbf{o}; \vartheta)[1 - h(y_t|\mathbf{o}; \vartheta)]}{1 - h(y_t|\mathbf{o}; \vartheta)} \frac{\partial}{\partial \vartheta} \ln P_m(y_t|\mathbf{o}; \vartheta)$$

$$= -h(y_t|\mathbf{o}; \vartheta)\frac{\partial}{\partial \vartheta} \ln P_m (y_t|\mathbf{o}; \vartheta)$$

$$= -\frac{P_m (y_t|\mathbf{o}; \vartheta)}{P_m (y_t|\mathbf{o}; \vartheta) + v P_n (y_t|\mathbf{o})} \frac{\partial}{\partial \vartheta} \ln P_m (y_t|\mathbf{o}; \vartheta) \quad (8.38)$$

we get

$$\frac{\partial}{\partial \vartheta} J_T (\mathbf{o}, \vartheta) = \frac{v P_n (s|\mathbf{o})}{P_m (s|\mathbf{o}; \vartheta) + v P_n (s|\mathbf{o})} \frac{\partial}{\partial \vartheta} \ln P_m (s|\mathbf{o}; \vartheta)$$

$$- \sum_{t=1}^{v} \left[\frac{P_m (y_t|\mathbf{o}; \vartheta)}{P_m (y_t|\mathbf{o}; \vartheta) + v P_n (y_t|\mathbf{o})} \frac{\partial}{\partial \vartheta} \ln P_m (y_t|\mathbf{o}; \vartheta) \right]. (8.39)$$

Note that the weights $\frac{P_m (y_t|\mathbf{o}; \vartheta)}{P_m (y_t|\mathbf{o}; \vartheta) + v P_n (y_t|\mathbf{o})}$ are always between 0 and 1 and so the NCE learning is very stable. In addition, since $P_m (s|\mathbf{o}; \vartheta)$ is an unnormalized model, the gradient $\frac{\partial}{\partial \vartheta} \ln P_m (.|\mathbf{o}; \vartheta)$ can be computed efficiently. However, in the unnormalized model there is a normalization factor c for each observation $\mathbf{o}$ which can become a problem when the AM training set is very large. Fortunately, experiments have shown that there is no or little performance degradation even if the same c is used for all observations or even when c is always set to 0 [4, 10]. Since it typically does not increase the computation a lot and often helps to boost the estimation accuracy using a shared c is recommended.

Since the conditional probability distributions for different observations are estimated with the same DNN, we cannot learn these distributions independently. Instead, we define a global NCE objective

$$J_T^G (\vartheta) = \sum_{t=1}^{T_d} J_T (\mathbf{o}_\text{t}, \vartheta) . \quad (8.40)$$

The above derivation can be easily extended to sequence-discriminative training. The only difference is that in the sequence-discriminative training there are significantly more classes than that in the frame-level training since each label sequence is considered as a different class. More specifically, for the mth utterance the distribution we need to estimate is

$$\log P \left(\mathbf{w}^m|\mathbf{o}^m; \theta\right) = \log p \left(\mathbf{o}^m|\mathbf{s}^m; \theta\right)^\kappa P \left(\mathbf{w}^m\right) + c^m. \quad (8.41)$$

Recall that here $\mathbf{o}^m = \mathbf{o}_1^m, \ldots, \mathbf{o}_t^m, \ldots, \mathbf{o}_{T_m}^m$ and $\mathbf{w}^m = \mathbf{w}_1^m, \ldots, \mathbf{w}_t^m, \ldots, \mathbf{w}_{N_m}^m$ are the observation sequence and the correct word transcription of the mth utterance, respectively, $\mathbf{s}^m = s_1^m, \ldots, s_t^m, \ldots, s_{T_m}^m$ is the sequence of states corresponding to $\mathbf{w}^m$, κ is the acoustic scaling factor, T_m is the total number of frames in utterance m, and N_m is the total number of words in the transcription of the same utterance. In the sequence-discriminative training, we can use uniform distributions over all possible state sequences or over sequences in the lattice as the noise distribution.

References

1. Bahl, L., Brown, P., De Souza, P., Mercer, R.: Maximum mutual information estimation of hidden markov model parameters for speech recognition. In: Proceedings of the International Conference on Acoustics, Speech and Signal Processing (ICASSP), vol. 11, pp. 49–52 (1986)
2. Goel, V., Byrne, W.J.: Minimum Bayes-risk automatic speech recognition. Comput. Speech Lang. **14**(2), 115–135 (2000)
3. Gutmann, M., Hyvärinen, A.: Noise-contrastive estimation: a new estimation principle for unnormalized statistical models. In: International Conference on Artificial Intelligence and Statistics, pp. 297–304 (2010)
4. Gutmann, M.U., Hyvärinen, A.: Noise-contrastive estimation of unnormalized statistical models, with applications to natural image statistics. J. Mach. Learn. Res. **13**, 307–361 (2012)
5. Hastie, T., Tibshirani, R., Friedman, J., Hastie, T., Friedman, J., Tibshirani, R.: The Elements of Statistical Learning, vol. 2. Springer, Heidelberg (2009)
6. Hennebert, J., Ris, C., Bourlard, H., Renals, S., Morgan, N.: Estimation of Global Posteriors and Forward-Backward Training of Hybrid HMM/ANN Systems (1997)
7. Kapadia, S., Valtchev, V., Young, S.: MMI training for continuous phoneme recognition on the TIMIT database. In: Proceedings of the International Conference on Acoustics, Speech and Signal Processing (ICASSP), vol. 2, pp. 491–494 (1993)
8. Kingsbury, B.: Lattice-based optimization of sequence classification criteria for neural-network acoustic modeling. In: Proceedings of the International Conference on Acoustics, Speech and Signal Processing (ICASSP), pp. 3761–3764 (2009)
9. Kingsbury, B., Sainath, T.N., Soltau, H.: Scalable minimum bayes risk training of deep neural network acoustic models using distributed hessian-free optimization. In: Proceedings of the Annual Conference of International Speech Communication Association (INTERSPEECH) (2012)
10. Mnih, A., Teh, Y.W.: A fast and simple algorithm for training neural probabilistic language models. arXiv preprint arXiv:1206.6426 (2012)
11. Mohamed, A.-R., Yu, D., Deng, L.: Investigation of full-sequence training of deep belief networks for speech recognition. In: Proceedings of the Annual Conference of International Speech Communication Association (INTERSPEECH), pp. 2846–2849 (2010)
12. Povey, D.: Discriminative Training for Large Vocabulary Speech Recognition. Ph.D. thesis, Cambridge University Engineering Department (2003)
13. Povey, D., Kanevsky, D., Kingsbury, B., Ramabhadran, B., Saon, G., Visweswariah, K.: Boosted MMI for model and feature-space discriminative training. In: Proceedings of the International Conference on Acoustics, Speech and Signal Processing (ICASSP), pp. 4057–4060 (2008)
14. Povey, D., Woodland, P.C.: Minimum phone error and I-smoothing for improved discriminative training. In: Proceedings of the International Conference on Acoustics, Speech and Signal Processing (ICASSP), vol. 1, pp. I–105 (2002)
15. Su, H., Li, G., Yu, D., Seide, F.: Error back propagation for sequence training of context-dependent deep networks for conversational speech transcription. In: Proceedings of the International Conference on Acoustics, Speech and Signal Processing (ICASSP) (2013)
16. Veselý, K., Ghoshal, A., Burget, L., Povey, D.: Sequence-discriminative training of deep neural networks. In: Proceedings of the Annual Conference of International Speech Communication Association (INTERSPEECH) (2013)
17. Yu, D., Yao, K., Su, H., Li, G., Seide, F.: Kl-divergence regularized deep neural network adaptation for improved large vocabulary speech recognition. In: Proceedings of the International Conference on Acoustics, Speech and Signal Processing (ICASSP), pp. 7893–7897 (2013)

Part IV
Representation Learning
in Deep Neural Networks

Chapter 9
Feature Representation Learning in Deep Neural Networks

Abstract In this chapter, we show that deep neural networks jointly learn the feature representation and the classifier. Through many layers of nonlinear processing, DNNs transform the raw input feature to a more invariant and discriminative representation that can be better classified by the log-linear model. In addition, DNNs learn a hierarchy of features. The lower-level features typically catch local patterns. These patterns are very sensitive to changes in the raw feature. The higher-level features, however, are built upon the low-level features and are more abstract and invariant to the variations in the raw feature. We demonstrate that the learned high-level features are robust to speaker and environment variations.

9.1 Joint Learning of Feature Representation and Classifier

Why do DNNs perform so much better than the conventional shallow models such as Gaussian mixture models (GMMs) and support vector machines (SVMs) in speech recognition? We believe it mainly attributes to the DNNs ability to learn complicated feature representations and classifiers jointly.

In the conventional shallow models, feature engineering is the key to the success of the system. Practitioner's main job is to construct features that perform well for a specific learning algorithm on a specific task. The improvement of the system's performance often comes from finding a better feature by someone who has great domain knowledge. Typical examples include the scale-invariant feature transform (SIFT) [17] widely used in image classification and the mel-frequency cepstrum coefficients (MFCC) [2] used in the speech recognition tasks.

Deep models such as DNNs, however, do not require hand-crafted high-level features.[1] Instead, they automatically learn the feature representations and classifiers jointly. Figure 9.1 depicts the general framework of a typical deep model, in which both the learnable representation and the learnable classifier are part of the model.

---

[1] Good raw features still help though since the existing DNN learning algorithms may generate an underperformed system even if a linear transformation such as discrete cosine transformation (DCT) is applied to the log filter-bank features.

© Springer-Verlag London 2015

D. Yu and L. Deng, *Automatic Speech Recognition*,

Signals and Communication Technology, DOI 10.1007/978-1-4471-5779-3_9

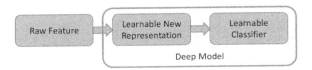

Fig. 9.1 Deep model jointly learns feature representation and classifier

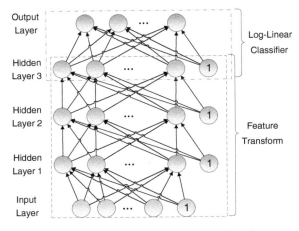

Fig. 9.2 DNN: A joint feature representation and classifier learning view

In the DNN, the combination of all hidden layers can be considered as a feature learning module as shown in Fig. 9.2. Although each hidden layer typically only employs simple nonlinear transformation, the composition of these simple nonlinear transformations results in very complicated nonlinear transformation. The last layer, which is the softmax layer, is essentially a simple log-linear classifier or sometimes referred as maximum entropy (MaxEnt) [25] model. Thus, in the DNN, the estimation of the posterior probability $p(y = s|\mathbf{o})$ can be considered as a two-step nonstochastic process: In the first step, the observation vector $\mathbf{o}$ is transformed into a feature vector $\mathbf{v}^{L-1}$ through $L-1$ layers of nonlinear transforms. In the second step, the posterior probability $p(y = s|\mathbf{o})$ is estimated using the log-linear model given the transformed feature $\mathbf{v}^{L-1}$. If we consider the first $L-1$ layers fixed, learning the parameters in the softmax layer is equivalent to training a MaxEnt model on feature $\mathbf{v}^{L-1}$. In the conventional MaxEnt model, features are manually designed, e.g., in most natural language processing tasks [25] and in speech recognition tasks [6, 31]. Manual feature construction works fine for tasks that people can easily inspect and know what feature to use but not for tasks whose raw features are highly variable. In DNNs, however, the features are defined by the first $L-1$ layers and are jointly learned with the MaxEnt model from the data. This not only eliminates the tedious and erroneous process of manual feature construction but also has the potential of extracting invariant and discriminative features, which are impossible to construct manually, through many layers of nonlinear transforms.

9.2 Feature Hierarchy

DNNs not only learn feature representations that are suitable for the classifier but also learn a feature hierarchy. Since each hidden layer is a nonlinear transformation of the input feature, it can be considered as a new representation of the raw input feature. The hidden layers that are closer to the input layer represent low-level features. Those that are closer to the softmax layer represent higher-level features. The lower-level features typically catch local patterns. These patterns are very sensitive to changes in the input feature. The higher-level features, however, are built upon the low-level features and are more abstract and invariant to the input feature variations. Figure 9.3, extracted from [34], depicts the feature hierarchy learned from imageNet dataset [9]. As we can see that the higher-level features are more abstract and invariant.

This property can also be observed in Fig. 9.4 in which the percentage of saturated neurons, defined as neurons whose activation value is either >0.99 or <0.01, at each layer are shown. The lower layers typically have smaller percentage of saturated neurons while the higher layers, that are close to the softmax layer, have large percentage of saturated neurons. Note that majority of the saturated neurons are deactivated (whose activation value is <0.01), which indicates that the associated features are sparse. This is because the training label, which is 1 for the correct class and 0 for all other classes, is parse.

In this hierarchy of features, higher layer features are more invariant and discriminative. This is because many layers of simple nonlinear processing can generate a complicated nonlinear transform. To show that this nonlinear transform is robust to small variations in the input features, let us assume the output of layer l, or equivalently the input to layer $l + 1$ is changed from $\mathbf{v}^\ell$ to $\mathbf{v}^\ell + \delta^\ell$, where δ^ℓ is

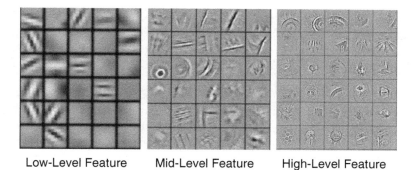

Low-Level Feature   Mid-Level Feature   High-Level Feature

Fig. 9.3 Feature hierarchy learned by a network with many layers on ImageNet. (Figure extracted from Zeiler and Fergusfrom [34], permitted to use by Zeiler.)

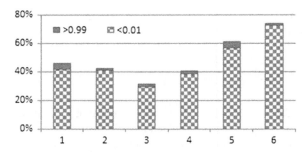

Fig. 9.4 Percentage of saturated neurons at each layer. (A similar figure has also been shown in Yu et al. [33])

a small change. This change will cause the output of layer $l + 1$, or equivalently the input to layer $\ell + 2$ to change by

$$
\begin{aligned}
\delta^{\ell+1} &= \sigma(\mathbf{z}^{\ell+1}(\mathbf{v}^{\ell} + \delta^{\ell})) - \sigma(\mathbf{z}^{\ell+1}(\mathbf{v}^{\ell})) \\
&\approx \text{diag}\left(\sigma'(\mathbf{z}^{\ell+1}(\mathbf{v}^{\ell}))\right)(\mathbf{W}^{\ell+1})^T \delta^{\ell}.
\end{aligned}
\tag{9.1}
$$

where

$$
\mathbf{z}^{\ell+1}(\mathbf{v}^{\ell}) = \mathbf{W}^{\ell+1}\mathbf{v}^{\ell} + \mathbf{b}^{\ell+1}
\tag{9.2}
$$

is the excitation, and $\sigma(\mathbf{z})$ is the sigmoid activation function. The norm of the change $\delta^{\ell+1}$ is

$$
\begin{aligned}
\|\delta^{\ell+1}\| &\approx \|\text{diag}\left(\sigma'(\mathbf{z}^{\ell+1}(\mathbf{v}^{\ell}))\right)(\mathbf{W}^{\ell+1})^T \delta^{\ell}\| \\
&\leq \|\text{diag}\left(\sigma'(\mathbf{z}^{\ell+1}(\mathbf{v}^{\ell}))\right)(\mathbf{W}^{\ell+1})^T\|\|\delta^{\ell}\| \\
&= \|\text{diag}(\mathbf{v}^{\ell+1} \bullet (1 - \mathbf{v}^{\ell+1}))(\mathbf{W}^{\ell+1})^T\|\|\delta^{\ell}\|
\end{aligned}
\tag{9.3}
$$

where $\bullet$ refers to an element-wise product.

In the DNN, the magnitude of the majority of the weights is typically very small if the size of the hidden layer is large as shown in Fig. 9.5. For example, in a $6 \times 2k$ DNN trained using 30 h of SWB data, the magnitude of 98 % of the weights in all layers except the input layer is less than 0.5.

While each element in $\mathbf{v}^{\ell+1} \bullet (1 - \mathbf{v}^{\ell+1})$ is less than or equal to 0.25, the actual value is typically much smaller. This is because a large percentage of hidden neurons are inactive, as shown in Fig. 9.4. As a result, the average norm $\|\text{diag}(\mathbf{v}^{\ell+1} \bullet (1 - \mathbf{v}^{\ell+1}))(\mathbf{W}^{\ell+1})^T\|_2$ in Eq. 9.3 across a 6 h SWB development set is smaller than one in all layers, as indicated in Fig. 9.6. Since all hidden layer values are bounded in the same range of $(0, 1)$, this indicates that when there is a small perturbation on the input, the perturbation shrinks at each higher layer. In other words, features generated by higher layers are more invariant to variations than those represented

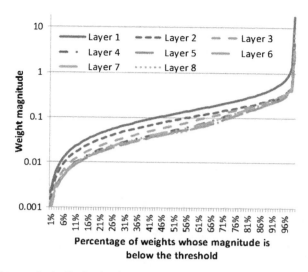

Fig. 9.5 Weight magnitude distribution in a typical DNN

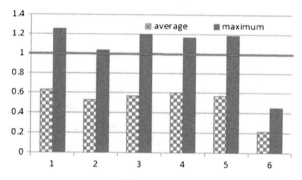

Fig. 9.6 Average and maximum $\|\text{diag}(\mathbf{v}^{\ell+1} \bullet (1 - \mathbf{v}^{\ell+1}))(\mathbf{W}^{\ell+1})^T\|_2$ across layers on a $6 \times 2\,\text{k}$ DNN. (A similar figure has also been shown in Yu et al. [33])

by lower layers. Note that the maximum norm over the same development set is larger than one, as shown in Fig. 9.6. This is necessary since the differences need to be enlarged around the class boundaries to have discrimination ability. These large norm cases also cause noncontinual points on the objective function, as indicated by the fact that you can almost always find inputs from which a small deviation would change the label predicted by the DNN [29].[2]

In general, features are processed in stages in the hierarchical deep model as shown in Fig. 9.7. Each stage can be consisted of several optional steps: normalization, filter-bank processing, nonlinear processing, and pooling [10]. Typical

---

[2] This behavior can be alleviated by adding small random noises to each training sample dynamically during the training time.

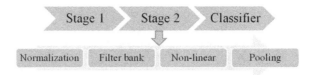

Fig. 9.7 A general framework of feature processing. In the figure, three stages are shown, each of the stage involves four optional steps

normalization techniques include average removal, local contrast normalization, and variance normalization, some of which have been discussed in Sect. 4.3.1. The purpose of the filter-bank processing is to project the feature to a higher dimensional space in which classification can be more easily done. This can be achieved by dimension expansion or feature projection. The nonlinear processing step is critical in the deep model since the combination of linear transformations is just another linear transformation. Often used nonlinear functions include sparsification, saturation, lateral inhibition, tanh, sigmoid, and winner-takes-all. The purpose of the pooling step, which involves aggregation and clustering, is to extract invariant feature and reduce the dimension.

9.3 Flexibility in Using Arbitrary Input Features

Note that GMMs typically require each dimension in the input feature to be statistically independent so that a diagonal covariance matrix may be used to reduce the number of parameters in the model. DNNs, however, are discriminative models and do not have similar constraints.

In the speech recognition applications, MFCC and perceptual linear predictive (PLP) features are two most often used raw features. However, both these two types of features are derived from the Mel-scaled log filter-bank features (MS-LFB). Although these features are more invariant than the Mel-scaled log filter-bank features, some information useful for classification may be lost during the manual feature transformation process. A natural thought is to use the MS-LFB features directly as the input to the DNNs [18]. Table 9.1, extracted from [16], compares the discriminatively trained CD-GMM-HMM baseline with the CD-DNN-HMMs using different raw input features on a voice search task. The 13-dimensional MFCC feature is extracted from the 24-dimensional Mel-scale log filter-bank feature with a truncated discrete cosine transform (DCT). All the input features are mean normalized and with dynamic features. The MFCC feature is with up to third-order derivatives, while the log filter-bank feature have up to the second-order derivatives. The HLDA [13] transform is only applied to the MFCC feature for the CD-GMM-HMM system. From this table, we can observe that switching from the MFCC feature to the 24 Mel-scale log filter-bank feature leads to 4.7 % relative word error rate

Table 9.1 Comparison of different raw input features for DNN

Combination of model and features	WER (rel. WERR)
CD-GMM-HMM (MFCC, fMPE+BMMI)	34.66 % (baseline)
CD-DNN-HMM (MFCC)	31.63 % (−8.7 %)
CD-DNN-HMM (24 MS-LFB)	30.11 % (−13.1 %)
CD-DNN-HMM (29 MS-LFB)	30.11 % (−13.1 %)
CD-DNN-HMM (40 MS-LFB)	29.86 % (−13.8 %)

All the input features are mean normalized and with dynamic features. Relative word error rate (WER) reduction in parentheses. (Summarized from Li et al. [16])

(WER) reduction. Increasing the number of filter-banks from 24 to 40 only provides less than 1 % relative WER reduction. Overall, CD-DNN-HMM outperforms CD-GMM-HMM trained using fMPE+BMMI by a relative WER reduction of 13.8 %. Note that this is achieved with much simpler training procedure than that is used to build the CD-GMM-HMM baseline. All the DNN models shown in this table were trained using the frame-level cross-entropy criterion. Further improvement can be obtained by using sequence-discriminative training discussed in Chap. 8.

Even the filter-banks can be automatically learned by the DNNs as reported in [26], in which FFT spectrum is used as the input to the DNNs. To suppress the dynamic range of the filter-bank output, the log function is used as the activation function in the filter-bank layer. In addition, the normalization step depicted in Fig. 9.7 is applied to the activations of the filter-bank layer before they are sent to the next layer. It is reported in [26] that by learning the filter-bank parameters directly 5 % relative WER reduction can be achieved over the baseline DNNs that use the manually designed Mel-scale filter-banks.

9.4 Robustness of Features

A key property of a good feature is its robustness to the variations. There are two main types of variations in speech signals: speaker variation and environment variation. In the conventional GMM-HMM systems, both types of variations need to be handled explicitly.

9.4.1 Robust to Speaker Variations

To deal with speaker variability, vocal tract length normalization (VTLN) [1] and feature-space maximum likelihood linear regression (fMLLR) [5] are critical in the GMM-HMM systems.

Table 9.2 Comparison of feature-transform-based speaker-adaptation techniques for GMM-HMMs, a shallow, and a deep NN

Adaptation technique	CD-GMM-HMM (40-mixture)	CD-MLP-HMM (1 × 2,048)	CD-DNN-HMM (7 × 2,048)
Speaker independent	23.6%	24.2%	17.1%
+ VTLN	21.5% (−9%)	22.5% (−7%)	16.8% (−2%)
+ fMLLR/fDLR×4	20.4% (−5%)	21.5% (−4%)	16.4% (−2%)

Word-error rates (WER) for Hub5'00-SWB (relative change in parentheses). (Summarized from Seide et al. [27])

VTLN warps the frequency axis of the filter-bank analysis to account for the fact that the locations of vocal-tract resonances vary roughly monotonically with the vocal tract length of the speaker. This is done in both training and testing with 20 quantized warping factors from 0.8 to 1.18. During the training, the optimal warping factor can be found using the expectation–maximization (EM) algorithm by repeatedly selecting the best factor given the current model and then updating the model using the selected factor. During the testing, the system can pick the best factor by running recognition for all factors and using the highest cumulative log probability.

On the other hand, fMLLR applies an affine transform to the feature vector so that the transformed feature better matches the model. It is typically applied to the testing utterance by first generating recognition results using the raw feature and then re-recognizing the speech with the transformed feature. This process can be iterated for several times. For GMM-HMMs, fMLLR transforms are estimated to maximize the likelihood of the adaptation data given the model. For DNNs, they are optimized to maximize cross entropy (with backpropagation), which is a discriminative criterion. This procedure is thus referred as feature-space discriminative linear regression (fDLR) [27]. The transformation may be applied to each input vector (which is typically a concatenation of multiple frames of features) in the DNN or applied to individual frames, prior to concatenation.

Table 9.2, extracted from [27], compares the effectiveness of VTLN and fMLLR/fDLR on GMMs, shallow multilayer perceptrons (MLPs), and DNNs. It can be observed that both VTLN and fMLLR are important for GMMs to reduce speaker variability. In fact, they provide 9 and 5% relative error rate reduction, respectively. These techniques are also important for shallow MLPs with 7 and 4% relative WER reduction. However, these techniques are less important on the DNN systems and provide only 2% relative error reduction over the speaker-independent baseline DNN system. This observation indicates that DNNs are more robust to the speaker variations than GMMs and shallow MLPs.

9.4.2 Robust to Environment Variations

Similarly, GMM-based acoustic models are highly sensitive to environmental mismatch. To deal with the issue several techniques, such as vector Taylor series (VTS) [12, 14, 15, 19] adaption and maximum likelihood linear regression (MLLR) [4], that normalize the input features or adapt the model parameters have been developed. In contrast, the analysis in the previous sections suggests that DNNs have the ability to generate internal representations that are robust to environmental variability seen in the training data.

In methods such as VTS adaptation, an estimated noise model is used to adapt the Gaussian parameters of the recognizer based on a physical model that defines how noise corrupts clean speech. The relationship between the clean speech $\mathbf{x}$, corrupted (or noisy) speech $\mathbf{y}$, and noise $\mathbf{n}$ in the log spectral domain can be approximated as

$$\mathbf{y} = \mathbf{x} + \log(1 + \exp(\mathbf{n} - \mathbf{x})). \tag{9.4}$$

In GMMs, this nonlinear relationship is often approximated with the first-order VTS. DNNs, however, with many layers of nonlinear transformation, can directly model arbitrary nonlinear relationships, including that described by Eq. 9.4. Since we are interested in the nonlinear mapping from the noisy speech $\mathbf{y}$, and noise $\mathbf{n}$ to the clean speech $\mathbf{x}$, we may augment each observation input (noisy speech) to the network with an estimate of the noise $\hat{\mathbf{n}}_t$ present in the signal, i.e.,

$$\mathbf{v}_t^0 = [\mathbf{y}_{t-\tau}, \ldots, \mathbf{y}_{t-1}, \mathbf{y}_t, \mathbf{y}_{t+1}, \ldots, \mathbf{y}_{t+\tau}, \hat{\mathbf{n}}_t], \tag{9.5}$$

where a window of $2\tau + 1$ frames of noisy speech and a frame of noise estimation is used as the input to the network. This is done in both training and decoding and thus is analogous to noise adaptive training (NAT) [11] without an explicit mismatch function. Since the DNN is being given noise estimation in order to automatically learn the mapping from the noisy speech and noise to the senone labels, implicitly through a clean speech estimation, this technique is referred as noise-aware training (NaT) [28, 33].

The robustness of the DNNs on environment distortions can be clearly observed in the experiments conducted on the Aurora 4 corpus [20], a 5,000-word vocabulary task based on the Wall Street Journal (WSJ0) corpus. The models were trained with the 16 kHz multi-condition training set consisting of 7,137 utterances from 83 speakers. One half of the utterances was recorded by a high-quality close-talking microphone and the other half was recorded using one of 18 different secondary microphones. Both halves include a combination of clean speech and speech corrupted by one of six different types of noise (street traffic, train station, car, babble, restaurant, airport) at a range of signal-to-noise ratios (SNR) between 10–20 dB.

The evaluation was conducted on the test set consisting of 330 utterances from 8 speakers. This test set was recorded by the primary microphone and a number of secondary microphones. These two sets were then each corrupted by the same six

Table 9.3 A comparison of several GMM systems in the literature to a DNN system on the Aurora 4 task

Systems	Distortion				AVG (%)
	None (clean) (%)	Noise (%)	Channel (%)	Noise+ channel (%)	
GMM baseline	14.3	17.9	20.2	31.3	23.6
MPE + NAT + VTS	7.2	12.8	11.5	19.7	15.3
NAT + Derivative kernels	7.4	12.6	10.7	19.0	14.8
NAT + Joint MLLR/VTS	5.6	11.0	8.8	17.8	13.4
DNN (7× 2,048)	5.6	8.8	8.9	20.0	13.4
DNN + NaT + dropout	5.4	8.3	7.6	18.5	12.4

Summarized from [28, 33]

noises used in the training set at SNRs between 5 and 15 dB, creating a total of 14 test sets. These 14 test sets can then be grouped into 4 subsets, based on the type of distortion: none (clean speech), additive noise only, channel distortion only, and noise + channel. Notice that the types of noise are common across training and test sets but the SNRs of the data are not.

The DNN was trained using 24-dimensional log mel-filter-bank features with utterance-level mean normalization. The first- and second-order derivative features were appended to the static feature vectors. The input layer was formed from a context window of 11 frames creating an input layer of 792 input neurons. The DNN had 7 hidden layers each with 2,048 neurons and the softmax output layer had 3,206 neurons, corresponding to the senones of the baseline HMM system. The network was initialized using layer-by-layer generative pretraining and then discriminatively trained using backpropagation. To reduce the overfitting, dropout [7] discussed in Sect. 4.3.4 was used in one of the DNN setups.

In Table 9.3, summarized from [28, 33], the performance obtained by the DNN acoustic model is compared to that obtained by several GMM systems. The first system is a baseline GMM-HMM system, while the remaining systems are representative of the state-of-the-art GMM systems in acoustic modeling and noise and speaker adaptation. All used the same training set.

The "MPE+NAT+VTS" system combines minimum phone error (MPE) discriminative training [23] and noise adaptive training (NAT) using VTS adaptation to compensate for noise and channel mismatch [3]. The "NAT+Derivative Kernels" system uses a multi-pass hybrid generative/discriminative classifier [24]. It first uses an adaptively trained HMM with VTS adaptation to generate features based on state likelihoods and their derivatives. These features are then input to a discriminative log-linear model to obtain the final hypothesis. The "NAT+Joint MLLR/VTS" system uses an HMM trained with NAT and combines VTS adaptation for environment compensation and MLLR for speaker adaptation [30] . The last two rows of the table show the performance of the two DNN-HMM systems. The "DNN (7 × 2 K)" system is simply a direct application of the CD-DNN-HMM with 7 hidden layers each with 2 K neurons. Nevertheless, it outperforms all but the "NAT+Joint MLLR/VTS" system. Finally, the "DNN+NaT+dropout" system that uses the noise-aware training and dropout has the best performance. In addition, all the DNN-HMM results were obtained in the first pass, while the other three systems required two or more recognition passes for noise, channel, or speaker adaptation. These results clearly demonstrate the inherent robustness of the DNN to unwanted variability from noise and channel mismatch.

9.5 Robustness Across All Conditions

The superior robustness results reported in Sect. 9.4 seem to suggest that DNNs provide significantly higher error rate reduction over GMM systems for noisier speech than for cleaner speech. This is, however, untrue. In fact, these results only indicate that the DNN systems are more robust than GMM systems to speaker and environmental distortions. The perturbation shrinking property in the higher layers as we discussed in Sect. 9.2 applies equally to all conditions. In this section, we show, using results from [8], that DNNs provide similar gains over GMM systems across different noise levels and speaking rates.

In [8], Huang et al. conducted a series of studies comparing GMMs and DNNs on the mobile voice search (VS) and short message dictation (SMD) datasets collected through the real-world applications that are used by millions of users with distinct speaking styles in diverse acoustic environments. These datasets were chosen because they cover almost all key LVCSR acoustic model challenges, each with enough data to ensure the statistical significance. In the study, Huang et al. trained a pair of GMM and DNN models using 400 h VS/SMD data. The GMM system, which used the 39-dimensional MFCC feature with up to the third-order derivatives, is a state-of-the-art model trained with the feature-space minimum phone error rate (fMPE) [22] and boosted MMI (bMMI) [21] criteria. The DNN system, which used the 29-dimensional log filter-bank (LFB) feature (and up to the second-order derivatives) with a context window of 11 frames, was trained using the cross-entropy (CE) criteria. The two models shared the same training data and decision tree. The same maximum likelihood estimation (MLE) GMM seed model was used for the lattice generation in the GMM and the senone state alignment in the DNN. The analytic study was conducted on a 100 h VS/SMD test data randomly sampled from the datasets and roughly follow the same distribution as the training data.

9.5.1 Robustness Across Noise Levels

Figures 9.8 and 9.9, provided in [8], compare the error pattern of the GMM-HMM and CD-DNN-HMM models under different signal-to-noise ratios (SNRs) for the VS and SMD datasets respectively. As we can observe from these tables, the CD-DNN-HMM

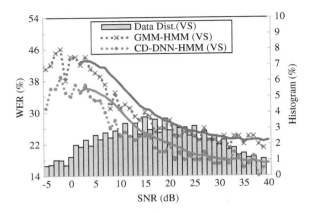

Fig. 9.8 Performance comparison of GMM-HMM and CD-DNN-HMM at different SNR levels for the VS task. The *solid lines* are the regression curves. (Figure from Huang et al. [8], permitted to use by ISCA.)

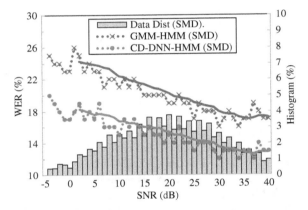

Fig. 9.9 Performance comparison of GMM-HMM and CD-DNN-HMM at different SNR levels for the SMD task. The *solid lines* are the regression curves. (Figure from Huang et al. [8], permitted to use by ISCA.)

significantly outperforms the GMM-HMM at all SNR levels, including both the clean and very noisy speech. But more interestingly, we can observe that the CD-DNN-HMM yields almost the uniform performance gain across different SNR levels over the GMM-HMM on both the VS and SMD datasets.

We can measure the noise robustness of the DNN in a different way by calculating the performance degradation per 1 dB SNR drop. For the VS task, each 1 dB SNR drop introduces about 0.40 % absolute (or 2.2 % relative) WER increment since when the SNR drops from 40 to 0 dB the WERs increase from 18 to 34 %. For the SMD task, the same 1 dB SNR drop results in 0.15 % absolute (or 1.3 % relative) WER increment because within the same SNR range the SMD WERs increase from 12 to 18 %. The quantitative difference of the sensitivity to the noise level between these tasks is likely due to the fact that the SMD has much lower LM perplexity. The same 1 dB SNR drop, however, introduces 0.6 % (2.6 % relative) and 0.20 % absolute (or 1.2 % relative) WER increment on the VS and SMD datasets, respectively, when the GMM system is used.

These results suggest that CD-DNN-HMM is more robust than GMM systems with less WER increment per 1 dB SNR drop on average and slightly more so at the low SNR range as indicated by the flatter slope compared to GMM systems. However, the difference is very small. The speech recognition performance of the DNN still drops quite a lot as the noise level increases within the normal range of the mobile speech applications. This indicates that the noise robustness remains an important research area and techniques such as speech enhancement, noise robust acoustic features, or other multi-condition learning technologies need to be explored to bridge the performance gap and further improve the overall performance of the deep learning-based acoustic model.

9.5.2 Robustness Across Speaking Rates

Speaking rate variation is another well known factor that would affect the speech intelligibility and thus the speech recognition accuracy. The speaking rate change can be due to different speakers, speaking modes, and speaking styles. There are several reasons that speaking rate change may result in speech recognition accuracy degradation. First, it may change the acoustic score dynamic range since the AM score of a phone is the sum of all the frames in the same phone segment. Second, the fixed frame rate, frame length, and context window size may be inadequate to capture the dynamics in transient speech events for fast or slow speech and therefore result in suboptimal modeling. Third, variable speaking rates may result in slight formant shift due to the human vocal instrumentation limitation. Last, extremely fast speech may cause formant target missing and phone deletion.

Figures 9.10 and 9.11, originally appear in [8], illustrate the WER difference across different speaking rates, measured as the number of phones per second,[3] on the VS and SMD datasets respectively. From these figures, we can notice that the CD-DNN-HMM system consistently outperforms the GMM-HMM system with almost uniform WER reduction across all speaking rates. Unlike in the noise robustness case, here we observe a U-shaped pattern on both VS and SMD datasets. On the VS dataset, the best WER is achieved around 10–12 phones per second. When the speaking rate deviates, either speeds up or slows down, 30 % from the sweet spot, 30 % relative WER increment is observed. On the SMD dataset, 15 % relative WER increment can be observed when the speaking rate deviates 30 % from the sweet spot.

---

[3] Huang et al. [8] also tried some of the variations such as the number of vowels per second and the speaking rate normalized by the average duration of different phonemes. It was reported that no matter which definition is used the WER pattern is very similar.

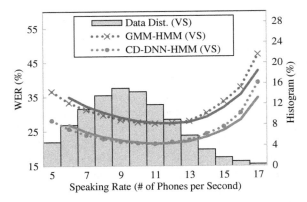

Fig. 9.10 Performance comparison of the GMM-HMM and the CD-DNN-HMM at different speaking rate for the VS task. (Figure from Huang et al. [8], permitted to use by ISCA.)

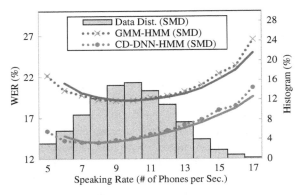

Fig. 9.11 Performance comparison of the GMM-HMM and the CD-DNN-HMM at different speaking rate for the SMD task. (Figure from Huang et al. [8], permitted to use by ISCA.)

To compensate for the speaking rate difference, additional modeling techniques need to be developed.

9.6 Lack of Generalization Over Large Distortions

In Sect. 9.2, we have shown that small perturbations in the input will be gradually shrunk as we move to the internal representation in the higher layers. This property leads to robustness of the DNN systems to the speaker and environment variations as shown in Sect. 9.4. In Sect. 9.5, we showed that this property applies across different SNR levels and speaking rates. In this section, we point out that the above result is only applicable to small perturbations around the training samples. When the test

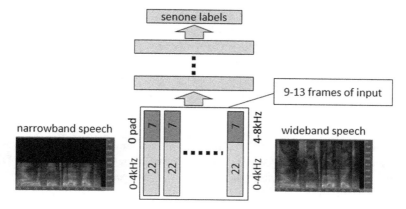

Fig. 9.12 Illustration of mixed-bandwidth speech recognition using a DNN. (Figure from Yu et al. [33], permitted to use by Yu.)

samples deviate significantly from the training samples, DNNs cannot accurately classify them. In other words, DNNs must see examples of representative variations in the data during training in order to generalize to similar variations in the test data. This is no different from other machine learning models.

This behavior can be demonstrated using a mixed-bandwidth ASR study. Typical speech recognizers are trained on either narrowband speech signals, recorded at 8 kHz, or wideband speech signals, recorded at 16 kHz. It would be advantageous if a single system could recognize both narrowband and wideband speech, i.e., mixed-bandwidth ASR. One such system, depicted in Fig. 9.12, was recently proposed on the CD-DNN-HMM framework [16]. In this mixed-bandwidth ASR system, the input to the DNN is the 29 mel-scale log filter-bank outputs together with dynamic features across an 11-frame context window. The DNN has 7 hidden layers, each with 2,048 nodes. The output layer has 1,803 neurons, corresponding to the number of senones determined by the GMM system.

The 29-dimensional filter-bank has two parts: the first 22 filters span 0–4 kHz and the last 7 filters span 4–8 kHz, with the center frequency of the first filter in the higher filter-bank at 4 kHz. When the speech is wideband, all 29 filters have observed values. However, when the speech is narrowband, the high-frequency information was not captured so the final 7 filters are set to 0.

Experiments were conducted on a mobile voice search (VS) corpus. This task consists of internet search queries made by voice on a smartphone [32]. There are two training sets, VS-1 and VS-2, consisting of 72 and 197 h of wideband audio data, respectively. These sets were collected during different times of year. The test set, called VS-T, has 26,757 words in 9,562 utterances. The narrow band training and test data were obtained by downsampling the wideband data.

Table 9.4, extracted from [16], summarizes the WER on the wideband and narrowband test sets when the DNN is trained with and without narrowband speech. From this table, we can observe that if all training data are wideband, the DNN performs well on the wideband test set (27.5 % WER) but very poorly on the

Table 9.4 Word error rate (WER) on wideband (16k) and narrowband (8k) test sets with and without narrowband training data

Training data	16 kHz VS-T (%)	8 kHz VS-T (%)
16 kHz VS-1 + 16 kHz VS-2	27.5	53.5
16 kHz VS-1 + 8 kHz VS-2	28.3	29.3

Table from Yu et al. [33] based on results extracted from Li et al. [16]

narrowband test set (53.5 % WER). However, if we convert VS-2 to narrowband speech and train the same DNN using mixed-bandwidth data (second row), the DNN performs very well on both wideband and narrowband speech.

To understand the difference between these two scenarios, we can measure the Euclidean distance

$$d_l(x_{\text{wb}}, x_{\text{nb}}) = \sqrt{\sum_{j=1}^{N^\ell} \left(v_j^\ell(x_{\text{wb}}) - v_j^\ell(x_{\text{nb}}) \right)^2} \qquad (9.6)$$

between the activation vectors at each layer for the wideband and narrowband input feature pairs, $v^\ell(x_{\text{wb}})$ and $v^\ell(x_{\text{nb}})$, where the hidden units are shown as a function of the wideband features, x_{wb}, or the narrowband features, x_{nb}. For the top layer, whose output is the senone posterior probability, we calculate the KL divergence in nats between $p(s_j|x_{\text{wb}})$ and $p(s_j|x_{\text{nb}})$.

$$d_y(x_{\text{wb}}, x_{\text{nb}}) = \sum_{j=1}^{N^L} p(s_j|x_{\text{wb}}) \log \frac{p(s_j|x_{\text{wb}})}{p(s_j|x_{\text{nb}})}, \qquad (9.7)$$

Table 9.5 Euclidean distance for the activation vectors at each hidden layer (L1–L7) and the KL divergence (nats) for the posteriors at the softmax layer between the narrowband (8 kHz) and wideband (16 kHz) input features, measured using the wideband DNN or the mixed-bandwidth DNN

Layer	Distance	Wideband DNN	Mixed-band DNN
1	Euclidean	13.28	7.32
2		10.38	5.39
3		8.04	4.49
4		8.53	4.74
5		9.01	5.39
6		8.46	4.75
7		5.27	3.12
Output	KL divergence	2.03	0.22

Table extracted from Li et al. [16] and Yu et al. [33]

where N^L is the number of senones, and s_j is the senone id. Table 9.5, extracted from [16], shows the statistics of d_l and d_y over 40,000 frames randomly sampled from the test set for the DNN trained using wideband speech only and that trained using mixed-bandwidth speech.

From Table 9.5, we can observe that the average distances in the data-mixed DNN are consistently smaller than those in the DNN trained on wideband speech only. This indicates that by using mixed-bandwidth training data, the DNN learns to consider the differences in the wideband and narrowband input features as irrelevant variations. These variations are suppressed after many layers of nonlinear transformation. The final representation is thus more invariant to this variation and yet still has the ability to distinguish between different class labels. This behavior is even more obvious at the output layer since the KL divergence between the paired outputs is only 0.22 nats in the mixed-bandwidth DNN, much smaller than the 2.03 nats observed in the wideband DNN.

References

1. Andreou, A., Kamm, T., Cohen, J.: Experiments in vocal tract normalization. In: Proceedings of the CAIP Workshop: Frontiers in Speech Recognition II (1994)
2. Davis, S., Mermelstein, P.: Comparison of parametric representations for monosyllabic word recognition in continuously spoken sentences. IEEE Trans. Acoust., Speech Signal Process. 28(4), 357–366 (1980)
3. Flego, F., Gales, M.J.: Discriminative adaptive training with VTS and JUD. In: Proceedings of the IEEE Workshop on Automatic Speech Recognition and Understanding (ASRU), pp. 170–175 (2009)
4. Gales, M.J., Woodland, P.: Mean and variance adaptation within the mllr framework. Comput. Speech Lang. 10(4), 249–264 (1996)
5. Gales, M.J.: Maximum likelihood linear transformations for HMM-based speech recognition. Comput. Speech Lang. 12(2), 75–98 (1998)
6. Gunawardana, A., Mahajan, M., Acero, A., Platt, J.C.: Hidden conditional random fields for phone classification. In: Proceedings of the Annual Conference of International Speech Communication Association (INTERSPEECH), pp. 1117–1120 (2005)
7. Hinton, G.E., Srivastava, N., Krizhevsky, A., Sutskever, I., Salakhutdinov, R.R.: Improving neural networks by preventing co-adaptation of feature detectors. arXiv preprint arXiv:1207.0580 (2012)
8. Huang, Y., Yu, D., Liu, C., Gong, Y.: A comparative analytic study on the gaussian mixture and context dependent deep neural network hidden markov models. In: Proceedings of the Annual Conference of International Speech Communication Association (INTERSPEECH) (2014)
9. Imagenet. http://www.image-net.org/
10. Jarrett, K., Kavukcuoglu, K., Ranzato, M., LeCun, Y.: What is the best multi-stage architecture for object recognition? In: Proceedings of the IEEE International Conference on Computer Vision (ICCV), pp. 2146–2153 (2009)
11. Kalinli, O., Seltzer, M.L., Droppo, J., Acero, A.: Noise adaptive training for robust automatic speech recognition. IEEE Trans. Audio, Speech Lang. Process. 18(8), 1889–1901 (2010)
12. Kim, D.Y., Kwan Un, C., Kim, N.S.: Speech recognition in noisy environments using first-order vector Taylor series. Speech Commun. 24(1), 39–49 (1998)
13. Kumar, N., Andreou, A.G.: Heteroscedastic discriminant analysis and reduced rank HMMs for improved speech recognition. Speech Commun. 26(4), 283–297 (1998)

14. Li, J., Deng, L., Yu, D., Gong, Y., Acero, A.: High-performance HMM adaptation with joint compensation of additive and convolutive distortions via vector Taylor series. In: Proceedings of the IEEE Workshop on Automatic Speech Recognition and Understanding (ASRU), pp. 65–70 (2007)
15. Li, J., Deng, L., Yu, D., Gong, Y., Acero, A.: HMM adaptation using a phase-sensitive acoustic distortion model for environment-robust speech recognition. In: Proceedings of the International Conference on Acoustics, Speech and Signal Processing (ICASSP), pp. 4069–4072 (2008)
16. Li, J., Yu, D., Huang, J.T., Gong, Y.: Improving wideband speech recognition using mixed-bandwidth training data in CD-DNN-HMM. In: Proceedings of the IEEE Spoken Language Technology Workshop (SLT), pp. 131–136 (2012)
17. Lowe, D.G.: Object recognition from local scale-invariant features. In: Proceedings of the Seventh IEEE International Conference on Computer vision, vol. 2, IEEE, pp. 1150–1157. (1999)
18. Mohamed, A.R., Hinton, G., Penn, G.: Understanding how deep belief networks perform acoustic modelling. In: Proceedings of the International Conference on Acoustics, Speech and Signal Processing (ICASSP), pp. 4273–4276 (2012)
19. Moreno, P.J., Raj, B., Stern, R.M.: A vector Taylor series approach for environment-independent speech recognition. In: Proceedings of the International Conference on Acoustics, Speech and Signal Processing (ICASSP), vol. 2, pp. 733–736 (1996)
20. Parihar, N., Picone, J.: Aurora working group: DSR front end LVCSR evaluation AU/384/02. Institute for Signal and Information Process, Mississippi State University, Technical Report (2002)
21. Povey, D., Kanevsky, D., Kingsbury, B., Ramabhadran, B., Saon, G., Visweswariah, K.: Boosted MMI for model and feature-space discriminative training. In: Proceedings of the International Conference on Acoustics, Speech and Signal Processing (ICASSP), pp. 4057–4060 (2008)
22. Povey, D., Kingsbury, B., Mangu, L., Saon, G., Soltau, H., Zweig, G.: fMPE: Discriminatively trained features for speech recognition. In: Proceedings of the International Conference on Acoustics, Speech and Signal Processing (ICASSP), vol. 1, pp. 961–964 (2005)
23. Povey, D., Woodland, P.C.: Minimum phone error and I-smoothing for improved discriminative training. In: Proceedings of the International Conference on Acoustics, Speech and Signal Processing (ICASSP), vol. 1, pp. I–105 (2002)
24. Ragni, A., Gales, M.: Derivative kernels for noise robust ASR. In: Proceedings of the IEEE Workshop on Automatic Speech Recognition and Understanding (ASRU), pp. 119–124 (2011)
25. Ratnaparkhi, A.: A simple introduction to maximum entropy models for natural language processing. IRCS Technical Reports Series, p. 81 (1997)
26. Sainath, T.N., Kingsbury, B., Mohamed, A.r., Ramabhadran, B.: Learning filter banks within a deep neural network framework. In: Proceedings of the IEEE Workshop on Automatic Speech Recognition and Understanding (ASRU) (2013)
27. Seide, F., Li, G., Chen, X., Yu, D.: Feature engineering in context-dependent deep neural networks for conversational speech transcription. In: Proceedings of the IEEE Workshop on Automfatic Speech Recognition and Understanding (ASRU), pp. 24–29 (2011)
28. Seltzer, M., Yu, D., Wang, Y.: An investigation of deep neural networks for noise robust speech recognition. In: Proceedings of the International Conference on Acoustics, Speech and Signal Processing (ICASSP) (2013)
29. Szegedy, C., Zaremba, W., Sutskever, I., Bruna, J., Erhan, D., Goodfellow, I., Fergus, R.: Intriguing properties of neural networks. arXiv preprint arXiv:1312.6199 (2013)
30. Wang, Y., Gales, M.J.: Speaker and noise factorization for robust speech recognition. IEEE Trans. Audio, Speech Lang. Process. **20**(7), 2149–2158 (2012)
31. Yu, D., Deng, L., Acero, A.: Hidden conditional random field with distribution constraints for phone classification. In: Proceedings of the Annual Conference of International Speech Communication Association (INTERSPEECH), pp. 676–679 (2009)
32. Yu, D., Ju, Y.C., Wang, Y.Y., Zweig, G., Acero, A.: Automated directory assistance system-from theory to practice. In: Proceedings of the Annual Conference of International Speech Communication Association (INTERSPEECH), pp. 2709–2712 (2007)

33. Yu, D., Seltzer, M.L., Li, J., Huang, J.T., Seide, F.: Feature learning in deep neural networks—studies on speech recognition tasks. In: Proceedings of the ICLR (2013)
34. Zeiler, M.D., Fergus, R.: Visualizing and understanding convolutional neural networks. arXiv preprint arXiv:1311.2901 (2013)

Chapter 10
Fuse Deep Neural Network and Gaussian Mixture Model Systems

Abstract In this chapter, we introduce techniques that fuse deep neural networks (DNNs) and Gaussian mixture models (GMMs). We first describe the Tandem and bottleneck approach in which DNNs are used as feature extractors. The hidden layers, which are better representation than the raw input feature, are used as features in the GMM systems. We then introduce techniques that fuse the recognition results and frame-level scores of the DNN-HMM hybrid system with that of the GMM-HMM system.

10.1 Use DNN-Derived Features in GMM-HMM Systems

In Chap. 9, we have shown that in the deep neural network (DNN)-hidden Markov model (HMM) hybrid systems DNNs jointly learn the nonlinear feature transformation and the log-linear classifier. More importantly, the feature representation learned by DNNs is more robust to the speaker and environment variations than the original feature. A natural idea is to treat the hidden and output layers in DNNs as better features and use them in the conventional GMM-HMM systems.

10.1.1 GMM-HMM with Tandem and Bottleneck Features

The idea of treating the hidden and output layers as better features was first proposed for the shallow multilayer perceptron (MLP) under the name of Tandem approach [1]. The Tandem approach augments the input to a GMM-HMM system with features derived from the suitably transformed output of one or more neural networks. Since the size of the output layer is the same as that of the training target, Tandem features are typically trained to produce distributions over monophone targets to control the dimension of the augmented feature.

Alternatively, Grezl et al. [2, 3] proposed to use features derived from a *bottleneck* hidden layer, which has smaller size than that of other layers, instead of using the neural network outputs directly. Since the hidden layer size can be chosen independent of the output layer size, this provides flexibility in choosing the training

© Springer-Verlag London 2015
D. Yu and L. Deng, *Automatic Speech Recognition*,
Signals and Communication Technology, DOI 10.1007/978-1-4471-5779-3_10

targets and the size of the augmented feature. The bottleneck layer creates a constriction in the network and forces the information pertinent to classification into a low-dimensional representation. Note that bottleneck layers are also used in autoencoders (described in Chap. 5) in which the neural network is trained to predict the input features themselves. Because the activations at the bottleneck layer are a low-dimensional nonlinear function of the input features, an autoencoder can be viewed as a method of nonlinear dimensionality reduction. However, since bottleneck features learned from autoencoders are unaware of the recognition task, these features are typically not discriminative as that extracted from bottleneck layers in neural networks that are trained to predict phonemes, phoneme states, or triphone states.

Many recent works [4–6] that exploit neural network features on large vocabulary speech recognition tasks use variants of these approaches, either augmenting the input to a GMM-HMM system with features based on the neural network outputs or some earlier hidden layers. More recently, DNNs are in place of shallow MLPs to extract more robust features. These DNNs are often trained to classify senones instead of monophone labels. For this reason, a hidden layer, instead of the output layer, is often used as the feature to the GMM system.

Figure 10.1 illustrates the typical hidden layers in DNNs from which the features are extracted. In Fig. 10.1a, a DNN whose hidden layers are of the same size is trained. The last hidden layer is then used as the feature, which is typically concatenated with raw features such as MFCC and PLP. However, in this architecture the feature generated typically has very high dimension. To make it more manageable, we can reduce the dimension of the concatenated feature with principal component analysis (PCA). Alternatively, we can directly reduce the size of the last hidden layer and make it a bottleneck layer as shown in Fig. 10.1b. Since all the hidden layers can be considered as some nonlinear transformation of the raw feature, we can use an arbitrary bottleneck layer as the feature sent to GMMs as shown in Fig. 10.1c.

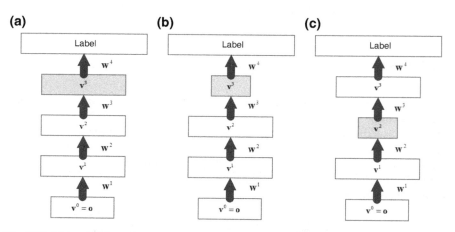

Fig. 10.1 Using DNNs as feature extractors in GMM-HMM systems. The *shaded layer* is the feature used by a GMM-HMM system

Since the hidden layer feature is to be modeled using GMMs, we should use excitations (i.e., before the nonlinear activation function is applied) instead of activations, especially if the sigmoid nonlinearity is used since sigmoid will generate features that are bounded within [0, 1] and concentrated in the two extreme values 0 and 1. Further more, even if we use bottleneck layers the dimension of the bottleneck feature may still be large and somewhat correlated between dimensions. For this reason it is often still useful to apply PCA or HLDA before it is used in the GMM-HMM system as shown in Fig. 10.2.

Note that since the Tandem (or bottleneck) features are trained independent of the GMM-HMM system, it is difficult to know which layer provides the best feature and whether adding more layers would perform better. For example, Yu and Seltzer [7] showed (in Table 10.1) that on the voice search dataset [8] (described in Sect. 6.2.1) a DNN with four hidden layers performs better than DNNs with either three or seven hidden layers. Table 10.1 also indicates that the generative pretraining (described in Chap. 5) helps in this case. In the experiment, they used 11-frames of 39-dimensioanl MFCC features as the input to the DNNs, 39 neurons in the bottleneck layer, and 2048 hidden neurons for nonbottleneck hidden layers. They used a learning rate

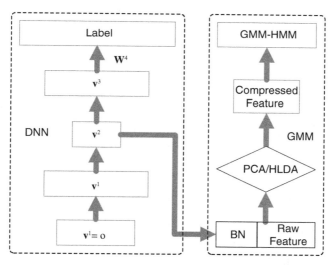

Fig. 10.2 Using Tandem (or bottleneck) features in the GMM-HMM. The DNN is trained to extract the Tandem or bottleneck feature, which is then combined with the raw feature. The combined feature is compressed and decorrelated with PCA or HLDA before it's modeled by the GMM-HMM

Table 10.1 Development set sentence error rate (SER) using bottleneck features when DNNs with different depths are used. (Summarized from Yu and Seltzer [7])

Hidden layers	3 (%)	5 (%)	7 (%)
Without DBN-pretraining	41.1	34.3	36.1
With DBN-pretraining	34.3	33.4	34.1

Table 10.2 Sentence error rate (SER) comparison of bottleneck features trained with different supervision labels

Labels used for bottleneck feature training	Dev SER (%)	Test SER (%)
None	39.4	42.1
Monophone states	35.2	37.0
Monophone states converted from senones	34.0	35.7
Senones	33.4	34.8

DBN-pretraining was applied under all conditions. (Summarized from Yu and Seltzer [7])

of $1.5e^{-5}$ per sample for all layers during pretraining, and used a learning rate of $3e^{-4}$ per sample for the first 6 epochs and a learning rate of $8e^{-6}$ per sample for the remaining 6 epochs during fine-tuning. The DNN training was carried out using stochastic mini-batch gradient descend with a minibatch size of 256 samples. The bottleneck features extracted from the DNNs are then used as alone or concatenated with the original MFCC features to train the GMM-HMMs. In both the bottleneck (BN) only and "BN+MFCC" configurations, the features were decorrelated using PCA and converted to 39 dimensions.

Yu and Seltzer also showed that using senone labels often leads to better performance than using monophone labels or using no label as shown in Table 10.2. The big performance gap between the bottleneck features that are learned without using labels and that with either monophone or senone labels clearly indicates that learning features with task-dependent information is important.

10.1.2 DNN-HMM Hybrid System Versus GMM-HMM System with DNN-Derived Features

The main difference between the DNN-HMM hybrid system and the GMM-HMM system that uses Tandem or bottleneck features is the classifier used. In the Tandem/bottleneck system, GMM is used in place of log-linear model (the softmax layer in DNNs) used in the DNN-HMM system. GMMs have better modeling power than the log-linear model when the same features are used. In fact, it is shown by Heigold et al. [9] that GMMs equal to the log-linear models when both the first and second order features are used in the log-linear models. Heigold et al.'s results also suggest that GMMs can be modeled with a single-hidden-layer neural network with a very wide hidden layer and sparse connections between the hidden and output layers. On the other hand, since the hidden layers are trained jointly with the log-linear classifier, they fit better to the log-linear model (i.e., in the DNN-HMM hybrid system) than the Tandem/bottleneck system. The net effect of these two factors cancels out and the two types of systems perform almost equally well when averaged over different tasks. However, the integrated CD-DNN-HMM system is simpler conceptually and in practice.

The main benefit of using DNN-derived features in GMM-HMM systems is the possibility of taking advantage of existing tools already available to train and to adapt GMM-HMM systems. It also allows DNNs used to derive features being trained with a subset of the training data and the GMM-HMM system that uses the DNN-derived features being trained with the full training set.

In [10], Yan et al. conducted a series of experiments to compare the CD-DNN-HMM system with the GMM-HMM system that uses the DNN-derived features. In their study they used the excitation of the last hidden layer depicted in Fig. 10.1a as the DNN-derived features. The DNN-derived features are then compressed with PCA and concatenated with the raw spectral features. The augmented features are again compressed with HLDA [11] so that the dimension is appropriate for GMM-HMM acoustic modeling. The experiments were conducted on the Switchboard (SWB) dataset described in Sect. 6.2.1. In the GMM-HMM system that uses the DNN-derived features (referred as DNN-GMM-HMM) an acoustic downscaling factor of 0.5 (simply multiply the acoustic log-likelihoods by 0.5) is used in decoding and word lattice-based sequence training. This was found essential to obtain the best recognition accuracy. The same PLP feature and training and test setups were used in the experiments so that the results can be compared with other works (e.g., [12–14]).

Table 10.3, summarized from [10, 14], compares the CD-DNN-HMM system with the DNN-GMM-HMM system on the SWB Hub5'00 evaluation set when 309 h training set is used. It can be observed that even though the region dependent linear transformation (RDLT) [15, 16], that is shown to improve the performance of the DNN-GMM-HMM system (from 17.8 to 16.1 %), is applied to the DNN-GMM-HMM, The MMI-trained DNN-GMM-HMM system still underperforms the CD-DNN-HMM trained with the same MMI training criterion (described in Chap. 8).

Table 10.4 compares CD-DNN-HMM with DNN-GMM-HMM when 2,000 h training set is used. From this table, we can observe that with RDLT the MMI-trained

Table 10.3 Word error rate (WER) on the SWB Hub5'00 evaluation set when 309 h training set is used

CD-DNN-HMM		DNN-GMM-HMM		
CE	MMI	ML	RDLT	+MMI
16.4 %	13.7 %	17.8 %	16.1 %	15.3 %

DNN has 7 hidden layers each with 2 K neurons. The output layer has 9.3 K senones. (Summarized from [10] and [14])

Table 10.4 Word error rate (WER) on the SWB Hub5'00 evaluation set when 2,000 h training set is used

CD-DNN-HMM		DNN-GMM-HMM		
CE	MMI	ML	RDLT	+MMI
14.6 %	13.3 %	15.6 %	14.5 %	13.0 %

DNN has seven hidden layers each with 2 K neurons. The output layer has 18 K senones. (Summarized from [10] and [14])

Table 10.5 Compare AE-BN system with baseline GMM-HMM system

Training method	Baseline GMM-HMM (%)	GMM-HMM with AE-BN (%)
FSA	20.2	17.6
+fBMMI	17.7	16.6
+BMMI	16.5	15.8
+MLLR	16.0	15.5

WER on English broadcast news dataset with 430 h training data. (Summarized from [17])

DNN-GMM-HMM slightly outperforms the MMI trained CD-DNN-HMM system. Combining the results in these two tables we can see that DNN-GMM-HMM, despite the increased complexity, does not have clear advantage over CD-DNN-HMM.

In the previous discussion, the DNN features are derived directly from the hidden layers. In [17] Sainath et al. investigated a less direct way of producing DNN-derived features for the GMM. In their setup, a DNN with six hidden layers of 1,024 neurons each was trained to classify 384 HMM states. Same as that in [10] the DNN did not have a bottleneck layer and so that it can classify the HMM states better than a DNN with a bottleneck layer. Different from [10], however, they used the excitation (before the softmax function is applied) of the output layer instead of the last hidden layer to derive the features. The 384 excitation values were compressed down to 40 values using a 384-128-40-384 autoencoder. This method of producing the DNN-derived features is referred as AE-BN because the bottleneck is in the autoencoder rather than in the DNN that is trained to classify HMM states.

They compared the results (shown in Table 10.5) with and without using AE-BN features in the GMM-HMM system on the English broadcast news task with 430 h training data. From the table, we can observe that with the same training methods, which include feature-space speaker adaptation (FSA), feature-space BMMI (fBMMI), BMMI [18], and maximum likelihood linear regression (MLLR) [19], the system with AE-BN feature always performs better than that without. They also compared the GMM-HMM system that uses the AE-BN features with the plain CD-DNN-HMM on a smaller task. By comparing the results in [17] and [20], we can observe that CD-DNN-HMM slightly outperforms the AE-BN system if the same training criterion is used.

10.2 Fuse Recognition Results

The errors made by the conventional GMM-HMMs and that by DNN-HMMs are different. This provides the possibility of improving the overall performance by fusing the complementary recognition results of the GMM-HMM and DNN-HMM systems. The mostly widely used system combination techniques are recognizer output voting error reduction (ROVER) [21], segmental conditional random field (SCARF) [22], and minimum Bayesian risk (MBR) based lattice combination [23].

10.2.1 ROVER

ROVER [21] is a two-step procedure comprised of alignment and voting as shown in Fig. 10.4. In the alignment step, an example of which is depicted in Fig. 10.3, the recognition results from two or more ASR systems are combined into a single word transition network (WTN). To align and combine three or more ASR results we first create a linear WTN for each of the ASR system outputs. In Fig. 10.3, for example, linear WTNs are created for three ASR results in step 1. By restricting the WTNs to linear topology, we can significantly simplify the combination process. To achieve best results these linear WTNs are ordered by increasing WER. The first WTN (WTN-1 in Fig. 10.3), which has the lowest WER, is designated as the base WTN from which the composite WTN is developed. The second WTN is then aligned to the base WTN using the dynamic programming (DP) alignment protocol. We then augment the base WTN with word transition arcs from the second WTN as

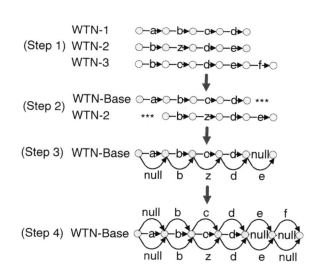

Fig. 10.3 Illustration of the word transition network (WTN) composition procedure. In *Step 1*, a linear WTN is generated for each ASR result. In *Step 2*, *WTN-1* is selected as the base WTN to which *WTN-2* is aligned. *WTN-2* is merged into the base WTN in *Step 3*. In *Step 4*, *WTN-3* is further merged into the base WTN

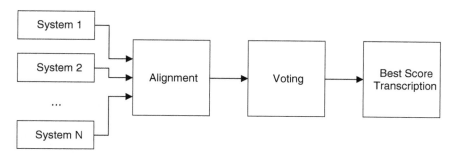

Fig. 10.4 Processing steps in ROVER

Table 10.6 The effect of system combination using ROVER

Method	50 h (%)	430 h (%)
Baseline GMM-HMM	18.8	16.0
GMM-HMM with AE-BN	17.5	15.5
ROVER over two	16.4	15.0

WER on the English broadcast news task. (Summarized from [17])

appropriate as shown in Step 3. With this new base WTN, the third WTN is merged into the base WTN as shown in Step 4. The process is repeated until all linear WTNs are merged into the base WTN.

Once the combined WTN is generated, the voting module evaluates each branching point using a voting scheme, which selects the best scoring word (with the highest number of votes) for the new transcription. There can be many different voting schemes, for example, based on frequency of occurrence, frequency of occurrence and average word confidence, or frequency of occurrence and maximum confidence. The general scoring formula is

$$\text{score}(w, i) = \frac{1}{N} \sum_{n=1}^{N} \left[\alpha \delta \left(w, w_{n,i} \right) + (1 - \alpha) \lambda_n \text{conf}_n (w, i) \right], \qquad (10.1)$$

where λ_n is the system dependent weight, δ is the Kronecker-δ, i denotes the position in the alignment, N is the number of systems, and conf(w, i) is the confidence score of word w at position i. Majority vote and averaged confidence score are smoothly interpolated via α, which is trained on a development set.

It has been shown that ROVER over different systems almost always provide additional improvement on the recognition accuracy. For example, Sainath et al. [17] reported that on the English broadcast news task, they can achieve additional 0.9 and 0.5 % WER reduction over the AE-BN system, respectively, trained with 50 and 430 h of training data, by combining the AE-BN system and the baseline GMM-HMM system as shown in Table 10.6.

10.2.2 SCARF

In the SCARF [22] framework, the conditional probability of a state sequence $\mathbf{s}$ given an observation sequence $\mathbf{o}$ is given by

$$p(\mathbf{s}|\mathbf{o}) = \frac{\sum_{\mathbf{q}:|\mathbf{q}|=|\mathbf{s}|} \exp\left(\sum_{e \in \mathbf{q},k} \lambda_k f_k \left(\mathbf{s}_l^e, \mathbf{s}_r^e, o(e) \right) \right)}{\sum_{\mathbf{s}'} \sum_{\mathbf{q}:|\mathbf{q}|=|\mathbf{s}'|} \exp\left(\sum_{e \in \mathbf{q},k} \lambda_k f_k \left(\mathbf{s}_l^{'e}, \mathbf{s}_r^{'e}, o(e) \right) \right)}, \qquad (10.2)$$

whereas $\mathbf{s}_l^e$ and $\mathbf{s}_r^e$ are the left and right states associated with an edge e in the recognition lattice, $\mathbf{q}$ is a segmentation of the observation sequence which induces a

Fig. 10.5 An example of SCARF. Shown in figure are three hypothesized states aligned with seven observations. s_1 equals to s_l^e, the *left* state of edge e, and s_2 equals to s_r^e, the *right* state of edge e. $o(e) = \{o_3, o_4\}$

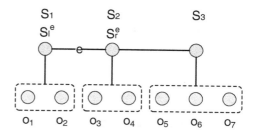

set of edges $e \in \mathbf{q}$ between the states, $o(e)$ is the segment associated with the right-hand state s_r^e and spans a block of observations from some start time to some end time, $f_k\left(s_l^e, s_r^e, o\left(e\right)\right)$ is a feature defined over the edge and associated segment, and λ_k is the weight associated with the feature. Figure 10.5 depicts an example SCARF, in which three states are hypothesized and aligned with seven observations. Weights λ_k are optimized to maximize the sequence conditional log-likelihood over a training set.

The key to the success of the SCARF model is the features extracted from recognition lattices generated by different ASR systems. Typical features used are [22]:

- Expectation features: defined with reference to a dictionary that specifies the spelling of each word in terms of the units.
- Levenshtein features: computed by aligning the observed unit sequence in a hypothesized span with that expected based on the dictionary entry for the word.
- Existence features: indicate the simple association between a unit in a detection stream, and a hypothesized word.
- Language model features: derived directly from LM.
- Baseline features: extracted from the baseline one-best sequence. The baseline feature for a segment is $+1$ when the hypothesized segment spans exactly one baseline word, and the label of the segment matches the baseline word. Otherwise it is -1.

In [24], Jaitly et al. applied the SCARF technique to combine the GMM-HMM system with the CD-DNN-HMM system. They reduced WER by 0.4 % (from 12.2 to 11.8 %) and 0.9 % (from 47.1 to 46.2 %) over the MMI trained CD-DNN-HMM system on the voice search and YouTube tasks, respectively.

10.2.3 MBR Lattice Combination

The MBR combination [23] finds the word sequence that minimizes the expected word error rate across the different systems being combined as

$$\mathbf{w}^* = \arg\min_{\mathbf{w}} \left\{ \sum_{n=1}^{N} \lambda_n \sum_{\mathbf{w}'} P_n\left(\mathbf{w}|\mathbf{o}\right) L\left(\mathbf{w}, \mathbf{w}'\right) \right\}, \quad (10.3)$$

where $L\left(\mathbf{w}, \mathbf{w}'\right)$ is the Levenshtein distance between two word sequences $\mathbf{w}$ and $\mathbf{w}'$ and $P_n\left(\mathbf{w}|\mathbf{o}\right)$ is the posterior probability of the word sequence $\mathbf{w}$ given the acoustic observation sequence $\mathbf{o}$ as computed by the n-th model. $P_n\left(\mathbf{w}|\mathbf{o}\right)$ can be estimated as

$$P_n\left(\mathbf{w}|\mathbf{o}\right) = \frac{p_n\left(\mathbf{o}|\mathbf{w}\right)^\kappa P\left(\mathbf{w}\right)}{\sum_\mathbf{w} p_n\left(\mathbf{o}|\mathbf{w}\right)^\kappa P\left(\mathbf{w}\right)}, \tag{10.4}$$

where κ is the acoustic scaling factor.

In [25], Swietojanski et al. reported that by combining the GMM-HMM and DNN-HMM systems with the MBR lattice combination technique they can achieve 1–8 % relative WER reduction over the DNN-HMM system across different setups. However, MBR lattice combination is less robust than ROVER as it sometimes increases the error rates.

10.3 Fuse Frame-Level Acoustic Scores

The systems can also be fused at the frame or state acoustic score level. The simplest yet effective approach is to perform frame-synchronous combination using a linear interpolation of the observation log-likelihoods of multiple systems as

$$\log p\left(\mathbf{o}|s\right) = \sum_{n=1}^{N} \alpha_n \log p_n\left(\mathbf{o}|s\right), \tag{10.5}$$

where α_n is the interpolation weight for system n, $p\left(\mathbf{o}|s\right)$ is the combined observation score of observation $\mathbf{o}$ given the state s, and $p_n\left(\mathbf{o}|s\right)$ is the score from the nth system. For the GMM-HMM system, this is simply the observation probability. For DNN-HMM hybrid systems, this is the scaled likelihood. Alternatively, we can model the state posterior probability

$$\log p\left(s|\mathbf{o}\right) = \sum_{n=1}^{N} \alpha_n \log p_n\left(s|\mathbf{o}\right) \tag{10.6}$$

instead. Note that this is a log-linear model with frame posterior scores as features and thus can be easily implemented with a neural network with no hidden layer. Additional improvement can be achieved by including hidden layers. The benefit of this formulation is that the frame cross-entropy (CE) and sequence discriminative training technique described in Chaps. 6 and 8 can be easily applied to train the interpolation weights or the fusing network.

In [25], Swietojanski et al. reported that by combining the frame likelihoods of GMM-HMM and DNN-HMM systems they can achieve 1–8 % relative WER reduction across different setups similar to the gains they achieved with the MBR lattice combination. Note, however, since GMM-HMM systems typically perform significantly worse than the CD-DNN-HMMs, the possible improvement is very limited.

A better approach is to combine the DNN-GMM-HMM that uses the DNN derived features with the CD-DNN-HMM system. This combination comes with two benefits. First, the DNN-GMM-HMM system performs very close to the CD-DNN-HMM system yet their results are still complementary. Second, since the same DNN is used in both the DNN-GMM-HMM and the CD-DNN-HMM systems the additional cost introduced during the decoding phase is very limited, especially when the architecture in Fig. 10.1a is used. By combining these two systems, we typically can obtain 5–10 % relative WER reduction over the CD-DNN-HMM system.

10.4 Multistream Speech Recognition

It is well known that the fixed-resolution (in both time and frequency domain) feature processing front-end used in the state-of-the-art speech recognition systems is a result of trade-off which does not model many phenomena well. For example, Huang et al. [26] showed that both CD-GMM-HMM and CD-DNN-HMM systems perform significantly worse when the speaking rate is very high or very low. One potential solution to this problem is a multistream system [27, 28] that can accommodate multiple time and/or frequency resolutions. The key design question of a multistream speech recognition system is how to combine streams. Figures 10.6, 10.7 and 10.8 illustrates three popular architectures used in the multistream speech recognition:

- Early integration: In the early integration architecture, the features are directly combined (e.g., concatenated) and processed by a single DNN-HMM to generate the decoding result.
- Intermediate integration: In the intermediate integration architecture, features in each stream are processed independently (e.g., with separate DNNs) first and then integrated at an intermediate stage. The integrated intermediate representations are then further processed by a single DNN-HMM for generating the final decoding result.
- Late integration: In the late integration architecture, features in each stream is processed independently with a separate DNN-HMM. The decoding results of each stream is then combined (e.g., through ROVER) to generate the final result.

There are many different kinds of streams. For example, we can use each narrow frequency sub-band as a stream. Such a system is often referred as multiband speech recognition system. Alternatively, we cause different feature extraction pipelines such as PLP and MFCC as streams. Such a system is sometimes called multichannel speech recognition system. Another popular way to construct streams is to use features extracted with different sampling rate, window size, or filter banks.

The work of Fletcher and his colleagues [29] suggests that in human speech perception narrow frequency sub-bands are processed independently, which favors the multiband system, and the decisions from these sub-bands are combined at some intermediate level, which favors the intermediate-integration framework. These

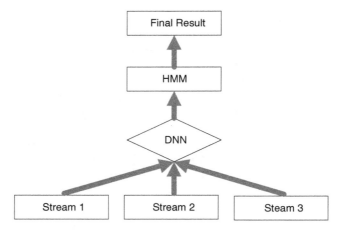

Fig. 10.6 An early-integration, multistream speech recognition architecture in which the features from all streams are combined and processed by a single DNN-HMM to generate the decoding result

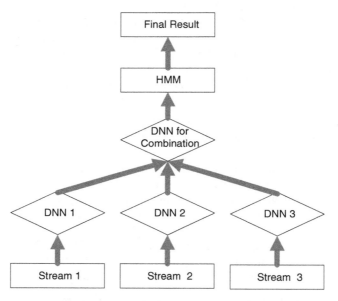

Fig. 10.7 An intermediate-integration, multistream speech recognition architecture in which features in each stream is processed independently using separate DNNs first and then integrated at an intermediate stage. The integrated intermediate representations are then further processed by a single DNN-HMM for generating the final decoding result

results are combined in such a way that the global error rate is equal to the product of error rates in the sub-bands. This product-of-error rule is very strong and essentially means that even if only one sub-band processing pipeline gives the correct

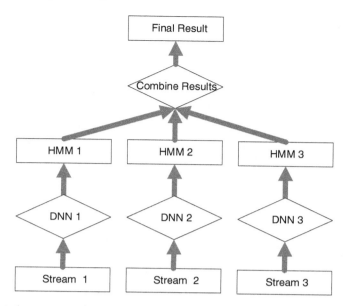

Fig. 10.8 A late-integration, multistream speech recognition architecture in which the decoding results of all the independently processed streams are combined to make the final decoding decision

result the system will recognize the utterance accurately. In [30], Zhou et al. compared early-integration, intermediate-integration, and late-integration architectures for multichannel speech recognition. Their results demonstrate that the intermediate-integration approach performs the best and reduces error by 6 % relative to the best singe-stream system on the TIMIT phone recognition task.

References

1. Hermansky, H., Ellis, D.P., Sharma, S.: Tandem connectionist feature extraction for conventional HMM systems. In: Proceedings of International Conference on Acoustics, Speech and Signal Processing (ICASSP), vol. 3, pp. 1635–1638 (2000)
2. Grézl, F., Fousek, P.: Optimizing bottle-neck features for LVCSR. In: Proceedings of International Conference on Acoustics, Speech and Signal Processing (ICASSP), pp. 4729–4732 (2008)
3. Grézl, F., Karafiát, M., Kontár, S., Černocký, J.: Probabilistic and bottle-neck features for LVCSR of meetings. In: Proceedings of International Conference on Acoustics, Speech and Signal Processing (ICASSP), pp. 757–760 (2007)
4. Fousek, P., Lamel, L., Gauvain, J.L.: Transcribing broadcast data using MLP features. In: Proceedings of Annual Conference of International Speech Communication Association (INTERSPEECH), pp. 1433–1436 (2008)
5. Valente, F., Doss, M.M., Plahl, C., Ravuri, S., Wang, W.: A comparative large scale study of MLP features for mandarin ASR. In: Proceedings of Annual Conference of International Speech Communication Association (INTERSPEECH), pp. 2630–2633 (2010)
6. Vergyri, D., Mandal, A., Wang, W., Stolcke, A., Zheng, J., Graciarena, M., Rybach, D., Gollan, C., Schlüter, R., Kirchhoff, K., et al.: Development of the SRI/nightingale Arabic ASR system.

In: Proceedings of Annual Conference of International Speech Communication Association (INTERSPEECH), pp. 1437–1440 (2008)

7. Yu, D., Seltzer, M.L.: Improved bottleneck features using pretrained deep neural networks. In: Proceedings of Annual Conference of International Speech Communication Association (INTERSPEECH), pp. 237–240 (2011)

8. Yu, D., Ju, Y.C., Wang, Y.Y., Zweig, G., Acero, A.: Automated directory assistance system-from theory to practice. In: Proceedings of Annual Conference of International Speech Communication Association (INTERSPEECH), pp. 2709–2712 (2007)

9. Heigold, G., Ney, H., Lehnen, P., Gass, T., Schluter, R.: Equivalence of generative and log-linear models. IEEE Trans. Audio, Speech Lang. Process. **19**(5), 1138–1148 (2011)

10. Yan, Z., Huo, Q., Xu, J.: A scalable approach to using DNN-derived features in GMM-HMM based acoustic modeling for LVCSR. In: Proceedings of Annual Conference of International Speech Communication Association (INTERSPEECH) (2013)

11. Kumar, N., Andreou, A.G.: Heteroscedastic discriminant analysis and reduced rank HMMs for improved speech recognition. Speech Commun. **26**(4), 283–297 (1998)

12. Seide, F., Li, G., Chen, X., Yu, D.: Feature engineering in context-dependent deep neural networks for conversational speech transcription. In: Proceedings of IEEE Workshop on Automfatic Speech Recognition and Understanding (ASRU), pp. 24–29 (2011)

13. Seide, F., Li, G., Yu, D.: Conversational speech transcription using context-dependent deep neural networks. In: Proceedings of Annual Conference of International Speech Communication Association (INTERSPEECH), pp. 437–440 (2011)

14. Su, H., Li, G., Yu, D., Seide, F.: Error back propagation for sequence training of context-dependent deep networks for conversational speech transcription. In: Proceedings of International Conference on Acoustics, Speech and Signal Processing (ICASSP) (2013)

15. Yan, Z.J., Huo, Q., Xu, J., Zhang, Y.: Tied-state based discriminative training of context-expanded region-dependent feature transforms for LVCSR. In: Proceedings of International Conference on Acoustics, Speech and Signal Processing (ICASSP), pp. 6940–6944 (2013)

16. Zhang, B., Matsoukas, S., Schwartz, R.: Discriminatively trained region dependent feature transforms for speech recognition. In: Proceedings of International Conference on Acoustics, Speech and Signal Processing (ICASSP), vol. 1, pp. I–I (2006)

17. Sainath, T.N., Kingsbury, B., Ramabhadran, B.: Auto-encoder bottleneck features using deep belief networks. In: Proceedings of International Conference on Acoustics, Speech and Signal Processing (ICASSP), pp. 4153–4156 (2012)

18. Povey, D., Kanevsky, D., Kingsbury, B., Ramabhadran, B., Saon, G., Visweswariah, K.: Boosted MMI for model and feature-space discriminative training. In: Proceedings of International Conference on Acoustics, Speech and Signal Processing (ICASSP), pp. 4057–4060 (2008)

19. Gales, M.J., Woodland, P.: Mean and variance adaptation within the mllr framework. Comput. Speech Lang. **10**(4), 249–264 (1996)

20. Sainath, T.N., Kingsbury, B., Ramabhadran, B., Fousek, P., Novak, P., Mohamed, A.r.: Making deep belief networks effective for large vocabulary continuous speech recognition. In: Proceedings of IEEE Workshop on Automfatic Speech Recognition and Understanding (ASRU), pp. 30–35 (2011)

21. Fiscus, J.G.: A post-processing system to yield reduced word error rates: Recognizer output voting error reduction (ROVER). In: Proceedings of IEEE Workshop on Automfatic Speech Recognition and Understanding (ASRU), pp. 347–354 (1997)

22. Zweig, G., Nguyen, P.: SCARF: a segmental conditional random field toolkit for speech recognition. In: Proceedings of Annual Conference of International Speech Communication Association (INTERSPEECH), pp. 2858–2861 (2010)

23. Xu, H., Povey, D., Mangu, L., Zhu, J.: Minimum Bayes risk decoding and system combination based on a recursion for edit distance. Comput. Speech Lang. **25**(4), 802–828 (2011)

24. Jaitly, N., Nguyen, P., Senior, A.W., Vanhoucke, V.: Application of pretrained deep neural networks to large vocabulary speech recognition. In: Proceedings of Annual Conference of International Speech Communication Association (INTERSPEECH) (2012)

25. Swietojanski, P., Ghoshal, A., Renals, S.: Revisiting hybrid and GMM-HMM system combination techniques. In: Proceedings of International Conference on Acoustics, Speech and Signal Processing (ICASSP) (2013)
26. Huang, Y., Yu, D., Liu, C., Gong, Y.: A comparative analytic study on the gaussian mixture and context dependent deep neural network hidden markov models. In: Proceedings of Annual Conference of International Speech Communication Association (INTERSPEECH) (2014)
27. Bourlard, H., Dupont, S., Martigny Valais Suisse C.R.: Multi stream speech recognition (1996)
28. Bourlard, H., et al.: Non-stationary multi-channel (multi-stream) processing towards robust and adaptive asr. In: Proceedings of Workshop on Robust Methods for Speech Recognition in Adverse Conditions, pp. 1–10 (1999)
29. Allen, J.B.: How do humans process and recognize speech? IEEE Trans. Speech Audio Process. **2**(4), 567–577 (1994)
30. Zhou, P., Dai, L., Liu, Q., Jiang, H.: Combining information from multi-stream features using deep neural network in speech recognition. In: IEEE International Conference on Signal Processing (ICSP) vol. 1, pp. 557–561 (2012)

Chapter 11
Adaptation of Deep Neural Networks

Abstract Adaptation techniques can compensate for the difference between the training and testing conditions and thus can further improve the speech recognition accuracy. Unlike Gaussian mixture models (GMMs), which are generative models, deep neural networks (DNNs) are discriminative models. For this reason, the adaptation techniques developed for GMMs cannot be directly applied to DNNs. In this chapter, we first introduce the concept of adaptation. We then describe the important adaptation techniques developed for DNNs, which are classified into the categories of linear transformation, conservative training, and subspace methods. We further show that adaptation in DNNs can bring significant error rate reduction at least for some speech recognition tasks and thus is as important as that in the GMM systems.

11.1 The Adaptation Problem for Deep Neural Networks

As in other machine learning techniques, in DNN systems it is assumed that the training and testing data follow the same probability distribution. In reality, however, this assumption is hardly satisfied. This is because before a speech application is deployed there is no matching training data. The application has to be bootstrapped from mismatched data. After the application is deployed some matched data can be collected under the real usage scenarios. However, at the early stage of the deployment the amount of matched data is often small and cannot cover many phenomena that may be observable in the subsequent usage. Even if enough matched training data are available after the application has been deployed for years, the mismatch problem may still exist. This is because DNN training optimizes for the average performance over the distribution of the training data. As a result, when targeting at a specific environment or speaker the training and testing conditions are still mismatched.

The training–testing mismatch problem can be solved with adaptation techniques, which either adapt the model to match better the testing condition or adapt the testing inputs to fit better the model. For example, in the conventional Gaussian mixture model (GMM)-hidden Markov model (HMM) speech recognition systems, speaker-adapted (SA) systems can cut errors by 5–30 % over the speaker-independent (SI) systems. Well known effective adaptation techniques in the GMM–HMM framework

© Springer-Verlag London 2015
D. Yu and L. Deng, *Automatic Speech Recognition*,
Signals and Communication Technology, DOI 10.1007/978-1-4471-5779-3_11

include maximum likelihood linear regression (MLLR) [11, 20, 34], constrained MLLR (cMLLR) [10] which is also known as the feature-domain MLLR (fMLLR), maximum a posteriori linear regression (MAP-LR) [7, 19], and vector Taylor series (VTS) expansion [18, 22–24, 26]. Many of these techniques can be applied to deal with both environment and speaker mismatches.

If the transcription is available for the adaptation data, it is called supervised adaptation; otherwise, it is called unsupervised adaptation, in which case the transcription needs to be inferred from the acoustic features. In most cases, the inferred transcription, often called pseudo-transcription, can be obtained by decoding the utterance using the speaker-independent models. This is the approach assumed in the following discussions. Under some strict conditions, the adaptation transcription may be inferred by exploiting the structures in the data, e.g., based on the distance between the feature vectors with and without labels. No matter which approach is used, the pseudo-transcription inevitably contains errors which will reduce the effect of adaptation. Furthermore, adaptation with the pseudo-transcription as the label reinforces what the model already can do well but limits what it can learn from what the SI model does not perform well. All these factors will limit the potential recognition accuracy improvement achievable from using the unsupervised adaptation.

Unlike GMMs, which are generative models, DNNs are discriminative models. For this reason, DNNs require different adaptation techniques than those developed for GMMs. For example, the model-space transform-based approaches such as MLLR that work well for GMMs cannot be directly ported to DNNs. This is because unlike in GMMs where Gaussian means or variances change in the same direction if they belong to the same phones or HMM states, there is no such structure in the model parameters of a DNN.

Note that DNN is a special case of the multilayer perceptron (MLP) and so some of the adaptation techniques developed for MLPs can be applied to DNNs directly. However, compared to the earlier ANN/HMM hybrid systems [27], the context-dependent (CD)-DNN-HMMs, that are typically used in the large vocabulary continuous speech recognition (LVCSR) systems, have significantly more parameters due to wider and deeper hidden layers used and the much larger output layer designed to model senones (tied-triphone states) directly. This difference casts additional challenges to adapting CD-DNN-HMMs, especially when the adaptation set is small.

Over the years many techniques have been developed to adapt DNNs. These techniques can be classified into three categories: linear transformation, conservative training, and subspace method. We will discuss these techniques in detail in the following sections.

11.2 Linear Transformations

The simplest and most popular approach to adapting DNNs is applying a linear transformation to either the input feature, the activation of a hidden layer, or the input to the softmax layer. No matter where the linear transformation is applied, it is typically trained from an identity weight matrix and zero bias to optimize either the cross-entropy (CE) training criterion discussed in Chap. 4 (Eq. (4.1)) or the sequence-discriminative training criterion discussed in Chap. 8 (Eqs. (8.1) and (8.8)), keeping fixed the weights of the original DNN.

11.2.1 Linear Input Networks

In the linear input network (LIN) [2, 4, 21, 28, 33, 35] and the very similar feature discriminative linear regression (fDLR) [30] adaptation techniques, the linear transformation is applied to the input features as shown in Fig. 11.1. The basic idea behind LIN is that the speaker-dependent (SD) feature can be linearly transformed to match that of the average speaker specified by the speaker-independent (SI) DNN model. In other words, we transform the SD feature $\mathbf{v}^0 \in \mathbb{R}^{N_0 \times 1}$ to $\mathbf{v}^0_{\text{LIN}} \in \mathbb{R}^{N_0 \times 1}$ by applying a linear transformation specified by the weight matrix $\mathbf{W}^{\text{LIN}} \in \mathbb{R}^{N_0 \times N_0}$ and bias vector $\mathbf{b}^{\text{LIN}} \in \mathbb{R}^{N_0 \times 1}$ as

$$\mathbf{v}^0_{\text{LIN}} = \mathbf{W}^{\text{LIN}} \mathbf{v}^0 + \mathbf{b}^{\text{LIN}}, \tag{11.1}$$

where N_0 is the size of the input layer.

Fig. 11.1 Illustration of the linear input network (LIN) and feature discriminative linear regression (fDLR) adaptation techniques. A linear layer is inserted right above and applied upon the input feature layer

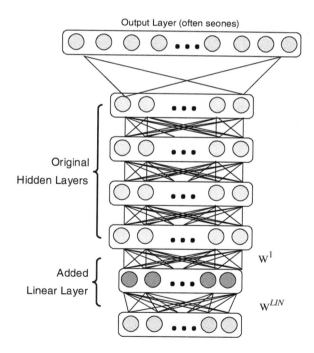

In speech recognition, the input feature vector $\mathbf{v}^0(t) = \mathbf{o}_t = \left[\mathbf{x}_{\max(0, t-\varpi)} \cdots \mathbf{x}_t \right.$ $\left. \cdots \mathbf{x}_{\min(T, t+\varpi)}\right]$ at time t for an utterance with T frames often covers $2\varpi + 1$ frames. When the adaptation set is small, it is desirable to apply a smaller per-frame transformation $\left[\mathbf{W}_f^{\mathrm{LIN}} \in \mathbb{R}^{D \times D}, \mathbf{b}_f^{\mathrm{LIN}} \in \mathbb{R}^{D \times 1}\right]$ as

$$\mathbf{x}^{\mathrm{LIN}} = \mathbf{W}_f^{\mathrm{LIN}} x + \mathbf{b}_f^{\mathrm{LIN}}, \tag{11.2}$$

and the transformed input feature vector can be constructed as $\mathbf{v}_{\mathrm{LIN}}^0(t) = \mathbf{o}_t^{\mathrm{LIN}} = \left[\mathbf{x}_{\max(0, t-\varpi)}^{\mathrm{LIN}} \cdots \mathbf{x}_t^{\mathrm{LIN}} \cdots \mathbf{x}_{\min(T, t+\varpi)}^{\mathrm{LIN}}\right]$, where D is the dimension of each feature frame and $N_0 = (2\varpi + 1) D$. Since W_f^{LIN} has only $\frac{1}{(2\varpi+1)^2}$ of parameters as that in $\mathbf{W}^{\mathrm{LIN}}$, it has less transformation power and is less effective than $\mathbf{W}^{\mathrm{LIN}}$. However, it can be more reliably estimated from a small adaptation set and thus may outperform $\mathbf{W}^{\mathrm{LIN}}$ overall.

11.2.2 Linear Output Networks

The linear transformation can also be applied to the softmax layer, in which case the adapted network is called linear output network (LON) [21] or output-feature discriminative linear regression (oDLR) [40, 43]. This is because all the hidden layers in the DNN can be considered as a complicated nonlinear feature transformation module and the last hidden layer can be considered as the transformed feature as we have discussed in Chap. 9. It is thus reasonable to apply a linear transformation on the last hidden layer for a specific speaker so that after the linear transformation it matches better to the average speaker. Different from the LIN/fDLR though, there are two ways to apply the linear transformation in LON/oDLR as shown in Fig. 11.2.

In Fig. 11.2a, the linear transformation is applied after the softmax layer weights. In other words,

$$\begin{aligned}
\mathbf{z}_{\mathrm{LONa}}^L &= \mathbf{W}_a^{\mathrm{LON}} z^L + \mathbf{b}_a^{\mathrm{LON}} \\
&= \mathbf{W}_a^{\mathrm{LON}} \left(\mathbf{W}^L v^{L-1} + b^L\right) + \mathbf{b}_a^{\mathrm{LON}} \\
&= \left(\mathbf{W}_a^{\mathrm{LON}} \mathbf{W}^L\right) v^{L-1} + \left(\mathbf{W}_a^{\mathrm{LON}} b^L + \mathbf{b}_a^{\mathrm{LON}}\right),
\end{aligned} \tag{11.3}$$

where the DNN is assumed to have L layers, N_{L-1} and N_L are the number of neurons at layer $L-1$ (last hidden layer) and L (softmax layer), respectively, $\mathbf{v}^{L-1} \in \mathbb{R}^{N_{L-1} \times 1}$ is the speaker-independent feature at the last hidden layer, $\mathbf{z}^L \in \mathbb{R}^{N_L \times 1}$ is the excitation of the softmax layer without adaptation, $\mathbf{W}^L$ and $\mathbf{b}^L$ are the softmax layer weight matrix and bias vector, respectively on the speaker-independent DNN, $\mathbf{W}_a^{\mathrm{LON}} \in \mathbb{R}^{N_L \times N_L}$ and $\mathbf{b}_a^{\mathrm{LON}} \in \mathbb{R}^{N_L \times 1}$ are the transformation matrix and the bias

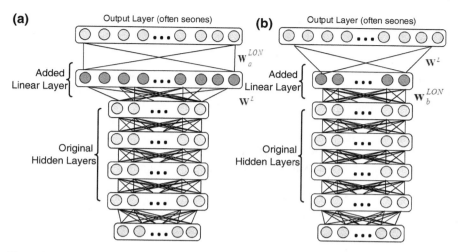

Fig. 11.2 Illustration of the linear output network (LON). The linear transformation can be applied after (**a**) or before (**b**) the original weight matrix $\mathbf{W}^L$ is applied

vector, respectively, in the LON(a), and $\mathbf{z}_{\mathrm{LON}a}^L \in \mathbb{R}^{N_L \times 1}$ is the excitation after the linear transformation.

In Fig. 11.2b, the linear transformation is applied before the softmax layer weights, or

$$
\begin{aligned}
\mathbf{z}_{\mathrm{LON}b}^L &= \mathbf{W}^L v_{\mathrm{LON}b}^{L-1} + \mathbf{b}^L \\
&= \mathbf{W}^L \left(\mathbf{W}_b^{\mathrm{LON}} v^{L-1} + \mathbf{b}_b^{\mathrm{LON}} \right) + \mathbf{b}^L \\
&= \left(\mathbf{W}^L \mathbf{W}_b^{\mathrm{LON}} \right) v^{L-1} + \left(\mathbf{W}^L \mathbf{b}_b^{\mathrm{LON}} + \mathbf{b}^L \right),
\end{aligned}
\tag{11.4}
$$

where $v_{\mathrm{LON}b}^{L-1} \in \mathbb{R}^{N_{L-1} \times 1}$ is the transformed feature at the last hidden layer, and $\mathbf{W}_b^{\mathrm{LON}} \in \mathbb{R}^{N_{L-1} \times N_{L-1}}$ and $\mathbf{b}_b^{\mathrm{LON}} \in \mathbb{R}^{N_{L-1} \times 1}$ are the transformation matrix and the bias vector, respectively, in the LON(b).

It is clear that these two approaches are equally powerful since the linear transformation of a linear transformation equals to a single linear transformation as shown in Eqs. (11.3) and (11.4). However, the number of parameters in these two approaches can be significantly different. If the number of output neurons is smaller than that of the last hidden layer, as in the case of monophone systems, $\mathbf{W}_a^{\mathrm{LON}}$ is smaller than $\mathbf{W}_b^{\mathrm{LON}}$. In the CD-DNN-HMM systems, however, the output layer size is significantly larger than that of the last hidden layers. As a result $\mathbf{W}_b^{\mathrm{LON}}$ is significantly smaller than $\mathbf{W}_a^{\mathrm{LON}}$ and can be more reliably estimated from the adaptation data.

11.3 Linear Hidden Networks

In the linear hidden network (LHN) [12], the linear transformation is applied to the hidden layers. This is because, as discussed in Chap. 9, a DNN can be separated into two parts at any hidden layer. The part that contains the input layer can be considered as a feature transformation module and the hidden layer that separates the DNN can be considered as a transformed feature. The part that contains the output layer can be considered as a classifier that operates on the hidden layer feature.

Similar to that in LON, in LHN there are also two ways to apply the linear transformation as shown in Fig. 11.3. For the same reason, we just discussed these two adaptation approaches that have the same adaptation power. Unlike in LON where the size of $\mathbf{W}_a^{LON}$ and $\mathbf{W}_b^{LON}$ can be significantly different, though, in LHN $\mathbf{W}_a^{LHN}$ and $\mathbf{W}_b^{LHN}$ often have the same size because in many systems the hidden layer sizes are the same.

The effectiveness of LIN, LON, and LHN is task dependent. Although they are very similar, there are subtle differences with regard to the number of parameters and the variability of the feature adapted as we just discussed. These factors together with the size of the adaptation set determine which technique is best for a specific task.

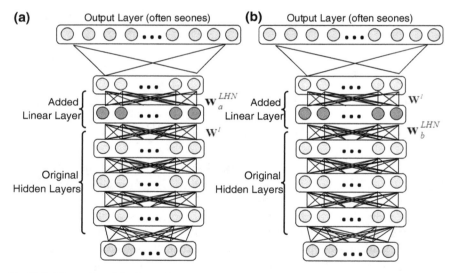

Fig. 11.3 Illustration of the linear hidden network (LHN). The linear transformation can be applied after (**a**) or before (**b**) the original weight matrix $\mathbf{W}^\ell$ at hidden layer ℓ is applied

11.4 Conservative Training

Although the linear transformation adaptation techniques can be useful under some conditions, their effectiveness is highly restricted by the linear transformation nature. A potentially more effective approach is to adapt all the parameters in the DNN to optimize the adaptation criterion $J(\mathbf{W}, \mathbf{b}; \mathbb{S})$ over the adaptation set $\mathbb{S} = \{(\mathbf{o}^m, \mathbf{y}^m) \mid 0 \leq m < M\}$, where $J(\mathbf{W}, \mathbf{b}; \mathbb{S})$ can be either the cross-entropy (CE) training criterion discussed in Chap. 4 (Eq. (4.11)) or the sequence-discriminative training criterion discussed in Chap. 8 (Eqs. (8.1) and (8.8)).

Unfortunately, this seemingly simple approach may destroy previously learned information and thus is not reliable especially since the adaptation set size is typically very small compared to the number of parameters in the DNN. To prevent this from happening, some conservative training (CT) [3, 25, 32, 43] strategy is needed. Conservative training (CT) can be achieved by adding regularization to the adaptation criterion. A simple heuristic is to adapt only selected weights. For example, in [32], only weights connected to the hidden nodes with maximum variance computed on the adaptation data are adapted. Alternatively, we can only adapt the large weights in the DNN. Adaptation with very small learning rate and early stopping can also be considered as CT.

In this section, we describe the two most popular explicit regularization techniques used in CT: L_2 regularization [25] and Kullback–Leibler divergence (KLD) regularization [43]. We also discuss techniques that can be used to reduce the footprint of the adapted models.

11.4.1 L_2 Regularization

The basic idea of the L_2 regularized CT is to add the L_2 norm of the model parameter difference

$$R_2(\mathbf{W}_{\text{SI}} - \mathbf{W}) = \|\text{vec}(\mathbf{W}_{\text{SI}} - \mathbf{W})\|_2^2$$

$$= \sum_{\ell=1}^{L} \left\|\text{vec}\left(\mathbf{W}_{\text{SI}}^\ell - \mathbf{W}^\ell\right)\right\|_2^2 \tag{11.5}$$

between the speaker-independent model $\mathbf{W}_{\text{SI}}$ and the adapted model $\mathbf{W}$ to the adaptation criterion $J(\mathbf{W}, \mathbf{b}; \mathbb{S})$, where $\text{vec}(\mathbf{W}^\ell) \in \mathbb{R}^{[N_\ell \times N_{\ell-1}] \times 1}$ is the vector generated by concatenating all the columns in the matrix $\mathbf{W}^\ell$, and $\|\text{vec}(\mathbf{W}^\ell)\|_2$ equals to $\|\mathbf{W}^\ell\|_F$—the Frobenious norm of the matrix $\mathbf{W}^\ell$.

When this L_2 regularization term is included, the adaptation criterion becomes

$$J_{L_2}(\mathbf{W}, \mathbf{b}; \mathbb{S}) = J(\mathbf{W}, \mathbf{b}; \mathbb{S}) + \lambda R_2(\mathbf{W}_{\text{SI}}, \mathbf{W}), \tag{11.6}$$

where λ is a regularization weight that controls the relative contribution of the two terms in the adaptation criterion. The L_2 regularized CT aims to constrain the change of the model parameter of the adapted model with regard to the speaker-independent model. Since the training criterion Eq. (11.6) is very similar to the weight decay we discussed in Sect. 4.3.3 of Chap. 4, the same training algorithm can be used directly in the L_2 regularized CT adaptation.

11.4.2 KL-Divergence Regularization

The intuition behind the KLD regularization approach is that the senone posterior distribution estimated from the adapted model should not deviate too far away from that estimated from the unadapted model. Since the DNN outputs are probability distributions, a natural choice in measuring the deviation is the Kullback–Leibler divergence (KLD). By adding this divergence as a regularization term to the adaptation criterion and removing the terms unrelated to the model parameters, we get the regularized optimization criterion

$$J_{\text{KLD}}\left(\mathbf{W}, \mathbf{b}; \mathbb{S}\right) = (1 - \lambda) J\left(\mathbf{W}, \mathbf{b}; \mathbb{S}\right) + \lambda R_{\text{KLD}}\left(\mathbf{W}_{\text{SI}}, \mathbf{b}_{\text{SI}}; W, \mathbf{b}; \mathbb{S}\right), \quad (11.7)$$

where λ is a regularization weight,

$$R_{\text{KLD}}\left(\mathbf{W}_{\text{SI}}, \mathbf{b}_{\text{SI}}; \mathbf{W}, \mathbf{b}; \mathbb{S}\right) = \frac{1}{M} \sum_{m=1}^{M} \sum_{i=1}^{C} P_{\text{SI}}\left(i | \mathbf{o}_m; \mathbf{W}_{\text{SI}}, \mathbf{b}_{\text{SI}}\right) \log P\left(i | \mathbf{o}_m; \mathbf{W}, \mathbf{b}\right),$$
$$(11.8)$$

$P_{\text{SI}}\left(i | \mathbf{o}_m; \mathbf{W}_{\text{SI}}, \mathbf{b}_{\text{SI}}\right)$ and $P\left(i | \mathbf{o}_m; \mathbf{W}, \mathbf{b}\right)$ are the probability that the mth observation $\mathbf{o}_m$ belongs to class i, estimated from the speaker-independent and the adapted DNNs, respectively, and will be simplified as $P_{\text{SI}}\left(i | \mathbf{o}_m\right)$ and $P\left(i | \mathbf{o}_m\right)$ in the following discussion. If the cross-entropy (CE) criterion

$$J_{\text{CE}}\left(\mathbf{W}, \mathbf{b}; \mathbb{S}\right) = \frac{1}{M} \sum_{m=1}^{M} J_{\text{CE}}\left(\mathbf{W}, \mathbf{b}; \mathbf{o}^m, \mathbf{y}^m\right) \quad (11.9)$$

is used, where

$$J_{\text{CE}}\left(\mathbf{W}, \mathbf{b}; \mathbf{o}, \mathbf{y}\right) = -\sum_{i=1}^{C} P_{\text{emp}}\left(i | \mathbf{o}_m\right) \log P\left(i | \mathbf{o}_m\right), \quad (11.10)$$

and $P_{\text{emp}}\left(i | \mathbf{o}\right)$ is the empirical (observed in the adaptation set) probability that the observation $\mathbf{o}$ belongs to class i, the regularized adaptation criterion can be converted to

$$J_{\text{KLD-CE}}\left(\mathbf{W}, \mathbf{b}; \mathbb{S}\right) = (1 - \lambda)\, J_{\text{CE}}\left(\mathbf{W}, \mathbf{b}; \mathbb{S}\right) + \lambda R_{\text{KLD}}\left(\mathbf{W}_{\text{SI}}, \mathbf{b}_{\text{SI}}; \mathbf{W}, \mathbf{b}; \mathbb{S}\right)$$

$$= -\frac{1}{M} \sum_{m=1}^{M} \sum_{i=1}^{C} \left((1 - \lambda)\, P_{\text{emp}}\left(i|\mathbf{o}_m\right) + \lambda P_{\text{SI}}\left(i|\mathbf{o}_m\right)\right) \log P\left(i|\mathbf{o}_m\right)$$

$$= -\frac{1}{M} \sum_{m=1}^{M} \sum_{i=1}^{C} \ddot{P}\left(i|\mathbf{o}_m\right) \log P\left(i|\mathbf{o}_m\right), \tag{11.11}$$

where we defined

$$\ddot{P}\left(i|\mathbf{o}_m\right) = (1 - \lambda)\, P_{\text{emp}}\left(i|\mathbf{o}_m\right) + \lambda P_{\text{SI}}\left(i|\mathbf{o}_m\right). \tag{11.12}$$

Note that Eq. (11.11) has the same form as the CE criterion except that the target distribution is now an interpolated value between the empirical probability $P_{\text{emp}}\left(i|\mathbf{o}_m\right)$ and the probability $P_{\text{SI}}\left(i|\mathbf{o}_m\right)$ estimated from the SI model. This interpolation prevents overtraining by keeping the adapted model from straying too far away from the SI model. It also indicates that the normal backpropagation (BP) algorithm can be directly used to adapt the DNN. The only thing needs to be changed is the error signal at the output layer, which is now defined based on $\ddot{P}\left(i|\mathbf{o}_m\right)$ instead of $P_{\text{emp}}\left(i|\mathbf{o}_m\right)$.

Note that the KLD regularization differs from the L_2 regularization, which constrains the model parameters themselves rather than the output probabilities. Since what we care about is the output probability instead of the model parameters themselves, KLD regularization is more attractive and often performs better than the L_2 regularization.

The interpolation weight, which is directly derived from the regularization weight λ, can be adjusted, typically using a development set, based on the size of the adaptation set, the learning rate used, and whether the adaptation is supervised or unsupervised. When $\lambda = 1$, we trust completely the SI model and ignore all new information from the adaptation data. When $\lambda = 0$, we adapt the model solely on the adaptation set, ignoring information from the SI model except using it as the starting point. Intuitively, we should use a large λ for a small adaptation set and a small λ for a large adaptation set.

The KLD regularized adaptation technique can be easily extended to the sequence-discriminative training. As discussed in Sect. 8.2.3, to prevent overfitting the frame smoothing is often used in the sequence-discriminative training which leads to the interpolated training criterion

$$J_{\text{FS-SEQ}}\left(\mathbf{W}, \mathbf{b}; \mathbb{S}\right) = (1 - H)\, J_{\text{CE}}\left(\mathbf{W}, \mathbf{b}; \mathbb{S}\right) + H J_{\text{SEQ}}\left(\mathbf{W}, \mathbf{b}; \mathbb{S}\right), \tag{11.13}$$

where H is the frame smoothing factor often set empirically. By adding the KLD regularization term, we get the adaptation criterion

$$J_{\text{KLD-FS-SEQ}}\left(\mathbf{W}, \mathbf{b}; \mathbb{S}\right) = J_{\text{FS-SEQ}}\left(\mathbf{W}, \mathbf{b}; \mathbb{S}\right) + \lambda_s R_{\text{KLD}}\left(\mathbf{W}_{\text{SI}}, \mathbf{b}_{\text{SI}}; \mathbf{W}, \mathbf{b}; \mathbb{S}\right), \tag{11.14}$$

where λ_s is the regularization weight for the sequence-discriminative training. $J_{\text{KLD-FS-SEQ}}$ can be converted to

$$J_{\text{KLD-FS-SEQ}}\left(\mathbf{W},\mathbf{b};\mathbb{S}\right) = H\,J_{\text{SEQ}}\left(\mathbf{W},\mathbf{b};\mathbb{S}\right) + \left(1 - H + \lambda_s\right)J_{\text{KLD-CE}}\left(\mathbf{W},\mathbf{b};\mathbb{S}\right)$$

$$\tag{11.15}$$

following the similar derivation by defining $\lambda = \frac{\lambda_s}{1-H+\lambda_s}$.

11.4.3 Reducing Per-Speaker Footprint

Conservative training can alleviate the overfitting problem in the adaptation process. However, it cannot solve the problem that a huge adapted model needs to be saved for each speaker. Since the DNN model typically has huge number of parameters the adapted model may be too large to be stored at either the client side (e.g., on an intelligent watch) or the server side (especially when the number of users is large).

The simplest approach to reduce the footprint is to adapt only part of the model. For example, we can adapt only the input layer, the output layer, or a specific hidden layer. Experiments conducted, however, indicate that adapting all layers in the DNN often achieves better performance than adapting only a subset of the layers.

Fortunately, techniques have been developed to reduce the per-speaker footprint while keeping the benefit achievable from adapting all the layers. Here, we describe two of such approaches proposed in [36].

The first approach is to store the compressed parameter difference between the speaker-independent model and the speaker-adapted model. Since the adapted model is very close to the SI model it is reasonable to believe that the delta matrices can be approximated with low-rank matrices. In other words, we can apply the singular value decomposition (SVD) technique to the delta matrices $\triangle\mathbf{W}_{m\times n} = \mathbf{W}_{m\times n}^{ADP} - \mathbf{W}_{m\times n}^{SI}$ as

$$\begin{aligned}
\triangle\mathbf{W}_{m\times n} &= \mathbf{U}_{m\times n}\Sigma_{n\times n}\mathbf{V}_{n\times n}^{T} \\
&\approx \widetilde{\mathbf{U}}_{m\times k}\widetilde{\Sigma}_{k\times k}\widetilde{\mathbf{V}}_{k\times n}^{T} \\
&= \widetilde{\mathbf{U}}_{m\times k}\widetilde{\mathbf{W}}_{k\times n}^{T},
\end{aligned}\tag{11.16}$$

where $\triangle\mathbf{W}_{m\times n} \in \mathbb{R}^{m\times n}$, $\Sigma_{n\times n}$ is a diagonal matrix that contains all the singular values, $k < n$ is the number of singular values kept, $\mathbf{U}$ and $\mathbf{V}^{T}$ are unitary, the columns of which form a set of orthonormal vectors, which can be regarded as basis vectors, and $\widetilde{\mathbf{W}}_{k\times n}^{T} = \widetilde{\Sigma}_{k\times k}\widetilde{\mathbf{V}}_{k\times n}^{T}$. Here, we only need to store $\widetilde{\mathbf{U}}_{m\times k}$ and $\widetilde{\mathbf{W}}_{k\times n}^{T}$ which contain $(m+n)\,k$ parameters instead of $m\times n$ parameters. Experiments in [36] have shown that no or little accuracy loss is observed even if only less than 10 % of delta parameters are stored.

The second approach is applied on top of the low-rank model approximation technique we have discussed in Chap. 7 as shown in Fig. 11.4. In Chap. 7, we have shown that a fully connected weight matrix $\mathbf{W}_{m\times n} \in \mathbb{R}^{m\times n}$ in the original DNN

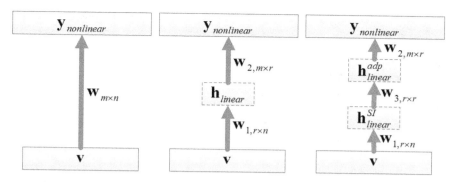

Fig. 11.4 Illustration of the SVD bottleneck adaptation technique

(Fig. 11.4a) can be approximated with two smaller matrices $\mathbf{W}_{1,r\times n} \in \mathbb{R}^{r\times n}$ and $\mathbf{W}_{2,m\times r} \in \mathbb{R}^{m\times r}$ as shown in Fig. 11.4b. To adapt this model, we insert another layer specified by the matrix $\mathbf{W}_{3,r\times r} \in \mathbb{R}^{r\times r}$ as shown in Fig. 11.4c. For the SI model, this matrix is set to the identity matrix. For the adapted model, this matrix is adapted to a specific speaker while keeping matrices $\mathbf{W}_{1,r\times n}$ and $\mathbf{W}_{2,m\times r}$ at all layers fixed. In this approach, the speaker-specific information is stored inside $\mathbf{W}_{3,r\times r}$ which contains only $r \times r$ instead of $m \times n$ parameters. Since r is significantly smaller than both m and n, this approach can reduce the per-speaker footprint significantly. For this same reason, $h_{\text{linear}}^{\text{SI}}$ and $h_{\text{linear}}^{\text{ADP}}$ are bottleneck layers in the model and thus this approach is called SVD bottleneck adaptation technique in [36]. This technique allows us to adapt all layers in the DNN while keeping the adapted matrix small for each speaker. This dramatically reduces the deployment cost for speaker personalization, and, at the same time can potentially reduce the amount of adaptation data needed for each new speaker. Experiments conducted by Xue et al. [36] indicate that this approach can reduce the per-speaker cost to 1 % of the original DNN without affecting the adaptation quality. Since in real-world deployment the low-rank approximation based model compression technique is often used and the SVD bottleneck adaptation technique has extremely low footprint for each speaker, it is one of the desired techniques for DNN adaptation.

Note that the SVD factorization of the model parameters also suggests another model adaptation technique [38]. Recall that at each layer the model weights $\mathbf{W}$ can be factorized into three components using SVD

$$\mathbf{W}_{m\times n} = \mathbf{U}_{m\times n}\Sigma_{n\times n}\mathbf{V}_{n\times n}^{T}, \tag{11.17}$$

where Σ is a diagonal matrix consisted of nonnegative singular values in the decreasing order, $\mathbf{U}$ and $\mathbf{V}^{T}$ are unitary, the columns of which form a set of orthonormal vectors. Here, Σ plays an important role and since it's a diagonal matrix the number of parameters in Σ is very small and equals to n. We can, instead of adapting the whole weight matrix $\mathbf{W}$, only adapt this diagonal matrix. If the adaptation set is small, we may even only adapt the top $k\%$ of the singular values.

11.5 Subspace Methods

Subspace methods aim to find a speaker subspace and then construct adapted ANN weights or transformations as a point in the subspace. Promising techniques in this category include principal component analysis (PCA) based approach [9], noise-aware [31] (which we have discussed in Chap. 9) and speaker-aware training [29], and tensor-based adaptation [41, 42]. Techniques in this category can adapt the SI model to a specific speaker very quickly.

11.5.1 Subspace Construction Through Principal Component Analysis

In [9], a fast adaptation technique was proposed. In this technique, subspace is used to estimate an affine transformation matrix by considering it as a random variable. Principal component analysis (PCA) was performed on a set of adaptation matrices to obtain the principal directions (i.e., eigenvectors) in the speaker space. Each new speaker adaptation model is then approximated by a linear combination of the retained eigenvectors.

The above mentioned technique can be extended to a more general condition. Given a set of S speakers, we can estimate a speaker-specific matrix $\mathbf{W}^{\mathrm{ADP}} \in \mathbb{R}^{m \times n}$ for each speaker. Here, the speaker-specific matrix can be the linear transformation in LIN, LHN, or LON, or the adapted weights or delta weights in the conservative training. Denote $\mathbf{a} = vec\left(\mathbf{W}^{\mathrm{ADP}}\right)$, the vectorization of the matrix, each matrix can be considered as an observation of a random variable in a speaker space of dimension $m \times n$. PCA can then be performed on the set of S vectors in the speaker space. The eigenvectors obtained from the PCA define principal adaptation matrices.

This approach assumes that new speakers can be represented as a point in the space spanned by the S speakers. In other words, S is big enough to cover the speaker space. Since each new speaker is represented by the linear combination of the eigenvectors, when S is large, the total number of linear interpolation weights can also be large. Fortunately, the dimensionality of the speaker space can be reduced by discarding the eigenvectors that correspond to small variances. In this way, each speaker-specific matrix can be efficiently represented by a reduced number of parameters.

For each new speaker,

$$\mathbf{a} = \overline{\mathbf{a}} + \mathbf{U}\mathbf{g_a} \approx \overline{\mathbf{a}} + \widetilde{\mathbf{U}}\widetilde{\mathbf{g}}_{\mathbf{a}} \tag{11.18}$$

where $\mathbf{U} = (\mathbf{u}_1, \ldots, \mathbf{u}_S)$ is the eigenvectors matrix, $\widetilde{\mathbf{U}} = (\mathbf{u}_1, \ldots, \mathbf{u}_k)$ is the reduced eigenvectors matrix, k is the number of retained eigenvectors, $\mathbf{g_a}$ and $\widetilde{\mathbf{g}}_{\mathbf{a}}$ are the full and reduced projection of the adaptation parameters vector onto the principal directions, and $\overline{\mathbf{a}}$ is the mean (across speakers) of the adaptation parameters. $\overline{\mathbf{a}}$ and $\widetilde{\mathbf{U}}$ are estimated from the training set consisted of S speakers. $\widetilde{\mathbf{g}}_{\mathbf{a}}$ is estimated for each new speaker from the adaptation set.

Fig. 11.5 Illustration of speaker-aware training. There are two sets of time-synchronous inputs: one set of acoustic features for phonetic discrimination and another set of features that characterizes the speaker

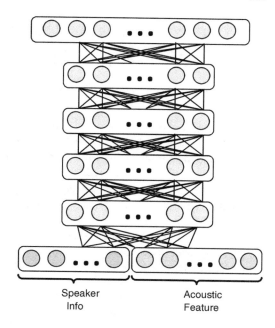

Speaker
Info

Acoustic
Feature

11.5.2 Noise-Aware, Speaker-Aware, and Device-Aware Training

Another set of subspace approaches explicitly estimate the noise or speaker information from the utterance and provide this information to the network in hope that the DNN training algorithm can automatically figure out how to adjust the model parameters to exploit the noise, speaker, or device information. We call such approaches noise-aware training (NaT) when the noise information is used, speaker-aware training (SaT) when the speaker information is exploited, and device-aware training (DaT) when the device information is used. Since NaT, SaT, and DaT are very similar and we have already discussed NaT in Chap. 9, we focus on SaT in this subsection by noting that NaT and DaT can be carried out similarly. Figure 11.5 illustrates the architecture of SaT, in which the input to the DNN has two parts: the acoustic feature and the speaker information (or noise information if NaT is used).

The reason SaT helps to boost the performance of DNNs can be easily understood with the following analysis. Without the speaker information, the activation of the first hidden layer is

$$\mathbf{v}^1 = f\left(\mathbf{z}^1\right) = f\left(\mathbf{W}^1 v^0 + \mathbf{b}^1\right), \tag{11.19}$$

where $\mathbf{v}^0$ is the acoustic feature vector, and $\mathbf{W}^1$ and $\mathbf{b}^1$ are the corresponding weight matrix and bias vector, respectively. When speaker information is used, it becomes

$$
\begin{aligned}
\mathbf{v}_{\text{SaT}}^1 = f\left(\mathbf{z}_{\text{SaT}}^1\right) &= f\left(\left[\,\mathbf{W}_v^1\ \mathbf{W}_s^1\,\right]\begin{bmatrix} v^0 \\ s \end{bmatrix} + \mathbf{b}_{\text{SaT}}^1\right) \\
&= f\left(\mathbf{W}_v^1 \mathbf{v}^0 + \mathbf{W}_s^1 \mathbf{s} + \mathbf{b}_{\text{SaT}}^1\right) \\
&= f\left(\mathbf{W}_v^1 \mathbf{v}^0 + \left(\mathbf{W}_s^1 \mathbf{s} + \mathbf{b}_{\text{SaT}}^1\right)\right),
\end{aligned}
\tag{11.20}
$$

where $\mathbf{s}$ is the vector that characterize the speaker, and $\mathbf{W}_v^1$ and $\mathbf{W}_s^1$ are the weight matrices associated with the acoustic feature and speaker information, respectively. Compared to the normal DNN, which uses a fixed bias vector $\mathbf{b}^1$, SaT uses a speaker-dependent bias $\mathbf{b}_s^1 = \mathbf{W}_s^1 \mathbf{s} + \mathbf{b}_{\text{SaT}}^1$. One benefit of SaT is that the adaptation is implicit and efficient and does not require a separate adaptation step. If the speaker information can be reliably estimated, SaT is a great candidate for speaker adaptation in the DNN framework.

The speaker information can be derived in many different ways. For example, in [1, 37], a speaker code is used as the speaker information. During the training, this code is jointly learned with the rest of the model parameters for each speaker. During the decoding process, we first learn a single speaker code for the new speaker using all the utterances spoken by the speaker. This can be done by treating the speaker code as part of the model parameters and estimating it using the backpropagation algorithm keeping rest of the parameters fixed. The speaker code is then used as part of the input to the DNN to compute the state likelihood.

The speaker information can also be estimated completely independent of the DNN training. For example, it may be learned from a separate DNN from which either the output node or the last hidden layer can be used to represent the speaker. In [29] i-vectors [8, 13] are used instead. I-vector is a popular technique for speaker verification and recognition. It encapsulates the most important information about a speaker's identity in a low-dimensional fixed-length representation and thus is an attractive tool for speaker adaptation techniques for ASR. I-vector has been used not only in DNN adaptation, but also in GMM adaptation. For example, it has been used in [16] for discriminative speaker adaptation with region dependent linear transforms and in [5, 39] for clustering speakers or utterances for more efficient adaptation. Since a single low-dimensional i-vector is estimated from all the utterances spoken by the same speaker, i-vector can be reliably estimated from less data than other approaches. Since i-vector is an important technique, we summarize its computation steps here.

11.5.2.1 Computation of I-Vector

We denote $\mathbf{x}_t \in \mathbb{R}^{D \times 1}$ the acoustic feature vectors generated from a universal background model (UBM) represented as a GMM with K diagonal covariance Gaussians

$$
\mathbf{x}_t \sim \sum_{k=1}^{K} c_k \mathcal{N}\left(\mathbf{x}; \mu_k, \Sigma_k\right),
\tag{11.21}
$$

where c_k, μ_k, and Σ_k are the mixture weight, Gaussian mean, and diagonal covariance of the kth Gaussian mixture. We assume the acoustic feature $\mathbf{x}_t(s)$ specific to speaker s are drawn from the distribution

$$\mathbf{x}_t(s) \sim \sum_{k=1}^{K} c_k \mathcal{N}(\mathbf{x}; \mu_k(s), \Sigma_k), \tag{11.22}$$

where $\mu_k(s)$ are the speaker-specific means of the GMM adapted from the UBM. We further assume that there is a linear dependence

$$\mu_k(s) = \mu_k + \mathbf{T}_k \mathbf{w}(s), \quad 1 \leq k \leq K, \tag{11.23}$$

between the speaker-adapted means $\mu_k(s)$ and the speaker-independent means μ_k, where $\mathbf{T}_k \in \mathbb{R}^{D \times M}$ is the factor loading submatrix, which contains M bases which span the subspace with important variability in the component mean vector space, corresponding to component k, and $\mathbf{w}(s)$ is the speaker identity vector (i-vector) corresponding to speaker s.

Note that the i-vector $\mathbf{w}$ is a latent variable. If we assume its prior follows a Gaussian distribution with a 0-mean and identity covariance, each frame is aligned to a fixed mixture component, and the factor loading submatrix $\mathbf{T}_k$ is known, we can estimate the posterior distribution [17]

$$p(\mathbf{w} | \{\mathbf{x}_t(s)\}) = \mathcal{N}\left(\mathbf{w}; \mathbf{L}^{-1}(s) \sum_{k=1}^{K} \mathbf{T}_k \mathcal{T} \Sigma_k^{-1} \theta_k(s), \mathbf{L}^{-1}(s)\right), \tag{11.24}$$

where the precision matrix $\mathbf{L}(s) \in \mathbb{R}^{M \times M}$ is

$$\mathbf{L}(s) = \mathbf{I} + \sum_{k=1}^{K} \gamma_k(s) \mathbf{T}_k \mathcal{T} \Sigma_k^{-1} \mathbf{T}_k, \tag{11.25}$$

and the zeroth-order and first-order statistics are

$$\gamma_k(s) = \sum_{t=1}^{T} \gamma_{tk}(s), \tag{11.26}$$

$$\theta_k(s) = \sum_{t=1}^{T} \gamma_{tk}(s) (\mathbf{x}_t(s) - \mu_k(s)). \tag{11.27}$$

with $\gamma_{tk}(s)$ being the posterior probability of mixture component k given $\mathbf{x}_t(s)$. The i-vector is simply the MAP point estimate of the variable $\mathbf{w}$

$$\mathbf{w}(s) = \mathbf{L}^{-1}(s) \sum_{k=1}^{K} \mathbf{T}_k \mathcal{T} \Sigma_k^{-1} \theta_k(s), \tag{11.28}$$

which is just the mean of the posterior distribution 11.24.

Note that $\{\mathbf{T}_k | 1 \le k \le K\}$ are unknown and need to be estimated from the speaker-specific acoustic features $\{\mathbf{x}_t(s)\}$ using the expectation–maximization (EM) algorithm to maximize the maximum likelihood (ML) training criterion [6]

$$Q(\mathbf{T}_1, \ldots, \mathbf{T}_K) = -\frac{1}{2} \sum_{s,t,k} \gamma_{tk}(s) \left[\log |\mathbf{L}(s)| + (\mathbf{x}_t(s) - \mu_k(s))^T \Sigma_k^{-1} (\mathbf{x}_t(s) - \mu_k(s)) \right],$$
$$\tag{11.29}$$

or equivalently

$$Q(\mathbf{T}_1, \ldots, \mathbf{T}_K) = -\frac{1}{2} \sum_{s,k} \left[\gamma_k(s) \log |\mathbf{L}(s)| + \gamma_k(s) \operatorname{Tr} \left\{ \Sigma_k^{-1} \mathbf{T}_k \mathbf{w}(s) \mathbf{w}^T(s) \mathbf{T}_k^T \right\} \right.$$
$$\left. -2 \operatorname{Tr} \left\{ \Sigma_k^{-1} \mathbf{T}_k \mathbf{w}(s) \theta_k^T(s) \right\} \right] + C, \tag{11.30}$$

Taking the derivative of Eq. 11.30 with respect to $\mathbf{T}_k$ and setting it to 0 we get the M-step

$$\mathbf{T}_k = \mathbf{C}_k \mathbf{A}_k^{-1}, \quad 1 \le k \le K \tag{11.31}$$

where

$$\mathbf{C}_k = \sum_s \theta_k(s) \mathbf{w}^T(s), \tag{11.32}$$

$$\mathbf{A}_k = \sum_s \gamma_k(s) \left[\mathbf{L}^{-1}(s) + \mathbf{w}(s) \mathbf{w}^T(s) \right] \tag{11.33}$$

are computed in the E-step.

Although we discussed SaT, NaT, and DaT separately, these techniques can be combined into a single network, in which the input has four segments: one for the speech feature and the rest are for the speaker, noise, and device codes, respectively. The speaker, noise, and device codes can be trained jointly and the learned codes can be carried over to form different combination of conditions.

11.5.3 Tensor

The speaker and speech subspaces can also be estimated and combined using three-way connections (or tensors). In [41], several such architectures are proposed. Figure 11.6 shows one of such architectures called disjoint factorized DNN. In this architecture, the speaker posterior probability $p(\mathbf{s}|\mathbf{x}_t)$ is estimated from the acoustic feature $\mathbf{x}_t$ using a DNN. The class posterior probability $p(\mathbf{y}_t = i|\mathbf{x}_t)$ is estimated as

$$p(\mathbf{y}_t = i|\mathbf{x}_t) = \sum_s p(\mathbf{y}_t = i|\mathbf{s},\mathbf{x}_t)\, p(\mathbf{s}|\mathbf{x}_t)$$

$$= \sum_s \frac{\exp\left(\mathbf{s}^T \mathbf{W}_i \mathbf{v}_t^{L-1}\right)}{\sum_j \exp\left(\mathbf{s}^T \mathbf{W}_j \mathbf{v}_t^{L-1}\right)}\, p(\mathbf{s}|\mathbf{x}_t), \qquad (11.34)$$

where $\mathbf{W} \in \mathbb{R}^{N_L \times S \times N_{L-1}}$ is a tensor, S is the number of neurons in the speaker identification DNN, N_{L-1} and N_L are the number of neurons in the last hidden layer and the output layer, respectively, of the senone classification DNN, and $\mathbf{W}_i \in$

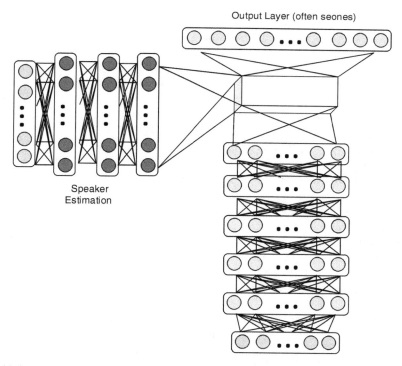

Fig. 11.6 Illustration of a typical architecture of the disjoint factorized DNN model for speaker adaptation

$\mathbb{R}^{S \times N_{L-1}}$ is a slice of the tensor. Unfortunately, the total number of parameters in the tensor networks is typically very large compared to other techniques we have discussed and thus is not practical for real-world applications.

11.6 Effectiveness of DNN Speaker Adaptation

As we have discussed in Chap. 9, DNN can extract features that are less sensitive to the perturbations in the acoustic features than GMM and other shallow models. In fact, it has been shown that much less improvement was observed with fDLR when the network goes from shallow to deep [30]. It thus raises the question on how much additional gain we can get from speaker adaptation on the CD-DNN-HMM systems. In this section, we show, using experimental results from [29, 36, 43] that speaker adaptation techniques are very important even to CD-DNN-HMM systems.

11.6.1 KL-Divergence Regularization Approach

The first set of experimental results, extracted from [43], was obtained on a short message dictation (SMD) task. The baseline SI models were trained using 300 h voice search and SMD data. The evaluation was conducted on data from 9 speakers, out of which 2 were used as the development set to determine the learning rate and 7 were used as the test set. The total number of test set words is 20,668.

The baseline SI CD-DNN-HMM system used 24 log-filter bank features with up to second derivatives and a context window size of 11, forming a vector of 792-dimension (72 x 11) input. On top of the input layer, there are 5 hidden layers with 2,048-neurons each. The output layer has a dimension of 5,976. The DNN system was trained using the senone alignments from the GMM–HMM system. The baseline SI CD-DNN-HMM system achieved 23.4 % WER on the 7-speaker test set. To evaluate how the size of the adaptation set affects the result, the number of adaptation utterances varies from 5 (32 s) to 200 (22 min).

Figures 11.7a,b summarize the WER on the SMD dataset using supervised and unsupervised KLD-Reg adaptation, respectively. From these figures, we can observe that with the optimal regularization weights determined by the development set, we get 30.3, 25.2, 18.6, 12.6, 8.8, and 5.6 % relative WER reduction using supervised adaptation, and 14.6, 11.7, 8.6, 5.8, 4.1, and 2.5 %, using unsupervised adaptation, respectively, with 200, 100, 50, 25, 10, and 5 utterances of adaptation data. These figures also show that larger regularization weights should be used for smaller adaptation set and smaller regularization weights should be used for larger adaptation set, although the error reductions are quite robust as long as the regularization weight is within [0.125 0.5] on this task. Compared to the supervised adaptation, the unsupervised setup can benefit from larger regularization weights. This is because the labels

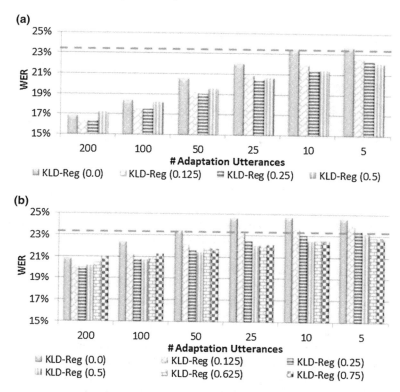

Fig. 11.7 WERs on the SMD dataset using KLD regularized adaptation for different regularization weights λ (*numbers in parentheses*). The *dashed line* is the SI DNN baseline. (Figure from Yu et al. [43]). **a** Supervised adaptation. **b** unsupervised adaptation

in the unsupervised adaptation setup are less reliable and thus we should trust the output from the SI model even more during the adaptation.

In [36], the SVD bottleneck adaptation technique described in Sect. 11.4.3 was used to reduce both the size of the SI DNN and the footprint of the adapted parameters. In the experiments conducted on the same SMD task, the full-rank DNN model, which has 30M parameters, was first converted to the low-rank model by keeping only 40 % of total singular values. This low-rank model is refined and the accuracy of the final model is the same as that achieved using the full-rank model. Table 11.1 summarizes the results obtained with SI models and with KLD regularized adaptation on the low-rank SI model. The regularization weight λ is set to 0.5 in this set of experiments. From the table, we can observe that even when the adaptation footprint is reduced from 7.4 M in the full low-rank DNN adaptation to 266 K in the SVD bottleneck adaptation we still can achieve over 18 % relative error reduction using the KLD regularized adaptation technique. Unpublished results on adapting the best SI DNN model we can built confirmed this result.

Table 11.1 Compare the full-rank adaptation and SVD bottleneck adaptation in WER on the SMD task, all supervised with a KLD regularization weight 0.5

	SI Low-Rank DNN	Adapt with 5 Utterances	Adapt with 100 Utterances
Full Model Adaptation (7.4 M)	25.12 % (baseline)	24.30 % (−3.2 %)	20.51 % (−18.4 %)
SVD Bottleneck Adaptation (266 K)	25.12 % (baseline)	24.23 % (−3.5 %)	19.95 % (−20.6 %)

Relative WER reduction in parenthesis. (Table based on results summarized from Xue et al. [36])

11.6.2 Speaker-Aware Training

In [29], the i-vector based adaptation approach was applied to the Switchboard (SWB) dataset [14, 15] described in Chap. 6 where 300 h of data was used to train the DNN model. In their experiments, each frame is represented by a feature vector of 13 perceptual linear prediction (PLP) cepstral coefficients which are mean- and variance-normalized per conversation side. Every 9 consecutive cepstral frames are spliced together and projected down to 40 dimensions using LDA, which is further diagonalized by means of a global semi-tied covariance transform.

They first trained a GMM UBM model with 2,048 40-dimensional Gaussian components with diagonal covariance using the maximum likelihood criteria. They then built a GMM of the same size, adapted from the UBM, for each speaker. The i-vector extraction matrices $T_1, \ldots, T_{2048}$ were initialized with values drawn randomly from the uniform distribution in $[-1, 1]$ and were estimated with 10 iterations of EM described in Eqs. (11.31)–(11.33). With these extraction matrices, they extracted M-dimensional i-vectors for all the training and test speakers.

The DNN input features cover a temporal context of 11 frames. In other words, the input layer has $40 \times 11 + M$ (for M ={40, 100, 200}) neurons. All DNNs have 6 hidden layers with sigmoid activation functions: the first 5 with 2,048 units and the last one with 256 units for parameter reduction and faster training. The output layer has 9,300 softmax units that correspond to the context-dependent HMM states. The decoding language model used is a 4 gram LM with 4M n-grams.

Table 11.2 compares the SI DNN and the DNN adapted with the i-vector based SaT in word error rate (WER) on the HUB5'00 and RT'03 test sets. From the table, we can observe that no matter whether the cross-entropy or the sequence-discriminative training criteria is used, the i-vector SaT adaptation can achieve on average over 10 % relative error reduction over the SI DNN.

In [37], Xue et al. reported results using the speaker-code-based approach on the Switchboard dataset. They showed that by using 10 adaptation utterances to learn the speaker code, a relative error rate reduction of 6.2 % (16.2 % → 15.2 %) and 4.3 % (14.0 % → 13.4 %) can be achieved on the frame-level and sequence-discriminative training baselines, respectively.

Table 11.2 Compare the SI DNN and the DNN adapted with the i-vector based speaker-aware training in WER on HUB5'00 and RT'03 test sets

Training	Model	Hub5'00	RT'03	
		SWB	FSH	SWB
Cross Entropy	SI	16.1%	18.9%	29.0%
	i-vector SaT	13.9% (−13.7%)	16.7% (−11.6%)	25.8% (−11.0%)
Sequence	SI	14.1%	16.9%	26.5%
	i-vector SaT	12.4% (−12.1%)	15.0% (−11.2%)	24.0% (−9.4%)

Relative WER reduction in parenthesis. (Summarized from Saon et al. [29])

References

1. Abdel-Hamid, O., Jiang, H.: Fast speaker adaptation of hybrid NN/HMM model for speech recognition based on discriminative learning of speaker code. In: Proceedings of the International Conference on Acoustics, Speech and Signal Processing (ICASSP), pp. 7942–7946 (2013)
2. Abrash, V., Franco, H., Sankar, A., Cohen, M.: Connectionist speaker normalization and adaptation. In: Proceedings of the European Conference on Speech Communication and Technology (EUROSPEECH) (1995)
3. Albesano, D., Gemello, R., Laface, P., Mana, F., Scanzio, S.: Adaptation of artificial neural networks avoiding catastrophic forgetting. In: Proceedings of the International Conference on Neural Networks (IJCNN), pp. 1554–1561 (2006)
4. Albesano, D., Gemello, R., Mana, F.: Hybrid HMM-NN modeling of stationary-transitional units for continuous speech recognition. Inf. Sci. **123**(1), 3–11 (2000)
5. Bacchiani, M.: Rapid adaptation for mobile speech applications. In: Proceedings of the International Conference on Acoustics, Speech and Signal Processing (ICASSP), pp. 7903–7907 (2013)
6. Brümmer, N.: The EM algorithm and minimum divergence. Agnitio Labs Technical Report. http://niko.brummer.googlepages (2009)
7. Chesta, C., Siohan, O., Lee, C.H.: Maximum a posteriori linear regression for hidden markov model adaptation. In: Eurospeech (1999)
8. Dehak, N., Kenny, P., Dehak, R., Dumouchel, P., Ouellet, P.: Front-end factor analysis for speaker verification. IEEE Trans. Audio, Speech Lang. Process. **19**(4), 788–798 (2011)
9. Dupont, S., Cheboub, L.: Fast speaker adaptation of artificial neural networks for automatic speech recognition. In: Proceedings of the International Conference on Acoustics, Speech and Signal Processing (ICASSP), vol. 3, pp. 1795–1798 (2000)
10. Gales, M.J.: Maximum likelihood linear transformations for HMM-based speech recognition. Comput. Speech Lang. **12**(2), 75–98 (1998)
11. Gales, M.J., Woodland, P.: Mean and variance adaptation within the mllr framework. Comput. Speech Lang. **10**(4), 249–264 (1996)
12. Gemello, R., Mana, F., Scanzio, S., Laface, P., De Mori, R.: Linear hidden transformations for adaptation of hybrid ANN/HMM models. Speech Commun. **49**(10), 827–835 (2007)
13. Glembek, O., Burget, L., Matejka, P., Karafiát, M., Kenny, P.: Simplification and optimization of i-vector extraction. In: Proceedings of the International Conference on Acoustics, Speech and Signal Processing (ICASSP), pp. 4516–4519 (2011)
14. Godfrey, J.J., Holliman, E.: Switchboard-1 release 2. Linguistic Data Consortium (1997)
15. Godfrey, J.J., Holliman, E.C., McDaniel, J.: Switchboard: telephone speech corpus for research and development. In: Proceeding of the International Conference on Acoustics, Speech and Signal Processing (ICASSP), vol. 1, pp. 517–520 (1992)

16. Karafiát, M., Burget, L., Matejka, P., Glembek, O., Cernocky, J.: iVector-based discriminative adaptation for automatic speech recognition. In: Proceedings of the IEEE Workshop on Automfatic Speech Recognition and Understanding (ASRU), pp. 152–157 (2011)
17. Kenny, P.: Joint factor analysis of speaker and session variability: theory and algorithms. CRIM, Montreal, (Report) CRIM-06/08-13 (2005)
18. Kim, D.Y., Kwan Un, C., Kim, N.S.: Speech recognition in noisy environments using first-order vector Taylor series. Speech Commun. 24(1), 39–49 (1998)
19. Lee, C.H., Huo, Q.: On adaptive decision rules and decision parameter adaptation for automatic speech recognition. Proc. IEEE 88(8), 1241–1269 (2000)
20. Leggetter, C.J., Woodland, P.: Maximum likelihood linear regression for speaker adaptation of continuous density hidden Markov models. Comput. Speech Lang. 9(2), 171–185 (1995)
21. Li, B., Sim, K.C.: Comparison of discriminative input and output transformations for speaker adaptation in the hybrid NN/HMM systems. In: Proceedings of the Annual Conference of International Speech Communication Association (INTERSPEECH), pp. 526–529 (2010)
22. Li, J., Deng, L., Yu, D., Gong, Y., Acero, A.: High-performance HMM adaptation with joint compensation of additive and convolutive distortions via vector Taylor series. In: Proceedings of the IEEE Workshop on Automatic Speech Recognition and Understanding (ASRU), pp. 65–70 (2007)
23. Li, J., Deng, L., Yu, D., Gong, Y., Acero, A.: HMM adaptation using a phase-sensitive acoustic distortion model for environment-robust speech recognition. In: Proceedings of the International Conference on Acoustics, Speech and Signal Processing (ICASSP), pp. 4069–4072 (2008)
24. Li, J., Deng, L., Yu, D., Gong, Y., Acero, A.: A unified framework of HMM adaptation with joint compensation of additive and convolutive distortions. Comput. Speech Lang. 23(3), 389–405 (2009)
25. Li, X., Bilmes, J.: Regularized adaptation of discriminative classifiers. In: Proceedings of the 2006 IEEE International Conference on Acoustics, Speech and Signal Processing (ICASSP), vol. 1. IEEE, pp. I–I (2006)
26. Moreno, P.J., Raj, B., Stern, R.M.: A vector Taylor series approach for environment-independent speech recognition. In: Proceedings of the International Conference on Acoustics, Speech and Signal Processing (ICASSP), vol. 2, pp. 733–736 (1996)
27. Morgan, N., Bourlard, H.: Continuous speech recognition using multilayer perceptrons with hidden Markov models. In: Proceedings of the International Conference on Acoustics, Speech and Signal Processing (ICASSP), pp. 413–416 (1990)
28. Neto, J., Almeida, L., Hochberg, M., Martins, C., Nunes, L., Renals, S., Robinson, T.: Speaker-adaptation for hybrid HMM-ANN continuous speech recognition system, pp. 2171–2174 (1995)
29. Saon, G., Soltau, H., Nahamoo, D., Picheny, M.: Speaker adaptation of neural network acoustic models using i-vectors. In: Proceedings of the IEEE Workshop on Automfatic Speech Recognition and Understanding (ASRU), pp. 55–59 (2013)
30. Seide, F., Li, G., Chen, X., Yu, D.: Feature engineering in context-dependent deep neural networks for conversational speech transcription. In: Proceedings of the IEEE Workshop on Automfatic Speech Recognition and Understanding (ASRU), pp. 24–29 (2011)
31. Seltzer, M., Yu, D., Wang, Y.: An investigation of deep neural networks for noise robust speech recognition. In: Proceedings of the International Conference on Acoustics, Speech and Signal Processing (ICASSP) (2013)
32. Stadermann, J., Rigoll, G.: Two-stage speaker adaptation of hybrid tied-posterior acoustic models. In: Proceedings of the International Conference on Acoustics, Speech and Signal Processing (ICASSP) (2005)
33. Trmal, J., Zelinka, J., Müller, L.: Adaptation of a feedforward artificial neural network using a linear transform. In: Text, Speech and Dialogue, pp. 423–430. Springer (2010)
34. Wang, Y., Gales, M.J.: Speaker and noise factorization for robust speech recognition. IEEE Trans. Audio, Speech Lang. Process. 20(7), 2149–2158 (2012)

35. Xiao, Y., Zhang, Z., Cai, S., Pan, J., Yan, Y.: A initial attempt on task-specific adaptation for deep neural network-based large vocabulary continuous speech recognition. In: Proceedings of the Annual Conference of International Speech Communication Association (INTERSPEECH) (2012)

36. Xue, J., Li, J., Yu, D., Seltzer, M., Gong, Y.: Singular value decomposition based low-footprint speaker adaptation and personalization for deep neural network. In: Proceedings of the International Conference on Acoustics, Speech and Signal Processing (ICASSP) (2014)

37. Xue, S., Abdel-Hamid, O., Jiang, H., Dai, L.: Direct adaptation of hybrid DNN/HMM model for fast speaker adaptation in LVCSR based on speaker code. In: Proceedings of the International Conference on Acoustics, Speech and Signal Processing (ICASSP), pp. 6389–6393 (2014)

38. Xue, S., Jiang, H., Dai, L.: Speaker Adaptation of Hybrid NN/HMM Model for Speech Recognition Based on Singular Value Decomposition. In: International Symposium on Chinese Spoken Language Processing (ISCSLP) (2014)

39. Yao, K., Gong, Y., Liu, C.: A feature space transformation method for personalization using generalized i-vector clustering. In: Proceedings of the Annual Conference of International Speech Communication Association (INTERSPEECH) (2012)

40. Yao, K., Yu, D., Seide, F., Su, H., Deng, L., Gong, Y.: Adaptation of context-dependent deep neural networks for automatic speech recognition. In: Proceedings of the IEEE Spoken Language Technology Workshop (SLT), pp. 366–369 (2012)

41. Yu, D., Chen, X., Deng, L.: Factorized deep neural networks for adaptive speech recognition. In: Proceedings of the International Workshop on Statistical Machine Learning for Speech Processing (2012)

42. Yu, D., Deng, L., Seide, F.: The deep tensor neural network with applications to large vocabulary speech recognition. IEEE Trans. Audio, Speech Lang. Process. **21**(3), 388–396 (2013)

43. Yu, D., Yao, K., Su, H., Li, G., Seide, F.: Kl-divergence regularized deep neural network adaptation for improved large vocabulary speech recognition. In: Proceedings of the International Conference on Acoustics, Speech and Signal Processing (ICASSP), pp. 7893–7897 (2013)

Part V
Advanced Deep Models

Chapter 12
Representation Sharing and Transfer in Deep Neural Networks

Abstract We have emphasized in the previous chapters that in deep neural networks (DNNs) each hidden layer is a new representation of the raw input to the DNN. The representation at higher layers is more abstract than that at lower layers. In this chapter, we show that these feature representations can be shared and transferred across related tasks through techniques such as multitask and transfer learning. We will use multilingual and crosslingual speech recognition as the main example, which uses a shared-hidden-layer DNN architecture, to demonstrate these techniques.

12.1 Multitask and Transfer Learning

12.1.1 Multitask Learning

Multitask learning (MTL) [3] is a machine learning technique that aims at improving the generalization performance of a learning task by jointly learning multiple-related tasks. The key to the successful application of MTL is that the tasks need to be related. Here *related* does not mean the tasks are similar. Instead, it means at some level of abstraction these tasks share part of the representation. If the tasks are indeed similar learning them together can help transfer knowledge among tasks since it effectively increases the amount of training data for each task. If the tasks are related yet not similar, learning them together can help constrain the possible functional space of each task and thus improve the generalization ability of each task. Multitask learning is mostly useful when the training set size is small compared to the model size.

Since each hidden layer in a deep neural network (DNN) is a new representation of the raw input to the DNN and the representation at higher layers is more abstract than that at lower layers, DNN is well suited to support MTL. Figure 12.1 illustrates a generic architecture of MTL in DNN. In this figure, three related tasks are shown. Note that these tasks process the raw input features independently at earlier processing stages (or layers). Part of the features, represented by the green layers inside the dotted rectangular box, however, are merged from and shared across all three tasks. The shared features are forked again for further task-dependent processing at the top of the network. Since each task has its own training criterion, a separate output layer is devoted to each task.

© Springer-Verlag London 2015
D. Yu and L. Deng, *Automatic Speech Recognition*,
Signals and Communication Technology, DOI 10.1007/978-1-4471-5779-3_12

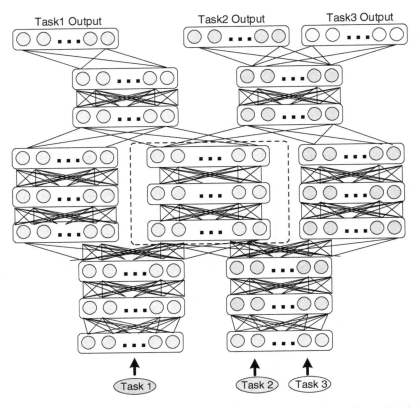

Fig. 12.1 A generic architecture of multitask learning in deep neural networks. Three related tasks are shown in the figure. The *green* layers inside the *dotted rectangular box* are shared across all tasks

This is a rather generic architecture. For a specific application, the decisions need to make include whether the raw input features are the same for tasks involved (i.e., similar to the task1 and task2 pair or the task2 and task3 pair in the figure), how many task-dependent layers are needed for each raw input feature before features are merged and shared, whether all or part of the features are shared (e.g., in Fig. 12.1 only part of the features are shared), and whether additional task-dependent hidden layers are needed for each task. These decisions are application specific and can significantly affect the effectiveness of MTL.

12.1.2 Transfer Learning

Transfer learning [24] aims at developing a reasonably performed system for a new task, domain, or distribution efficiently and effectively by retaining and leveraging the knowledge learned from one or more similar tasks, domains, or distributions. Different from multitask learning which focuses on improving performance on all or a primary task involved, transfer learning emphasizes on the performance of the target task by transferring knowledge across tasks, domains, and distributions that are similar but not the same.

Due to the more abstract and invariant features represented by the hidden layers, DNNs are well suited for transfer learning. In transfer learning, the main design questions are what and how to transfer. In the DNN framework, these questions are translated to how to represent knowledge to be transferred in DNNs (e.g., at which abstraction level) and how to leverage the knowledge from the other domains.

Transfer learning has great practical implications. In many real-world applications, due to the high cost of human manual labeling and/or environmental/societal restrictions, it is not uncommon that sufficient matching training data are not available. In such cases, transfer learning between task domains would be very desirable. Transfer learning has been successfully applied to many such machine learning tasks (refer to [24] for a recent survey). In these applications, feature transfer, which is very well suited in DNNs, is the main approach to transfer knowledge between tasks.

In the following sections, we will discuss how multitask and transfer learning can be applied to attack speech recognition problems. We focus on three key applications: multilingual and crosslingual speech recognition [11, 12, 16, 28, 33], multiobjective training of DNNs for speech recognition [4, 21, 30], and robust speech recognition exploiting audio-visual information [15].

12.2 Multilingual and Crosslingual Speech Recognition

In most traditional automatic speech recognition (ASR) systems, languages (and dialects) are considered independently, and a separate acoustic model (AM) is trained for each language from scratch. This introduces several problems. First, training an AM for a language from scratch requires a lot of manually labeled training data, which not only costs a lot but also takes time. This also results in considerable differences in quality between resource-rich and resource-scarce languages. This is because only small models with low complexity can be estimated for the resource-scarce languages. The requirement of having a large collection of labeled training data is also an unavoidable bottleneck for languages with low traffic and new launches where it is difficult to find large amounts of representative data. Second, training an AM for each language independently increases the accumulated training time. This is true especially for deep neural network (DNN)-based ASR systems since, as we mentioned in Chap. 7, training DNNs is substantially slower than training Gaussian mixture models (GMMs) due to the number of parameters in the DNNs and the backpropagation (BP) algorithm used. Third, building a separate AM for each language prevents smooth recognition and increases the cost of recognizing the mixed-lingual speech. To train accurate acoustic models efficiently and effectively for a large number of languages, to reduce cost incurred in training the AMs, and to support new usage scenarios where mixed-lingual ASR is crucial (e.g., English words are frequently inserted inside Chinese phrases in Hong Kong), there is an increasing interest in building multilingual ASR systems and reusing multilingual resources.

Although resource constraint (in both labeled data and computational power) is the very practical reason to look into the multilingual ASR problem, this is not the only reason. By investigating and engineering such techniques, we may also improve our understanding of the algorithms used and the relationship between different languages. There have been much research work on the multilingual and crosslingual ASR in the literature (e.g., [20, 34]). In this section, we focus only on those that use neural networks.

There can be many different architectures of multilingual ASR systems based on the DNNs as we will discuss in the next several subsections. However, these architectures share the same core idea: the hidden layers of a DNN can be considered as a cascade of feature extractors and only the output layer provides the direct correspondence to the classes of interest as we have emphasized in Chap. 9. These feature extractors can be shared across many different languages, jointly trained with data from multiple languages, and transferred to new (and typically resource-scarce) languages. By transferring the shared hidden layers to a new language, we can mitigate the requirement of having a significant amount of data to train large DNNs from scratch since only the weights of the language-specific output layer needs to be trained.

12.2.1 Tandem/Bottleneck-Based Crosslingual Speech Recognition

Most early studies on using neural networks in multilingual and crosslingual acoustic modeling focused on tandem and bottleneck approaches [25, 27, 28, 32, 33] because it is only after the publication of [7, 29] that the DNN-HMM hybrid system has become an important alternative acoustic model for large vocabulary continuous speech recognition (LVCSR). As we have described in Chap. 10, in the tandem/bottleneck approaches, a neural network is trained to classify either monophone states or tied triphone states. The neural network outputs or hidden layer excitations are then used as discriminative features for a GMM-HMM acoustic model.

Since both the hidden and output layers of the neural network contains information to classify phoneme states in a language and different languages share similar phonetic phenomena, it is possible to use the tandem/bottleneck features extracted from neural networks trained for one language (referred as source language) to be used as features of a GMM-HMM system for recognizing another language (referred as target language). Experiments show that these transferred features can improve on a competitive target language baseline when the amount of transcribed data in the target language is small.

The neural network used to extract tandem/bottleneck features can be multilingually trained [33] by using a different output layer (corresponding to context-independent phones) for each language similar to that depicted in Fig. 12.2. In addition, multiple neural networks can be trained each using different features, e.g., one using spectral (perceptual linear predictive or PLP [13]) features and the other using modulation (frequency domain linear prediction or FDLP [2]) features. Features extracted from these neural networks can be combined to further improve the recognition accuracy.

The tandem/bottleneck-based approaches are mainly used in the crosslingual ASR to improve the resource-scarce languages. They are seldom used in the multilingual ASR. This is because even if the same neural network is used to extract the tandem/bottleneck features, a completely different GMM-HMM system is often still needed for each language. This restriction, however, may be removed if multiple languages share the same set of phonemes (or context-dependent phone states) and decision trees as is done in [20]. The shared phone set may be determined with domain knowledge, e.g., using the international phonetic alphabet (IPA) [1], or through data driven approaches, e.g., by measuring the distance between phonemes and triphone states of different languages [34].

12.2.2 Shared-Hidden-Layer Multilingual DNN

The multilingual and crosslingual ASR can be more easily implemented in the CD-DNN-HMM framework. Figure 12.2 depicts the proposed architecture used for

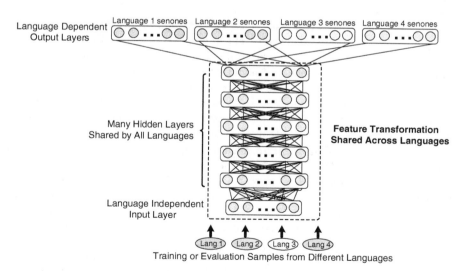

Fig. 12.2 Architecture of the shared-hidden-layer multilingual DNN (A similar figure has been shown in Huang et al. [16])

multilingual ASR. This architecture is referred as shared-hidden-layer multilingual DNN (SHL-MDNN) in [16] because the input and hidden layers are shared across all the languages the SHL-MDNN can recognize in this architecture. The softmax layers, however, are not shared. Instead, each language has its own softmax layer to estimate the posterior probabilities of the senones (tied triphone states) specific to that language. This same architecture was also independently proposed in [11, 12].

Note that the shared hidden layers in this architecture can be considered as a universal feature transformation or a special universal front end. The input to the SHL-MDNN covers a long contextual window of the acoustic feature (e.g., MFCC or Mel-scale log filter-bank) frames as in the single-language CD-DNN-HMM systems. However, since the shared hidden layers are to be used by many languages, language-specific transformations such as HLDA cannot be applied. Fortunately, this restriction does not limit the performance of the SHL-MDNN because any linear transformation can be subsumed by the DNN as described in Chap. 9.

The SHL-MDNN depicted in Fig. 12.2 is a special case of multitask learning [3] in which multiple tasks are learned in parallel with a shared representation. Multitask learning may be beneficial for learning DNNs for several reasons. First, it provides a representation bias in that local optima supported by all tasks are preferred. Second, it can alleviate the overfitting problem since the shared hidden layers (feature transformations) are now more reliably estimated with data from multiple languages. This is in particular helpful for resource-scarce tasks. Third, it helps to learn features that are easier to learn in a parallel task. Fourth, it helps improve model generalization since the model is now trained with noises from different datasets.

Although SHL-MDNNs can provide these benefits, we cannot achieve them if the SHL-MDNN is not correctly trained. The key to the successful training of the SHL-MDNN is to train the model for all the languages simultaneously. When batch training algorithms, such as L-BFGS or the Hessian-free [22] algorithm, are used, this is trivial since all the data will be used in each update of the model. However, if the minibatch-based stochastic gradient descent (SGD) training algorithm is used, it is desirable to include training data from all languages in each minibatch. This can be efficiently accomplished by randomizing the training utterance list across the languages before feeding it into the DNN training tool.

An alternative training procedure is proposed in [11]. In this procedure, all the hidden layers are first pretrained using the unsupervised DBN-pretraining procedure described in Chap. 5. A language is then selected and a randomly initialized softmax layer that is corresponding to the selected language is added. This softmax layer and the whole SHL-MDNN is fine-tuned on the chosen language. After it is refined, the softmax layer is replaced by another randomly initialized one corresponding to the next language, and fine-tuned on that language. This process is repeated for all the languages. A potential problem with this language-sequential training is that it will lead to more biased estimates and underperforms the simultaneous training procedure.

The SHL-MDNN can be pretrained using either generative or discriminative pre-training techniques described in Chap. 5. The fine-tuning of the SHL-MDNN can be carried out using the conventional backpropagation (BP) algorithm. However, since a different softmax layer is used for each different language, the algorithm needs to be adjusted slightly. When a training sample is presented to the SHL-MDNN trainer, only the shared hidden layers and the language-specific softmax layer are updated. Other softmax layers are kept intact.

After being trained, the SHL-MDNN can be used to recognize speech of any language used in the training process. Since multiple languages can be simultaneously decoded with its unified DNN structure, the SHL-MDNN makes multilingual LVCSR easy and efficient. Supporting a new language in SHL-MDNN is trivial as shown in Fig. 12.3. It only involves adding a new softmax layer in the existing SHL-MDNN and training the newly added softmax layer on the new language.

By sharing the hidden layers in the SHL-MDNN and by using the joint training strategy, we can improve the recognition accuracy of all the languages decode-able by the SHL-MDNN over the monolingual DNNs trained using data from individual languages only. Experiments have been conducted on a Microsoft internal speech recognition task to evaluate the SHL-MDNN [16]. The SHL-MDNN used in the experiment has 5 hidden layers, each with 2,048 neurons. The input to the DNN is 11 (5-1-5) frames of the 13-dim MFCC feature with its derivatives and accelerations. It is trained with 138-h French (FRA), 195-h German (DEU), 63-h Spanish (ESP), and 63-h Italian (ITA) speech data. For each language, the output layer has 1.8k senones determined by the GMM-HMM system trained with the maximum likelihood estimation (MLE) on the same training set. The SHL-MDNN was initialized using the unsupervised DBN-pretraining procedure, and then refined with BP using senone labels derived from the MLE model alignment. The trained DNNs are plugged in the CD-DNN-HMM framework described in Chap. 6.

Table 12.1 compares the word error rate (WER) obtained on the language-specific test sets using the monolingual DNN, trained using only the data from that language,

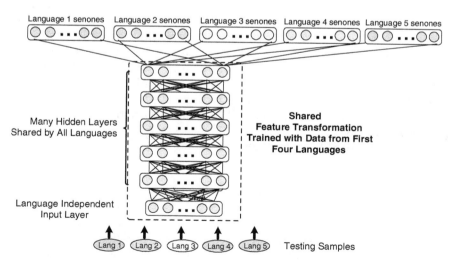

Fig. 12.3 Support a fifth language in the SHL-MDNN trained for four languages

Table 12.1 Compare monolingual DNN and shared-hidden-layer multilingual DNN in WER

	FRA	DEU	ESP	ITA
Test set size (words)	40k	37k	18k	31k
Monolingual DNN WER	28.1 %	24.0 %	30.6 %	24.3 %
SHL-MDNN WER	27.1 % (−3.6 %)	22.7 % (−5.4 %)	29.4 % (−3.9 %)	23.5 % (−3.3 %)

Relative WER reduction over the monolingual DNN is in parentheses (Summarized from Huang et al. [16])

and the SHL-MDNN, whose hidden layers are trained using data from all four languages. From the table, we can observe that the SHL-MDNN outperforms the monolingual DNN with a 3–5 % relative WER reduction across all the languages. We believe the gain obtained by the SHL-MDNN attributes to the cross-language knowledge. It is encouraging that even for FRA and DEU, which have more than 100 h of training data, SHL-MDNN can still provide improvements.

12.2.3 Crosslingual Model Transfer

The shared hidden layers extracted from the multilingual DNN can be considered as an intelligent feature extraction module jointly trained with data from multiple source languages. As such they carry rich information to distinguish phonetic classes in multiple languages and can be carried over to distinguish phonemes in new languages.

The procedure of crosslingual model transfer is simple. We just extract the shared hidden layers from the SHL-MDNN and add a new softmax layer on top of it as shown in Fig. 12.4. The softmax layer's output nodes correspond to the senones in the target language. We then fix the hidden layers and only train the softmax layer using training data from the target language. If enough training data are available, additional gains may be achieved by further tuning the entire network.

To evaluate the effectiveness of crosslingual model transfer, a series of experiments have been conducted in [16] and are used here. In these experiments, two different languages are used as the target languages: American English (ENU), which is phonetically close to the European languages used to train the SHL-MDNN described in Sect. 12.2.2, and Mandarin Chinese (CHN), which is far away from the European languages. The ENU test set consists of 2,286 utterances (or 18k words) and the CHN test set has 10,510 utterances (or 40k characters).

12.2.3.1 Hidden Layers Are Transferable

The first question to ask is whether the hidden layers are indeed transferable to other languages. To answer this question empirically it is assumed that a 9-h ENU training data (55,737 utterances) is available to build an ENU ASR system. Table 12.2 summarizes the experimental results. The baseline DNN is trained solely using the

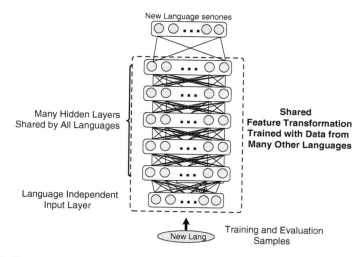

Fig. 12.4 Crosslingual transfer. The hidden layers are borrowed from the multilingual DNN while the softmax layer needs to be trained with data from the target language

Table 12.2 Compare WER on the ENU test set with and without using hidden layers (HLs) transferred from the FRA DNN

Setup	WER (%)
Baseline DNN (trained with 9-h US English data only)	30.9
French trained hidden layers + refine all layers	30.6
French trained hidden layers + refine Softmax layer only	27.3
Shared-hidden layers trained with four languages + refine Softmax layer only	25.3

(Summarized from Huang et al. [16])

9-h ENU training set. With this approach, a WER of 30.9 % is achieved on the ENU test set. An alternative approach is to leverage the hidden layers (feature transformation) learned from other languages. In this experiment, a monolingual DNN is trained with 138 h of FRA training data. The hidden layers of this DNN are then extracted and reused in the ENU DNN. If the hidden layers are fixed and only the ENU-specific softmax layer is trained using the 9-h ENU training data, a 2.6 % WER reduction (30.9 % → 27.3 %) over the baseline DNN can be obtained. If the whole FRA DNN is retrained using the 9-h ENU data, a WER of 30.6%, which is only slightly better than the 30.9 % baseline WER is obtained. These results indicate that the feature transformation represented by the hidden layers in the FRA DNN can be effectively transferred to recognize the ENU speech.

In addition, if the shared hidden layers extracted from the SHL-MDNN described in Sect. 12.2.2 are extracted and used in the ENU DNN an additional absolute 2.0 % WER reduction (27.3 % → 25.3 %) can be obtained. This indicates that the hidden layers extracted from the SHL-MDNN are more effective than that extracted from the FRA DNN when transferred to build the ENU DNN. Overall, by using the

Table 12.3 Compare the effect of target language training set size in WER when hidden layers are transferred from the SHL-MDNN

US English training data	3-h (%)	9-h (%)	36-h (%)
Baseline DNN (trained with English data only)	38.9	30.9	23.0
SHL-MDNN + refine Softmax layer only	28.0	25.3	22.4
SHL-MDNN + refine all layers	33.4	28.9	21.6
Best case relative WER reduction (%)	28.0	18.1	6.1

(Summarized from Huang et al. [16])

crosslingual model transfer, a 4.6 % (or 18.1 % relative) WER reduction was obtained over the baseline ENU DNN.

12.2.3.2 Size of Target Language Training Data Matters

The second question to ask is how the size of target language training data affect the performance of the multilingual DNN crosslingual model transfer. To answer this question, experiments were conducted by Huang et al. assuming 3, 9, and 36 h of English (target language) training data are available. Table 12.3 from [16] summarizes the results. From the table, we can observe that DNNs that exploit transferred hidden layers consistently outperform the baseline DNNs that do not use crosslingual model transfer. We can also observe that when different sizes of target languages are available, the best learning strategy is different. When less than 10 h of target language training data are available, the best strategy is to only train a new softmax layer. By doing so, we observe 28.0 and 18.1 % relative WER reduction over the baseline DNNs, when 3 and 9 h of ENU speech data are available, respectively. However, when the amount of training data is large enough, further adapting the whole DNN can provide additional error reduction. For example, when 36 h of ENU speech data are available, we observe additional 0.8 % WER reduction (22.4 % → 21.6 %) by adapting all layers.

12.2.3.3 Transferring from European Languages to Mandarin Chinese Is Effective

The third question to ask is whether the effectiveness of the crosslingual model transfer approach is sensitive to the language similarities between the source and the target languages. To answer this question, Huang et al. [16] used Mandarin Chinese (CHN) as the target language, which differs significantly to the European languages used to train the SHL-MDNN. Table 12.4 from [16] lists the character error rates (CERs) using both the baseline and the multilingual-boosted DNN when the size of Chinese training data varies. To achieve the results reported in this table, only the softmax layers were trained when less than 9 h of CHN data are available and

Table 12.4 Effectiveness of crosslingual model transfer on CHN measured in character error rate (CER) reduction

Chinese training set	3-h (%)	9-h (%)	36-h (%)	139-h (%)
Baseline DNN (trained with Chinese data only)	45.1	40.3	31.7	29.0
SHL-MDNN model transfer	35.6	33.9	28.4	26.6
	(−21.1)	(−15.9)	(−10.4)	(−8.3)

Relative CER reduction in parenthesis (Summarized from Huang et al. [16])

all layers are further refined when more than 10 h of CHN data are available. We can see that in all cases CER reduction is observed by using the transferred hidden layers. Even if 139 h of CHN training data are available, we can still benefit from the SHL-MDNN with 8.3 % relative CER reduction. Moreover, using only 36 h of CHN data, we can achieve 28.4 % CER on the test set by transferring the SHLs from the SHL-MDNN. This is better than the 29.0 % CER obtained with the baseline DNN trained using the 139 h of CHN training data, a save of over 100 h of CHN transcription effort.

12.2.3.4 Using Label Information Is Important

The fourth question to ask is whether the features extracted through unsupervised learning can perform classification tasks equally well as that through supervised training. It can provide significant advantage if the answer is true since it is much easier to obtain untranscribed speech data than transcribed ones for model training. In this subsection, we show that the label information is important for effectively learning the shared representation from the multilingual data. Table 12.5, based on results from [16], compares the systems with and without using the label information when training the shared hidden layers. We see from Table 12.5 that while there is a small gain by using pretrained only multilingual DNN and adapting the whole network with ENU data (30.9 % → 30.2 %), the gain is significantly smaller than that obtained when label information is used (30.9 % → 25.3 %). These results clearly indicate that labeled data are much more valuable than unlabeled data and using label information is critical in learning effective features from multilingual data.

Table 12.5 Compare features learned from multilingual data with and without using label information on ENU data

	SHL-MDNN trained with label?	WER (%)
Baseline DNN (trained with 9-h US English data only)	–	30.9
SHL-MDNN + refine softmax layer only	No	38.7
SHL-MDNN + refine all layers	Yes	30.2
SHL-MDNN + refine softmax layer only	Yes	25.3

(Summarized from Huang et al. [16])

12.3 Multiobjective Training of Deep Neural Networks for Speech Recognition

Since multitask learning can potentially improve the generalization ability of all the tasks involved, it has also been used in more general multiobjective training of DNNs for speech recognition. In this section, we sample three such applications. Note that in all these three tasks, the training set is small.

12.3.1 Robust Speech Recognition with Multitask Learning

In [21], Lu et al. proposed to improve the robustness of the noisy digit recognition task by using multitask learning. They used a single-hidden layer recurrent neural network as shown in Fig. 12.5 for digit classification. Different from previous work, they trained the neural network to simultaneously classify the digits, enhance the noisy speech, and recognize the gender of the speaker. In their experiments, 1,000 examples were chosen for training and 110 examples were used to control the stopping condition in training. Tests were performed on the isolated digit samples in Aurora test set A. They observed that by including both the enhancement and gender recognition tasks they can reduce error rate by close to 50 % relatively over the system trained to do digit classification alone.

12.3.2 Improved Phone Recognition with Multitask Learning

In [30], Seltzer and Droppo proposed to improve the recognition accuracy of the DNN-HMM system on the TIMIT phone recognition task [10] by adding a secondary task to the training of DNNs. They adopted a standard DNN used for phone recognition [14], which contains four 2,048-neuron hidden layers and uses 183 monophone states as the DNN training target, and investigated three different secondary tasks:

Fig. 12.5 Improve noisy digit recognition by training the neural network to simultaneously classify the digits, enhance the noisy speech, and recognize the gender of the speaker

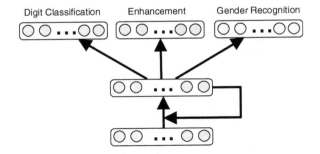

- Phone Label Task: They created the phone label for each training example by mapping the state symbol up to its corresponding phone label and used this phone label as the target of the secondary task. The intuition is that by doing so the DNN can know which states belong to the same phone and so do not need to separate these states too aggressively.
- State Context Task: Instead of classifying just the state label of the central frame, they added a secondary task to classify the previous and next frames' acoustic state labels. This secondary objective function measures the ability of the model to predict the current acoustic model state as well as the previous and next acoustic model states. The idea is by giving the model information about the time-evolution of the acoustic state the system can distinguish the states at the phone boundary from those in the middle.
- Phone Context Task: Since the primary task is to recognize the monophone state the phone context information such as that in the triphone state models is absent. To compensate for the missing of the context information, they added a secondary task to recognize the left and right context phone labels.

They conducted a series of experiments on the TIMIT corpus [10], which contains continuous speech from 630 native English speakers, with eight usable recorded utterances for each speaker. The core test set consists of twenty-four speakers who are not included in the training set consisted of 462 speakers. Recognition was performed using a set of 61 phoneme labels, each of which has three states, for a total of 183 possible monophone states. The secondary tasks were only used during the training and discarded during the testing. After decoding, the 61 phone labels were collapsed into a set of 39 phone classes for scoring, following [18].

Their results indicate that adding phone label classification as the secondary task does not affect the primary task's result. This is understandable since the phone label does not provide any additional information than the state label already used in the primary task. Using the phone context classification as the secondary task gave the highest phonetic error rate (PER) reduction ($21.63 \rightarrow 20.25\%$) on the core test set, and surpassing the best performance in the literature for a DNN that uses a standard feed-forward network architecture. As a conclusion, if suitable secondary tasks are chosen, the network can leverage the common structure in the different tasks to learn a model with better generalization capability.

12.3.3 Recognizing both Phonemes and Graphemes

In [4], Chen et al. proposed to improve the generalization performance of triphone models by jointly training DNNs with trigrapheme models of the same language under the MTL framework for low-resource languages. Triphone modeling and trigrapheme modeling are obviously related learning tasks for the same language. It is reasonable to believe that the features used to classify phonemes and graphemes can be shared.

In [4], triphone acoustic modeling was picked as the primary task and trigrapheme acoustic modeling as the secondary task. The architecture of their system is very similar to that used in the multilingual speech recognition, except that the two output layers are trained to model the posterior probabilities of triphone tied states and trigrapheme tied states respectively for a given input acoustic frame. All the hidden layers are shared across two tasks. Chen et al. evaluated this MTL system on three low-resource South African languages, namely Afrikaans, Sesotho, and Swati, each with about 1 h of training speech. They found that the MTL-DNN outperformed the single task learning (STL) DNN with a 5–13 % relative error reduction. More interestingly, the MTL-DNN even outperforms the integration of the triphone and trigrapheme STL DNNs by ROVER [9] with a 0.5–4.2 % relative error reduction.

12.4 Robust Speech Recognition Exploiting Audio-Visual Information

In some tasks, the goal is to improve the primary task's (e.g., speech recognition) accuracy by exploiting other sources of information (e.g., visual information). Such problems can be cast as a multitask learning problem if the tasks for which the additional information is designed are also optimized at the same time. In most cases, however, the secondary tasks are not optimized since the performance gain is mainly from using the additional information instead of multitasking learning. The key design question in such applications is how to exploit that additional information.

The mixed-bandwidth speech recognition we discussed in Chap. 9 is an example of such applications. As depicted in Fig. 9.12, there are two sources of information: output from the lower filter-banks and that from the higher filter-banks. Lower recognition accuracy was observed if only the information from the lower filter-banks is available, for example, when the signal is narrowband. If both sources of information are available as in the wideband signal case, additional error rate reduction can be achieved. Not only that, by training the DNN with mixed-bandwidth data, the performance on both the narrowband and wideband speech can be improved due to regularization imposed by the DNN architecture and the combination of the training data.

In [15], Huang and Kingsbury proposed a similar architecture. Their goal, however, is to improve the robust speech recognition accuracy by exploiting audio-visual information. The architecture of their system is illustrated in Fig. 12.6, which is a special case of Fig. 12.1.

Their work is motivated by the bimodality (auditory and visual) of human speech perception [31]. Because visual information is separate from the audio and invariant to acoustic noise, it can potentially improve over audio-only speech recognition in both clean and noisy conditions [5, 6, 8, 17, 19, 23, 26]. It is not surprising that the most successful systems extract visual features from the facial region of interest. In [15], Huang and Kingsbury investigated the use of DNNs to improve audio-visual

Fig. 12.6 A DNN architecture for improving noisy speech recognition by exploiting audio-visual information. The *green* hidden layers in the *dotted rectangular box* are shared by two modalities

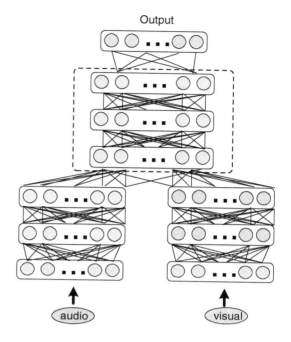

speech recognition with a focus on using the visual modality to supplement the audio. They investigated two techniques to achieve this goal. The first technique fuses the decisions made by two single-modality DNNs, one for audio and one for visual features. The second technique fuses the features at a mid-level hidden layer as shown in Fig. 12.6. On a continuously spoken digit recognition task, their experiments show that these methods can reduce word error rate by as much as 21 % relative over a baseline multistream audio-visual GMM/HMM system.

References

1. Association, I.P., et al.: Report on the 1989 Kiel convention. J. Int. Phonetic Assoc. **19**(2), 67–80 (1989)
2. Athineos, M., Ellis, D.P.: Frequency-domain linear prediction for temporal features. In: Proceedings of the IEEE Workshop on Automfatic Speech Recognition and Understanding (ASRU), pp. 261–266 (2003)
3. Caruana, R.: Multitask learning. Mac. Learn. **28**(1), 41–75 (1997)
4. Chen, D., Mak, B., Leung, C.C., Sivadas, S.: Joint acoustic modeling of triphones and trigraphemes by multi-task learning deep neural networks for low-resource speech recognition. In: Proceedings of the International Conference on Acoustics, Speech and Signal Processing (ICASSP) (2014)
5. Chen, T., Rao, R.R.: Audio-visual integration in multimodal communication. Proc. IEEE **86**(5), 837–852 (1998)

6. Chibelushi, C.C., Deravi, F., Mason, J.S.: A review of speech-based bimodal recognition. Multimedia IEEE Trans. **4**(1), 23–37 (2002)
7. Dahl, G.E., Yu, D., Deng, L., Acero, A.: Context-dependent pre-trained deep neural networks for large-vocabulary speech recognition. IEEE Trans. Audio Speech Lang. Process. **20**(1), 30–42 (2012)
8. Dupont, S., Luettin, J.: Audio-visual speech modeling for continuous speech recognition. Multimedia IEEE Trans. **2**(3), 141–151 (2000)
9. Fiscus, J.G.: A post-processing system to yield reduced word error rates: recognizer output voting error reduction (ROVER). In: Proceedings of the IEEE Workshop on Automatic Speech Recognition and Understanding (ASRU), pp. 347–354 (1997)
10. Garofolo, J.S.: Darpa Timit: Acoustic-Phonetic Continuous Speech Corps CD-ROM. US Department of Commerce, National Institute of Standards and Technology, Gaithersburg (1993)
11. Ghoshal, A., Swietojanski, P., Renals, S.: Multilingual training of deep-neural netowrks. In: Proceedings of the International Conference on Acoustics, Speech and Signal Processing (ICASSP) (2013)
12. Heigold, G., Vanhoucke, V., Senior, A., Nguyen, P., Ranzato, M., Devin, M., Dean, J.: Multilingual acoustic models using distributed deep neural networks. In: Proceedings of the International Conference on Acoustics, Speech and Signal Processing (ICASSP) (2013)
13. Hermansky, H.: Perceptual linear predictive (PLP) analysis of speech. J. Acoust. Soc. Am. **87**, 1738 (1990)
14. Hinton, G., Deng, L., Yu, D., Dahl, G.E.,Mohamed, A.r., Jaitly, N., Senior, A., Vanhoucke, V., Nguyen, P., Sainath, T.N., et al.: Deep neural networks for acoustic modeling in speech recognition: the shared views of four research groups. IEEE Signal Process. Mag. **29**(6), 82–97 (2012)
15. Huang, J., Kingsbury, B.: Audio-visual deep learning for noise robust speech recognition. In: Proceedings of the International Conference on Acoustics, Speech and Signal Processing (ICASSP), pp. 7596–7599 (2013)
16. Huang, J.T., Li, J., Yu, D., Deng, L., Gong, Y.: Cross-language knowledge transfer using multilingual deep neural network with shared hidden layers. In: Proceedings of the International Conference on Acoustics, Speech and Signal Processing (ICASSP) (2013)
17. Kim, M.W., Ryu, J.W., Kim, E.J.: Speech recognition by integrating audio, visual and contextual features based on neural networks. Advances in Natural Computation, pp. 155–164. Springer, Berlin (2005)
18. Lee, K.F., Hon, H.W.: Speaker-independent phone recognition using hidden Markov models. IEEE Trans. Speech Audio Process. **37**(11), 1641–1648 (1989)
19. Lewis, T.W., Powers, D.M.: Audio-visual speech recognition using red exclusion and neural networks. J. Res. Pract. Inf. Technol. **35**(1), 41–64 (2003)
20. Lin, H., Deng, L., Yu, D., Gong, Y.f., Acero, A., Lee, C.H.: A study on multilingual acoustic modeling for large vocabulary ASR. In: Proceedings of the International Conference on Acoustics, Speech and Signal Processing (ICASSP), pp. 4333–4336 (2009)
21. Lu, Y., Lu, F., Sehgal, S., Gupta, S., Du, J., Tham, C.H., Green, P., Wan, V.: Multitask learning in connectionist speech recognition. In: Proceedings of the Australian International Conference on Speech Science and Technology (2004)
22. Martens, J.: Deep learning via Hessian-free optimization. In: Proceedings of the International Conference on Machine Learning (ICML), pp. 735–742 (2010)
23. Ngiam, J., Khosla, A., Kim, M., Nam, J., Lee, H., Ng, A.Y.: Multimodal deep learning. In: Proceedings of the 28th International Conference on Machine Learning (ICML-11), pp. 689–696 (2011)
24. Pan, S.J., Yang, Q.: A survey on transfer learning. IEEE Trans. Knowl. Data Eng. **22**(10), 1345–1359 (2010)
25. Plahl, C., Schluter, R., Ney, H.: Cross-lingual portability of chinese and english neural network features for french and german LVCSR.In: Proceedings of the IEEE Workshop on Automfatic Speech Recognition and Understanding (ASRU), pp. 371–376 (2011)

26. Potamianos, G., Neti, C., Gravier, G., Garg, A., Senior, A.W.: Recent advances in the automatic recognition of audiovisual speech. Proc. IEEE **91**(9), 1306–1326 (2003)
27. Qian, Y., Liu, J.: Cross-lingual and ensemble MLPs strategies for low-resource speech recognition. In: Proceedings of the Annual Conference of International Speech Communication Association (INTERSPEECH) (2012)
28. Schultz, T., Waibel, A.: Multilingual and crosslingual speech recognition. In: Proceedings of the DARPA Workshop on Broadcast News Transcription and Understanding, pp. 259–262 (1998)
29. Seide, F., Li, G., Yu, D.: Conversational speech transcription using context-dependent deep neural networks. In: Proceedings of the Annual Conference of International Speech Communication Association (INTERSPEECH), pp. 437–440 (2011)
30. Seltzer, M.L., Droppo, J.: Multi-task learning in deep neural networks for improved phoneme recognition. In: Proceedings of the International Conference on Acoustics, Speech and Signal Processing (ICASSP), pp. 6965–6969 (2013)
31. Sumby, W.H., Pollack, I.: Visual contribution to speech intelligibility in noise. J. Acoust. Soc. Am. (JASA) **26**(2), 212–215 (1954)
32. Thomas, S., Ganapathy, S., Hermansky, H.: Cross-lingual and multi-stream posterior features for low resource LVCSR systems. In: Proceedings of the Annual Conference of International Speech Communication Association (INTERSPEECH), pp. 877–880 (2010)
33. Thomas, S., Ganapathy, S., Hermansky, H.: Multilingual MLP features for low-resource LVCSR systems. In: Proceedings of the International Conference on Acoustics, Speech and Signal Processing (ICASSP), pp. 4269–4272 (2012)
34. Yu, D., Deng, L., Liu, P., Wu, J., Gong, Y., Acero, A.: Cross-lingual speech recognition under runtime resource constraints. In: Proceedings of the International Conference on Acoustics, Speech and Signal Processing (ICASSP), pp. 4193–4196 (2009)

Chapter 13
Recurrent Neural Networks and Related Models

Abstract A recurrent neural network (RNN) is a class of neural network models where many connections among its neurons form a directed cycle. This gives rise to the structure of internal states or memory in the RNN, endowing it with the dynamic temporal behavior not exhibited by the DNN discussed in earlier chapters. In this chapter, we first present the state-space formulation of the basic RNN as a nonlinear dynamical system, where the recurrent matrix governing the system dynamics is largely unstructured. For such basic RNNs, we describe two algorithms for learning their parameters in some detail: (1) the most popular algorithm of back-propagation through time (BPTT); and (2) a more rigorous, primal-dual optimization technique, where constraints on the RNN's recurrent matrix are imposed to guarantee stability during RNN learning. Going beyond basic RNNs, we further study an advanced version of the RNN, which exploits the structure called long-short-term memory (LSTM), and analyzes its strengths over the basic RNN both in terms of model construction and of practical applications including some latest speech recognition results. Finally, we analyze the RNN as a bottom-up, discriminative, dynamic system model against the top-down, generative counterpart of dynamic system as discussed in Chap. 4. The analysis and discussion lead to potentially more effective and advanced RNN-like architectures and learning paradigm where the strengths of discriminative and generative modeling are integrated while their respective weaknesses are overcome.

13.1 Introduction

As discussed in previous chapters, for many years and until the recent rise of deep learning technology as we reviewed in the several preceding chapters, automatic speech recognition (ASR) technology had been dominated by a "shallow" architecture-hidden Markov models (HMMs) with each state characterized by a Gaussian mixture model (GMM), which we covered in some detail in Chaps. 2 and 3. While significant technological successes had been achieved using complex and carefully engineered variants of GMM-HMMs and acoustic features suitable for them, researchers had for long anticipated that the next generation of ASR would require solutions to many new technical challenges under diversified deployment

© Springer-Verlag London 2015
D. Yu and L. Deng, *Automatic Speech Recognition*,
Signals and Communication Technology, DOI 10.1007/978-1-4471-5779-3_13

environments and that overcoming these challenges would likely require *deep* architectures that can at least functionally emulate the human speech recognition system known to have dynamic and hierarchical structure in both speech production and speech perception [18, 29, 35, 94]. An attempt to incorporate a primitive level of understanding of this deep speech structure, initiated at the 2009 NIPS Workshop on Deep Learning for Speech Recognition and Related Applications [24], has helped create an impetus in the ASR community to pursue a deep representation learning approach based on the deep neural network (DNN) architecture, which was pioneered by the machine learning community only a few years earlier [51, 52] but rapidly evolved into the new state of the art in speech recognition with industry-wide adoption; e.g., [12, 13, 23, 24, 26, 47, 50, 60, 75, 84–87, 91, 92, 100, 101, 108].

In the meantime, however, it has been realized by many that the DNN-HMM approach has not modeled speech dynamics properly. This relates to the same type of limitations as we analyzed in Chap. 3 on the topic of HMM where several variants of the HMM were discussed aiming to overcome such limitations. The deep and temporally recurrent neural network (RNN), which is the focus of this chapter, has been developed in the past few years by deep learning and ASR researchers to overcome the dynamic-modeling challenge; e.g., [10, 21, 43–46, 66, 78, 88, 89, 96, 99]. In the RNN, the internal representation of dynamic speech features is discriminatively formed by feeding the low-level acoustic features into the hidden layer together with the recurrent hidden features from the past history. In contrast, there is no internal representation of speech dynamics in the DNN-HMM. The RNN is a class of neural network models where connections among many of its units form a directed cycle, hence the term *recurrent*. Such a cycle or recurrence is associated with the time-delay operation. The use of time-delayed recurrence over the temporal dimension gives rise to the *memory* structure, expressed as internal states, in the RNN, permitting it to exhibit the type of dynamic temporal behavior not exhibited by the DNN and DNN-HMM discussed in previous chapters.

Even without stacking RNNs one on top of another as carried out in [45, 46, 88, 89] or feeding DNN features into RNNs as explored in [10, 21], an RNN itself is a deep model since temporal unfolding of the RNN creates as many layers in the network as the length of the input speech utterance. Recent progress on speech recognition has seen excellent speech recognition accuracy achieved by RNNs, including the long-short-term memory (LSTM) version of the RNN which started in as early as 1997 by neural network researchers [39, 40, 44, 45, 53, 103]. One focus of this chapter is to present the background and mathematical formulation of the RNN (Sect. 13.2), as well as the learning methods including the most popular backpropagation through time (BPTT) technique (Sect. 13.3).

The use of RNNs or related neural predictive models for speech recognition dates back to late 1980s and early 1990s; e.g., [22, 82, 102], which achieved relatively low recognition accuracy. Since deep learning became popular in recent years, much more research has been devoted to the RNN, including the applications to both speech [45, 46, 88, 89] and languages [11, 67–73], and its stacked versions, also called deep RNNs [45, 46, 48, 77, 88]. Most work on RNNs made use of the method of BPTT to train their parameters, and empirical tricks need to be exploited (e.g., truncate gradients when they become too large [69, 70]) in order to make the training effective. It is not until recently that careful analysis was made to fully understand the source of difficulties in learning RNNs and somewhat more principled, but still rather heuristic, solutions were developed. For example, in [3, 6, 78] strategies of gradient norm clipping was proposed to deal with the gradient exploding problem during BPTT training. There are other solutions offered to improve learning methods for the RNN; e.g., [10, 56]. The method described in [10] is based on more principled optimization techniques than most other methods, and will be reviewed in Sect. 13.4. Further, the LSTM version of the RNN has recently been shown to perform extremely well in both small- and large-scale ASR, and its structure is well motivated. We will devote Sect. 13.5 to this topic.

It is important to note that before the recent rise of deep learning for speech modeling and recognition, a number of earlier attempts had been made, which we briefly discussed in Sect. 3.7 of Chap. 3 on HMM variants, to develop computational architectures that are "deeper" than the conventional GMM-HMM architecture. One prominent class of such models are hidden dynamic models where the internal representation of dynamic speech features is generated probabilistically from the higher levels in the overall deep speech model hierarchy [9, 15, 17, 30, 34, 62, 80, 97, 105, 107]. Despite separate developments of the RNNs and of the hidden dynamic or trajectory models, they share a very similar motivation-representing aspects of dynamic structure in human speech. Nevertheless, a number of different ways in which these two types of deep dynamic models are constructed endow them with distinct pros and cons. Careful analysis of the contrast between these two model types and of the similarity to each other will help provide insights into the strategies for developing new types of deep dynamic models with the hidden representations of speech features superior to both existing RNNs and hidden dynamic models. We will devote Sect. 13.6 of this chapter to a contrastive analysis between (discriminative) RNNs and (generative) hidden dynamic models. In this multifaceted analysis, we will focus mainly on the most prominent contrasts between the two types of dynamic models in terms of the opposing top-down versus bottom-up information flow, and in terms of the opposing local versus distributed representations adopted by the latent vectors in these two types of models.

13.2 State-Space Formulation of the Basic Recurrent Neural Network

An RNN is fundamentally different from the feed-forward DNN in that the RNN operates not only based on inputs, as for the DNN, but also on internal states. The internal states encode the past information in the temporal sequence that has already been processed by the RNN. In this sense, the RNN is a dynamic system, more general than the DNN which performs static input–output transformation. The use of the state space in the RNN enables its representation and learning of sequentially extended dependencies over a long time span, at least in principle.

Let us now formulate the simple one-hidden-layer RNN in terms of the (noise-free) nonlinear state-space model commonly used in signal processing. This formulation allows us to later compare the RNN with the same state-space formulation of non-linear dynamic systems used as generative models for speech acoustics. The contrast between the discriminative RNN and the use of the same mathematical model in the generative model will be made to shed light onto why one approach works better than another and how a combination of the two would be desirable.

At each time point t, let $\mathbf{x}_t$ be the $K \times 1$ vector of inputs, $\mathbf{h}_t$ be the $N \times 1$ vector of hidden state values, and $\mathbf{y}_t$ be the $L \times 1$ vector of outputs, the simple one-hidden-layer RNN can be described as

$$\mathbf{h}_t = f(\mathbf{W}_{xh}\mathbf{x}_t + \mathbf{W}_{hh}\mathbf{h}_{t-1}) \tag{13.1}$$

$$\mathbf{y}_t = g(\mathbf{W}_{hy}\mathbf{h}_t), \tag{13.2}$$

where $\mathbf{W}_{hy}$ is the $L \times N$ matrix of weights connecting the N hidden units to the L outputs, $\mathbf{W}_{xh}$ is the $N \times K$ matrix of weights connecting the K inputs to the N hidden units, and $\mathbf{W}_{hh}$ is the $N \times N$ matrix of weights connecting the N hidden units from time $t - 1$ to time t, $\mathbf{u}_t = \mathbf{W}_{xh}\mathbf{x}_t + \mathbf{W}_{hh}\mathbf{h}_{t-1}$ is the $N \times 1$ vector of hidden layer potentials, $\mathbf{v}_t = \mathbf{W}_{hy}\mathbf{h}_t$ is the $L \times 1$ vector of output layer potentials, $f(\mathbf{u}_t)$ is the hidden layer activation function, and $g(\mathbf{v}_t)$ is the output layer activation function. Typical hidden layer activation functions are sigmoid, tanh, and rectified linear units while the typical output layer activation functions are linear and softmax functions. Equations 13.1 and 13.2 are often called the observation and state equations, respectively.

Note that outputs from previous time frames can also be used to update the state vector, in which case the state equation becomes

$$\mathbf{h}_t = f(\mathbf{W}_{xh}\mathbf{x}_t + \mathbf{W}_{hh}\mathbf{h}_{t-1} + \mathbf{W}_{yh}\mathbf{y}_{t-1}), \tag{13.3}$$

where $\mathbf{W}_{yh}$ denotes the weight matrix connecting from output layer to the hidden layer. For simplicity and without loss of generality, we only consider the case without output feedback in this chapter.

13.3 The Backpropagation-Through-Time Learning Algorithm

The standard BPTT method, which was well explained in the tutorial material [7, 56] and in the original paper [83], for learning the weight matrices of an RNN *unfolds* the network in time and propagates error signals backwards through time. It is an extension of the classic backpropagation algorithm for feed-forward networks, where the stacked hidden layers for the same training frame, t, are replaced by the T same single hidden layers across time, $t = 1, 2, \ldots, T$.

Referring to Eqs. 13.1 and 13.2, we denote $h_t(j)$ the jth hidden unit where $j = 1, 2, \ldots, N$, $w_{hy}(i, j)$ as the weight connecting the jth hidden unit to the ith output unit for $i = 1, 2, \ldots, L$ and $j = 1, 2, \ldots, N$.

13.3.1 Objective Function for Minimization

As for the classic backpropagation, we begin by defining the cost function (or training criterion). In this section, we use the sum-square error

$$E = c \sum_{t=1}^{T} \| \mathbf{l}_t - \mathbf{y}_t \|^2 = c \sum_{t=1}^{T} \sum_{j=1}^{L} (l_t(j) - y_t(j))^2 \tag{13.4}$$

between the actual output, $\mathbf{y}_t$, and the target vector, $\mathbf{l}_t$, over all time frames as the cost function, where $l_t(j)$ and $y_t(j)$ are the jth units in the target and output vectors, respectively, and $c = 0.5$ is a conveniently chosen scale factor.

We seek to minimize this cost with respect to the weights using the gradient descent algorithm. For a specific weight, w, in the RNN, the update rule for gradient descent is

$$w^{new} = w - \gamma \frac{\partial E}{\partial w}, \tag{13.5}$$

where γ is the learning rate. To compute the gradient, we define the error terms

$$\delta_t^y(j) = -\frac{\partial E}{\partial v_t(j)}, \quad \delta_t^h(j) = -\frac{\partial E}{\partial u_t(j)} \tag{13.6}$$

as the gradient of the cost with respect to the unit's input potential. The error terms and gradients can be recursively computed as we will explain next.

13.3.2 Recursive Computation of Error Terms

In the error propagation part of the BPTT algorithm, all RNN weights are duplicated spatially for an arbitrary number of time steps. That is, they are tied over time. Therefore, the standard backpropagation algorithm for feed-forward neural networks needs to be modified by incorporating this tying constraint.

At the final time frame $t = T$, we can calculate the error terms at the output as

$$\delta_T^y(j) = -\frac{\partial E}{\partial y_T(j)} \frac{\partial y_T(j)}{\partial v_T(j)} = (l_T(j) - y_T(j)) g'(v_T(j)) \quad \text{for } j = 1, 2, \ldots, L$$

$$\text{or} \quad \delta_T^y = (\mathbf{l}_T - \mathbf{y}_T) \bullet g'(\mathbf{v}_T), \tag{13.7}$$

and that at the hidden layer as

$$\delta_T^h(j) = -\left(\sum_{i=1}^{L} \frac{\partial E}{\partial v_T(i)} \frac{\partial v_T(i)}{\partial h_T(j)} \frac{\partial h_T(j)}{\partial u_T(j)}\right) = \sum_{i=1}^{L} \delta_T^y(i) w_{hy}(i,j) f'(u_T(j))$$

$$\text{for } j = 1, 2, \ldots, N$$

$$\text{or} \quad \delta_T^h = \mathbf{W}_{hy}^T \delta_T^y \bullet f'(\mathbf{u}_T) \tag{13.8}$$

where $\bullet$ is the element-wise multiplication operator.

For all other time frames, $t = T - 1, T - 2, \ldots, 1$, we can compute the error terms as

$$\delta_t^y(j) = (l_t(j) - y_t(j)) g'(v_t(j)) \quad \text{for } j = 1, 2, \ldots, L$$

$$\text{or} \quad \delta_t^y = (\mathbf{l}_t - \mathbf{y}_t) \bullet g'(\mathbf{v}_t) \tag{13.9}$$

for the output units and

$$\delta_t^h(j) = -\left[\sum_{i=1}^{N} \frac{\partial E}{\partial u_{t+1}(i)} \frac{\partial u_{t+1}(i)}{\partial h_t(j)} + \sum_{i=1}^{L} \frac{\partial E}{\partial v_t(i)} \frac{\partial v_t(i)}{\partial h_t(j)}\right] \frac{\partial h_t(j)}{\partial u_t(j)}$$

$$= \left[\sum_{i=1}^{N} \delta_{t+1}^h(i) w_{hh}(i,j) + \sum_{i=1}^{L} \delta_t^y(i) w_{hy}(i,j)\right] f'(u_t(j))$$

$$\text{for } j = 1, 2, \ldots, N$$

$$\text{or} \quad \delta_t^h = \left[\mathbf{W}_{hh}^T \delta_{t+1}^h + \mathbf{W}_{hy}^T \delta_t^y\right] \bullet f'(\mathbf{u}_t) \tag{13.10}$$

for the hidden units, recursively, where the error term δ_t^y is propagated back from the output layer at time frame t, and δ_{t+1}^h is propagated back from the hidden layer at time frame $t + 1$.

13.3.3 Update of RNN Weights

Given all the error terms and gradients computed above, we can easily update the weights. For the output weight matrices, we have

$$w_{hy}^{new}(i,j) = w_{hy}(i,j) - \gamma \sum_{t=1}^{T} \frac{\partial E}{\partial v_t(i)} \frac{\partial v_t(i)}{\partial w_{hy}(i,j)} = w_{hy}(i,j) - \gamma \sum_{t=1}^{T} \delta_t^y(i) h_t(j)$$

$$\text{or} \quad \mathbf{W}_{hy}^{new} = \mathbf{W}_{hy} + \gamma \sum_{t=1}^{T} \delta_y^t \mathbf{h}_t^T. \tag{13.11}$$

For the input weight matrices, we get

$$w_{xh}^{new}(i,j) = w_{xh}(i,j) - \gamma \sum_{t=1}^{T} \frac{\partial E}{\partial u_t(i)} \frac{\partial u_t(i)}{\partial w_{xh}(i,j)} = w_{xh}(i,j) - \gamma \sum_{t=1}^{T} \delta_t^h(i) x_t(j)$$

$$\text{or} \quad \mathbf{W}_{xh}^{new} = \mathbf{W}_{xh} + \gamma \sum_{t=1}^{T} \delta_h^t \mathbf{x}_t^T. \tag{13.12}$$

For the recurrent weight matrices, we have

$$w_{hh}^{new}(i,j) = w_{hh}(i,j) - \gamma \sum_{t=1}^{T} \frac{\partial E}{\partial u_t(i)} \frac{\partial u_t(i)}{\partial w_{hh}(i,j)}$$

$$= w_{hh}(i,j) - \gamma \sum_{t=1}^{T} \delta_t^h(i) h_{t-1}(j)$$

$$\text{or} \quad \mathbf{W}_{hh}^{new} = \mathbf{W}_{hh} + \gamma \sum_{t=1}^{T} \delta_h^t \mathbf{h}_{t-1}^T. \tag{13.13}$$

Note that different from the BP algorithm used in the DNN system, here the gradients are summed over all the time frames since the same weight matrices are used across time. Algorithm 13.1 summarizes the BPTT algorithm for the single-hidden-layer RNN described above.

Algorithm 13.1 The Backpropagation Through Time Algorithm for the Single-Hidden-Layer RNN with the Sum of Squared Error Cost Function

1: **procedure** BPTT($\{\mathbf{x}_t, \mathbf{I_t}\} 1 \leq t \leq T$)

                                ▷ $\mathbf{x}_t$ is the input feature sequence
                                 ▷ $\mathbf{I_t}$ is the label sequence
                                 ▷ forward computation

2:     **for** $t \leftarrow 1; t \leq T; t \leftarrow t + 1$ **do**
3:         $\mathbf{u}_t \leftarrow \mathbf{W}_{xh}\mathbf{x}_t + \mathbf{W}_{hh}\mathbf{h}_{t-1}$
4:         $\mathbf{h}_t \leftarrow f(\mathbf{u}_t)$
5:         $\mathbf{v}_t \leftarrow \mathbf{W}_{hy}\mathbf{h}_t$
6:         $\mathbf{y}_t \leftarrow g(\mathbf{v}_t)$
7:     **end for**

                                  ▷ backpropagation through time
                                 ▷ •: element-wise multiplication

8:     $\delta_T^y \leftarrow (\mathbf{l}_T - \mathbf{y}_T) \bullet g'(\mathbf{v}_T)$
9:     $\delta_T^h \leftarrow \mathbf{W}_{hy}^T \delta_T^y \bullet f'(\mathbf{u}_T)$
10:    **for** $t \leftarrow T - 1; t \geq 1T; t \leftarrow t - 1$ **do**
11:        $\delta_t^y \leftarrow (\mathbf{l}_t - \mathbf{y}_t) \bullet g'(\mathbf{v}_t)$
12:        $\delta_t^h \leftarrow \left[\mathbf{W}_{hh}^T \delta_{t+1}^h + \mathbf{W}_{hy}^T \delta_t^y \right] \bullet f'(\mathbf{u}_t)$ ▷ propagate from δ_t^y and δ_{t+1}^h
13:    **end for**

                                  ▷ model update

14:    $\mathbf{W}_{hy} \leftarrow \mathbf{W}_{hy} + \gamma \sum_{t=1}^{T} \delta_y^t \mathbf{h}_t^T$
15:    $\mathbf{W}_{hh} \leftarrow \mathbf{W}_{hh} + \gamma \sum_{t=1}^{T} \delta_h^t \mathbf{h}_{t-1}^T$
16: **end procedure**

The computational complexity of the BPTT described above can be shown to be $O(M^2)$ per time step where $M = LN + NK + N^2$ is the total number of weight parameters that need to be learned. Compared to the classic feed-forward backpropagation, BPTT converges slower due to dependencies between frames, and is more likely to converge to a poor local optimum due to exploding and vanishing gradients [78, 96] and utterance-level (instead of frame-level) randomization. It is far from trivial to achieve good results without much experimentation and tuning. The training speed can be improved if we truncate the past history to no more than the last p time steps.

13.4 A Primal-Dual Technique for Learning Recurrent Neural Networks

13.4.1 Difficulties in Learning RNNs

It is well known that learning RNNs is difficult partly because of the exploding and vanishing gradient problems, as analyzed in [78]. A sufficient condition for the vanishing gradient problem to occur is

$$\|\mathbf{W}_{hh}\| < d \tag{13.14}$$

where $d = 4$ for sigmoidal hidden units and $d = 1$ for linear units. $\|\mathbf{W}_{hh}\|$ is the L_2-norm (the largest singular value) of the recurrent weight matrix $\mathbf{W}_{hh}$ of the RNN. On the other hand, a necessary condition for exploding gradient to occur is

$$\|\mathbf{W}_{hh}\| > d. \tag{13.15}$$

Therefore, the property of recurrent matrix $\mathbf{W}_{hh}$ is essential for learning an RNN. In [6, 78], the proposed method for solving the exploding gradient problem is to empirically clip the gradient so that the norm of the gradient cannot exceed certain threshold. The way to avoid the vanishing gradient is also empirical: either adding a regularization term to push up the gradient or exploiting the information about the curvature of the objective function [66]. Here we review the study described in [10], which proposed and successfully experimented a more rigorous and effective approach to learning RNNs by directly exploiting the constraints that need to be imposed on $\mathbf{W}_{hh}$.

13.4.2 Echo-State Property and Its Sufficient Condition

We now show that conditions described in Eqs. 13.14 and 13.15 are closely related to whether the RNN satisfies the echo-state property, which, following [56], states that "if the network has been run for a very long time, the current network state is uniquely determined by the history of the input and the (teacher-forced) output." It is also shown in [55] that this echo-state property is equivalent to the *state contracting* property. For networks with no feedback from the output, a network is state contracting if for all right-infinite input sequences $\{\mathbf{x}_t\}$, where $t = 0, 1, 2, \ldots$, there exists a null sequence $(\varepsilon_t)_{t \geq 0}$ such that for all starting states $\mathbf{h}_0$ and $\mathbf{h}_0'$ and for all $t > 0$ it holds that $\|\mathbf{h}_t - \mathbf{h}_t'\| < \varepsilon_t$, where $\mathbf{h}_t$ and $\mathbf{h}_t'$ are two hidden state vectors at time t obtained when the network is driven by $\mathbf{x}_t$ up to time t after having been stated in $\mathbf{x}_0$ and $\mathbf{x}_0'$, respectively. It is further shown that a sufficient condition for the *non-existence*, or a necessary condition for the *existence*, of echo-state property is that the spectral radius of the recurrent matrix $\mathbf{W}_{hh}$ is greater than one when tanh nonlinear units are used in the RNN's hidden layer.

In echo-state machines, the reservoir or recurrent weight matrix $\mathbf{W}_{hh}$ is randomly generated and normalized according to the rule above and will remain unchanged over time in the training. The input weight matrix $\mathbf{W}_{xh}$ is fixed as well. To improve the learning, we here learn both $\mathbf{W}_{hh}$ and $\mathbf{W}_{xh}$ subject to the constraint that the RNN satisfies the echo-state property. To this end, the following sufficient condition for the echo-state property was recently developed in [10], which can be more easily handled in the training procedure than the original definition:

Let $d = 1/\max_x |f'(x)|$. Then the RNN satisfies the echo-state property if

$$\|\mathbf{W}_{hh}\|_\infty < d \tag{13.16}$$

where $\|\mathbf{W}_{hh}\|_\infty$ denote the ∞-norm of matrix $\mathbf{W}_{hh}$ (i.e., maximum absolute row sum), $d = 1$ for tanh units, and $d = 4$ for sigmoid units.

An important consequence of condition 13.16 is that it naturally avoids the exploding gradient problem. If the condition 13.16 can be enforced in the training process, there is no need to clip the gradient in a heuristic way.

13.4.3 Learning RNNs as a Constrained Optimization Problem

Given the sufficient condition for the echo-state property, we can now formulate the problem of learning the RNN that preserves the echo-state property as the following constrained optimization problem:

$$\min_{\Theta} \quad E(\Theta) = E(\mathbf{W}_{hh}, \mathbf{W}_{xh}, \mathbf{W}_{hy}) \tag{13.17}$$

$$\text{subject to} \quad \|\mathbf{W}_{hh}\|_\infty \leq d \tag{13.18}$$

That is, we need to find the set of RNN parameters that best predict the target values on average while preserving the echo-state property. Recall that $\|\mathbf{W}_{hh}\|_\infty$ is defined as the maximum absolute row sum. Therefore, the above RNN learning problem is equivalent to the following constrained optimization problem:

$$\min_{\Theta} \quad E(\Theta) = E(\mathbf{W}_{hh}, \mathbf{W}_{xh}, \mathbf{W}_{hy}) \tag{13.19}$$

$$\text{subject to} \quad \sum_{j=1}^{N} |W_{ij}| \leq d, \quad i = 1, \dots, N \tag{13.20}$$

where W_{ij} denotes the (i, j)th entry of the matrix $\mathbf{W}_{hh}$. Next, we proceed to derive the learning algorithm that can achieve this objective.

13.4.4 A Primal-Dual Method for Learning RNNs

13.4.4.1 A Brief Introduction to Primal-Dual Method

Here let us solve the constrained optimization problem above by the primal-dual method, a popular technique in modern optimization literature; e.g., [8]. First, the Lagrangian of the problem can be written as

$$L(\Theta, \boldsymbol{\lambda}) = E(\mathbf{W}_{hh}, \mathbf{W}_{xh}, \mathbf{W}_{hy}) + \sum_{i=1}^{N} \lambda_i \left(\sum_{j=1}^{N} |W_{ij}| - d \right) \tag{13.21}$$

where λ_i denotes the ith entry of the Lagrange vector λ (i.e., dual variable) and is required to be non-negative. Let the dual function $q(\lambda)$ be defined as the following *unconstrained* optimization problem

$$q(\lambda) = \min_{\Theta} L(\Theta, \lambda) \tag{13.22}$$

The dual function $q(\lambda)$ in the above-unconstrained optimization problem is always concave, even when the original cost $E(\Theta)$ is nonconvex [8]. In addition, the dual function is always a lower bound of the original constrained optimization problem. That is,

$$q(\lambda) \leq E(\Theta^\star) \tag{13.23}$$

Maximizing $q(\lambda)$ subject to the constraint $\lambda_i \geq 0$, $i = 1, \dots, N$ will be the best lower bound that can be obtained from the dual function [8]. This new problem is called the dual problem of the original optimization problem:

$$\max_{\lambda} \quad q(\lambda) \tag{13.24}$$

$$\text{subject to} \quad \lambda_i \geq 0, \quad i = 1, \ldots, N \tag{13.25}$$

which is a convex optimization problem since we are maximizing a concave objective with linear inequality constraints. After solving $\lambda^\star$ in Eqs. 13.24 and 13.21, we can substitute the corresponding $\lambda^\star$ into the Lagrangian 13.21 and then solve the corresponding set of parameters $\Theta^o = \{\mathbf{W}_{hh}^0, \mathbf{W}_{xh}^0, \mathbf{W}_{hy}^0\}$ that minimizes $L(\Theta, \lambda)$ for this given $\lambda^\star$:

$$\Theta^o = \arg\min_\Theta L(\Theta, \lambda^\star) \tag{13.26}$$

Then, the obtained $\Theta^o = \{\mathbf{W}_{hh}^0, \mathbf{W}_{xh}^0, \mathbf{W}_{hy}^0\}$ will be an approximation to an optimal solution to the original constrained optimization problem. For convex optimization problems, this approximate solution will be the same global optimal solution under some mild conditions [8]. This property is called strong duality. However, in general nonconvex problems, it will not be the exact solution. But since finding the globally optimal solution to the original problem of 13.24 and 13.21 is not realistic, it would be satisfactory if it can provide a good approximation.

Now let us return to the problem of 13.24 and 13.21. We are indeed solving the following problem

$$\max_{\lambda \succeq 0} \min_\Theta L(\Theta, \lambda) \tag{13.27}$$

where the notation $\lambda \succeq 0$ denotes that each entry of the vector λ is greater than or equal to zero. The preceding analysis shows that in order to solve the problem, we need to first minimize the Lagrangian $L(\Theta, \lambda)$ with respect to Θ, while in the mean time, maximize the dual variable λ subjected to the constraint that $\lambda \succeq 0$. Therefore, as we will proceed next, updating the RNN parameters consists of two steps:

- *primal update*—minimization of $L(\Theta, \lambda^*)$ with respect to Θ, and
- *dual update*—maximization of $L(\Theta^*, \lambda)$ with respect to λ.

13.4.4.2 Primal-Dual Method Applied to RNN Learning: Primal Update

We may directly apply the gradient descent algorithm to the primal update rule to minimize the Lagrangian $L(\Theta, \lambda^*)$ with respect to Θ. However, it is better to exploit the structure in this objective function, which consists of two parts: $E(\Theta)$ that measures the prediction quality, and the part that penalizes the violation of the constraint expressed in 13.25. The latter part is a sum of many ℓ_1 regularization terms on the rows of the matrix $\mathbf{W}_{hh}$:

$$\sum_{j=1}^N |W_{ij}| = \|\mathbf{w}_i\|_1 \tag{13.28}$$

where $\mathbf{w}_i$ denotes the ith row vector of the matrix $\mathbf{W}_{hh}$. With this observation, the Lagrangian in 13.21 can be written in the equivalent form of

$$L(\Theta, \lambda) = E(\mathbf{W}_{hh}, \mathbf{W}_{xh}, \mathbf{W}_{hy}) + \sum_{i=1}^{N} \lambda_i (\|\mathbf{w}_i\|_1 - d) . \tag{13.29}$$

In order to minimize $L(\Theta, \lambda)$ with the structure above with respect to $\Theta = \{\mathbf{W}_{hh}, \mathbf{W}_{xh}, \mathbf{W}_{hy}\}$, we can use a technique similar to the one proposed in [2] to derive the following iterative soft-thresholding algorithm for the primal update of $\mathbf{W}_{hh}$:

$$\mathbf{W}_{hh}^{\{k\}} = S_{\lambda\mu_k} \left\{ \mathbf{W}_{hh}^{\{k-1\}} - \mu_k \frac{\partial E(\mathbf{W}_{hh}^{\{k-1\}}, \mathbf{W}_{xh}^{\{k-1\}}, \mathbf{W}_{hy}^{\{k-1\}})}{\partial \mathbf{W}_{hh}} \right\}, \tag{13.30}$$

where $S_{\lambda\mu_k}(\mathbf{X})$ denote a component-wise shrinkage (soft-thresholding) operator on a matrix $\mathbf{X}$, defined as

$$[S_{\lambda\mu_k}(\mathbf{X})]_{ij} = \begin{cases} X_{ij} - \lambda_i\mu_k & \mathbf{X}_{ij} \geq \lambda_i\mu_k \\ \mathbf{X}_{ij} + \lambda_i\mu_k & \mathbf{X}_{ij} \leq -\lambda_i\mu_k \\ 0 & \text{otherwise} \end{cases} \tag{13.31}$$

The above primal update 13.30 for $\mathbf{W}_{hh}$ is implemented by a standard stochastic gradient descent followed by a shrinkage operator. On the other hand, the primal updates for $\mathbf{W}_{xh}$ and $\mathbf{W}_{hy}$ will follow the standard stochastic gradient descent rule since there is no constraints on them. In order to accelerate the convergence of the algorithm, one can, for example, add momentum or use Nesterov method to replace the gradient descent steps, as was carried out in the experiments reported in [10, 21].

13.4.4.3 Primal-Dual Method Applied to RNN Learning: Dual Update

The dual update step is aimed to maximize $L(\Theta, \lambda)$ with respect to λ subject to the constraint that $\lambda \succeq 0$. To this end, we use the following rule of gradient ascent with projection, which increases the function value of $L(\Theta, \lambda)$ while enforcing the constraint:

$$\lambda_{i,k} = \left[\lambda_{i,k-1} + \mu_k \left(\|\mathbf{w}_{i,k-1}\|_1 - d \right) \right]_+ \tag{13.32}$$

where $[x]_+ = \max\{0, x\}$. Note that λ_i is a regularization factor in $L(\Theta, \lambda)$ that penalizes the violation of constraint for the ith row of $\mathbf{W}_{hh}$. The dual update can be interpreted as a rule to adjust the regularization factor in an adaptive manner. When the sum of the absolute values of the ith row of $\mathbf{W}_{hh}$ exceeds d, i.e., violating the constraint, the recursion 13.32 will increase the regularization factor $\lambda_{i,k}$ on the ith row in 13.21. On the other hand, if the constraint for a certain i is not violated, then the dual update 13.21 will decrease the value of the corresponding λ_i. The projection operator $[x]_+$ makes sure that once the regularization factor λ_i is decreased below zero, it will be set to zero and the constraint for the ith row in 13.25 will not be penalized in 13.21.

13.5 Recurrent Neural Networks Incorporating LSTM Cells

13.5.1 Motivations and Applications

The basic RNN described so far does not have sufficient structure in modeling sophisticated temporal dynamics, and thus in practice has been shown not capable of looking far back into the past in many types of input sequences. These problems were first carefully analyzed in the early work published in [53], and subsequently in [39, 40, 44], one solution is to impose a *memory* structure into the RNN, resulting in the so-called *long-short-term memory* cells in the RNN proposed first in the above early publications. Such LSTM version of the RNN was shown to overcome some fundamental problems of traditional RNNs, and be able to efficiently learn to solve a number of previously unlearnable tasks involving recognition of temporally extended patterns in noisy input sequences and of the temporal order of widely separated events in noisy input streams. Like the basic RNNs described so far in this chapter, the LSTM version can be shown to be an universal computing machine in the sense that given enough network units, it can compute anything a conventional computer can compute and if it has proper weight matrices. But unlike the basic RNNs, the LSTM version is better-suited to learn from input sequence data to classify, process, and predict time series when there are very long time lags of unknown lengths between important events.

Over the years since the early establishment of the LSTM-RNN, it has been shown to perform very well in many useful tasks such as handwriting recognition, phone recognition, keyword spotting, reinforcement learning for robot localization and control (in partially observable environments), online learning for protein structure prediction, learning music composition, and learning of grammars, etc. An overview of these progresses has been provided in [90]. Most recently, excellent results on large vocabulary speech recognition by the use of the LSTM-RNN have been reported in [88, 89]. A simpler version of the LSTM-RNN is also recently found effective for language identification [42], speech synthesis [36, 37], and for environment-robust speech recognition [38, 103]. In the latter study, it was shown that the LSTM architecture can be effectively used to exploit temporal context in learning the correspondences of noise-distorted and reverberant speech features and effectively handle highly nonstationary, convolutive noise involved in the task of 2013 Computational Hearing in Multisource Environments (CHiME) Challenge (track 2).

13.5.2 The Architecture of LSTM Cells

The basic idea behind the LSTM cell in the RNN is to use various types of gating (i.e., element-wise multiplication) structure to control the information flow in the network. An LSTM-RNN is an advanced version of the RNN that contains LSTM cells instead of, or in addition to, regular network units. An LSTM cell can be regarded as a complex and smart network unit capable of remembering information for a long length of time. This is accomplished by the gating structure that determines when the input is significant enough to remember, when it should continue to remember or forget the information, and when it should output the information.

Mathematically, a set of LSTM cells can be described by the following forward operations iteratively over time $t = 1, 2, ..., T$, following [39, 40, 44, 53, 88]:

$$\mathbf{i}_t = \sigma \left(\mathbf{W}^{(xi)} \mathbf{x}_t + \mathbf{W}^{(hi)} \mathbf{h}_{t-1} + \mathbf{W}^{(ci)} \mathbf{c}_{t-1} + \mathbf{b}^{(i)} \right) \tag{13.33}$$

$$\mathbf{f}_t = \sigma \left(\mathbf{W}^{(xf)} \mathbf{x}_t + \mathbf{W}^{(hf)} \mathbf{h}_{t-1} + \mathbf{W}^{(cf)} \mathbf{c}_{t-1} + \mathbf{b}^{(f)} \right) \tag{13.34}$$

$$\mathbf{c}_t = \mathbf{f}_t \bullet \mathbf{c}_{t-1} + \mathbf{i}_t \bullet \tanh \left(\mathbf{W}^{(xc)} \mathbf{x}_t + \mathbf{W}^{(hc)} \mathbf{h}_{t-1} + \mathbf{b}^{(c)} \right) \tag{13.35}$$

$$\mathbf{o}_t = \sigma \left(\mathbf{W}^{(xo)} \mathbf{x}_t + \mathbf{W}^{(ho)} \mathbf{h}_{t-1} + \mathbf{W}^{(co)} \mathbf{c}_t + \mathbf{b}^{(o)} \right) \tag{13.36}$$

$$\mathbf{h}_t = \mathbf{o}_t \bullet \tanh \left(\mathbf{c}_t \right), \tag{13.37}$$

where $\mathbf{i}_t$, $\mathbf{f}_t$, $\mathbf{c}_t$, $\mathbf{o}_t$, and $\mathbf{h}_t$ are vectors, all with the same dimensionality, which represent five different types of information at time t of the input gate, forget gate, cell activation, output gate, and hidden layer, respectively. $\sigma (.)$ is the logistic sigmoid function, $\mathbf{W}$'s are the weight matrices connecting different gates, and $\mathbf{b}$'s are the corresponding bias vectors. All the weight matrices are full except the weight matrix $\mathbf{W}^{(ci)}$ is diagonal. Note functionally, the above LSTM set, with input vector $\mathbf{i}_t$ and hidden vector $\mathbf{h}_t$ is akin to the input and hidden layers of the basic RNN as described by Eq. 13.1. An additional output layer need to be provided (not included in the above set of equations) on top of the LSTM-RNN's hidden layer. In [46], a straightforward linear mapping from LSTM-RNN's hidden layer to the output layer was exploited. In [88], two intermediate, linear projection layers were created to reduce the large dimensionality of the LSTM-RNN hidden layer's vector $\mathbf{h}_t$. They were then linearly combined to form the final output layer.

13.5.3 Training the LSTM-RNN

All LSTM-RNN parameters can be learned in a similar way to learning the basic RNN parameters using the BPTT as described in Sect. 13.3. That is, to minimize LSTM's total loss on a set of training sequences, stochastic gradient descent methods can be used to update each weight parameter where the error derivative with respect to the weight parameter can be computed by the BPTT. As we recall, one major difficulty associated with the BPTT for the basic RNN is that error gradients vanish exponentially quickly with the size of the time lag between important events, or the gradients explode equally fast, both making the learning ineffective unless heuristic rules are applied or principled constrained optimization is exploited as described in Sect. 13.4. With the use of LSTM cells that replaces the basic RNN's input-to-hidden layer mapping and hidden-to-hidden layer transition, this difficulty is largely overcome. The reason is that when error signals are backpropagated from the output, they become trapped in the memory portion of the LSTM cell. Therefore, meaningful error signals are continuously fed back to each of the gates until the RNN parameters are well trained. This makes the BPTT effective for training LSTM cells, which can remember information in the input sequence for a long time when such pattern is present in the input data and is needed to perform the required sequence-processing task.

Of course, due to the greater structural complexity of the LSTM cells than its counterpart in the basic RNN where the two nonlinear mappings (i.e., input-to-hidden and hidden-to-hidden) typically have no designed structure, gradient computation as required in the BPTT is expectedly more involved than that for the basic RNN. We refer readers to Chap. 14 on computational networks in this book for additional discussions on this topic.

13.6 Analyzing Recurrent Neural Networks—A Contrastive Approach

We devote this section to the important topic of analyzing the capabilities and limitations of the RNN, which is mathematically formulated as a state-space dynamic system model described in Sect. 13.2. We take a contrastive approach to the analysis by examining the similarities to and differences from the closely related state-space formulation of the hidden dynamic model (HDM) described in Sect. 3.7 of Chap. 3 on the HMM variants. The main goal of performing such a comparative analysis is to understand the respective strengths and weaknesses of these two types of speech models derived from quite distinct motivations yet sharing striking commonality in their mathematical formulation of the models. Based on this understanding, it is possible to construct potentially more effective RNN-like dynamic architectures as well as new learning paradigms to further advance the state of the art in acoustic modeling for ASR.

13.6.1 Direction of Information Flow: Top-Down versus Bottom-Up

The first aspect examined in our contrastive analysis between the basic RNN model and the HDM is the opposing direction in which information flows in these two types of models. In the HDM, the speech object is modeled by a generative process from the top linguistic symbol or label sequence to the bottom continuous-valued acoustic observation sequence, mediated by the intermediate latent dynamic process which is also continuous valued. That is, this top-down generative process starts with specification of the latent linguistic sequence at the top level. Then the label sequence generates the latent dynamic vector sequence, which in turn generates the visible acoustic sequence at the bottom level in the model hierarchy. In contrast, in bottom-up modeling paradigm as adopted by the RNN, information flow starts at the bottom level of acoustic observation, which activates the hidden layer of the RNN modeling temporal dynamics via the recurrent matrix. Then the output layer of the RNN computes the linguistic label or target sequence as a numerical-vectorial sequence at the top level of the model hierarchy. Since the top layer determines the speech-class distinction, this bottom-up processing approach adopted by the RNN can be appropriately called (direct) discriminative learning. See more detailed discussions on discriminative learning versus generative learning in [27]. Let us now further elaborate on the top-down versus bottom-up comparisons as connected to generative versus discriminative learning here.

13.6.1.1 The Hidden Dynamic Model as a Top-Down Generative Process

To facilitate the comparison with the RNN, let us rewrite the HDM state Eq. 3.51 and observation Eq. 3.52 both in Sect. 3.7.2 of Chap. 3, into the following form most compatible with the state-space formulation of the basic RNN described in Sect. 13.2 of this chapter:

$$\mathbf{h}_t = q(\mathbf{h}_{t-1}; \mathbf{W}_{l_t}, \Lambda_{l_t}) + \text{StateNoiseTerm} \tag{13.38}$$

$$\mathbf{x}_t = r(\mathbf{h}_t, \Omega_{l_t}) + \text{ObsNoiseTerm} \tag{13.39}$$

where $\mathbf{W}_{l_t}$ is the system matrix that shapes the (articulatory-like) state dynamics, which can be easily structured to follow physical constraint in speech production; e.g., [15, 16]. The parameter set Λ_{l_t} include those of phonetic targets as correlates with the components of the phonological units (e.g., symbolic articulatory features), which can also be interpreted as the input driving force derived from speech production's motor control to the articulatory state dynamics. Both sets of parameters, $\mathbf{W}_{l_t}$ and Λ_{l_t}, are dependent on the label l_t at time t with segmental properties; hence the model is also called a (segmental) switching dynamic system. The system matrix $\mathbf{W}_{l_t}$ is analogous to $\mathbf{W}_{hh}$ in the RNN. The parameter set Ω_{l_t}, on the other hand, governs the nonlinear mapping from the hidden (articulatory-like) states in speech production to continuous-valued acoustic features $\mathbf{x}_t$, which is the output of the HDM, on a frame-by-frame basis. In some early implementations, Ω_{l_t} took the form of shallow neural network weights [9, 28, 80, 97, 98]. In another implementation, Ω_{l_t} took the form of a set of matrices in a mixture of linear experts [63, 65].

The state equation in several previous implementations of the HDM of speech did not take nonlinear forms. Rather, the following linear form was used (e.g., [28, 63, 65]):

$$\mathbf{h}_t = \mathbf{W}_{hh}(l_t)\mathbf{h}_{t-1} + [\mathbf{I} - \mathbf{W}_{hh}(l_t)]\mathbf{t}_{l_t} + \text{StateNoiseTerm} \qquad (13.40)$$

which exhibits the target-directed property for articulatory-like dynamics. Here, the parameters $\mathbf{W}_{hh}$ are a function of the (phonetic) label l_t at a particular time frame t, and $\mathbf{t}_{l_t}$ is a mapping from the symbolic quantity l_t of a linguistic unit to the continuous-valued target vector with the segmental property. To make the comparisons easy with the RNN, let us keep the nonlinear form and remove both the state and observation noise, yielding the state-space generative model of

$$\mathbf{h}_t = q(\mathbf{h}_{t-1}; \mathbf{W}_{l_t}, \mathbf{t}_{l_t}) \qquad (13.41)$$
$$\mathbf{x}_t = r(\mathbf{h}_t, \Omega_{l_t}) \qquad (13.42)$$

13.6.1.2 The Recurrent Neural Net as a Bottom-Up Discriminative Classifier

Likewise, to facilitate the comparison with the HDM, let us rewrite the RNN's state and observation Eqs. 13.1 and 13.2 into a more general form:

$$\mathbf{h}_t = f(\mathbf{h}_{t-1}; \mathbf{W}_{hh}, \mathbf{W}_{xh,}, \mathbf{x}_t) \qquad (13.43)$$
$$\mathbf{y}_t = g(\mathbf{h}_t; \mathbf{W}_{hy}) \qquad (13.44)$$

where information flow starts from observation data $\mathbf{x}_t$, going to hidden vectors $\mathbf{h}_t$, and then going to the predicted target label vectors $\mathbf{y}_t$, often coded in a one-hot manner, in the bottom-up direction.

This contrasts with the HDM's corresponding state and observation Eqs. 13.41 and 13.42, which describe the information flow from the top-level label-indexed phonetic target vector $\mathbf{t}_{l_t}$ to hidden vectors $\mathbf{h}_t$ and then to observation data $\mathbf{x}_t$, where we clearly see top-down information flows, opposite to the RNN's bottom-up information flow.

To further examine other differences between the HDM and the RNN beyond the top-down versus bottom-up difference identified above, we can keep the same mathematical description of the RNN but exchange the variables of input $\mathbf{x}_t$ and output $\mathbf{y}_t$ in Eqs. 13.43 and 13.44. This yields

$$\mathbf{h}_t = f(\mathbf{h}_{t-1}; \mathbf{W}_{hh}, \mathbf{W}_{yh}, \mathbf{y}_t) \qquad (13.45)$$
$$\mathbf{x}_t = g(\mathbf{h}_t; \mathbf{W}_{hx}). \qquad (13.46)$$

After normalizing the RNN into the generative form of Eqs. 13.45 and 13.46 via input-output exchange, with the same direction of information flow as the HDM, we will analyze below the remaining contrasts between the RNN and HDM with respect to the different nature of the hidden-space representations (while keeping the same generative form of the models). We will then analyze other aspects of the contrast between them including different ways of exploiting model parameters.

13.6.2 The Nature of Representations: Localist or Distributed

Localist and distributed representations are important concepts in cognitive science as two distinct styles of information representation. In the localist representation, each neuron represents a single concept on a stand-alone basis. That is, localist units have their own meaning and interpretation, not so for the units in distributed representation. The latter pertains to an internal representation of concepts in such a way that they are modeled as being explained by the interactions of many hidden factors. A particular factor learned from configurations of other factors can often generalize well to new configurations, not so in localist representations.

Distributed representations, based on vectors consisting of many nonzero elements or units, naturally occur in a "connectionist" neural network, where a concept is represented by a pattern of activity across a number of many units and where at the same time a unit typically contributes to many concepts. One key advantage of such many-to-many correspondence is that they provide robustness in representing the internal structure of the data in terms of graceful degradation and damage resistance. Such robustness is enabled by redundant storage of information. Another advantage is that they facilitate automatic generalization of concepts and relations, thus enabling reasoning abilities. Further, distributed representations allow similar vectors to be associated with similar concepts and thus allow efficient use of representational resources. These attractive properties of distributed representations, however, come with a set of weaknesses—nonobviousness in interpreting the representations, difficulties with representing hierarchical structure, and inconvenience in representing variable-length sequences. Distributed representations are also not directly suitable for input and output to a network and some translation with localist representations are needed.

Local representations, on the other hand, have advantages of explicitness and ease of use; i.e., the explicit representation of the components of a task is simple and the design of representational schemes for structured objects is easy. But the weaknesses are many, including inefficiency for large sets of objects, highly redundant use of connections, and undesirable growth of units in networks which represent complex structure.

The HDM discussed above adopts "localist" representations for the symbolic linguistic units, and the RNN makes use of distributed representations. This can be seen directly from Eq. 13.41 for the HDM and from Eq. 13.45 for the RNN. In the former, symbolic linguistic units l_t as a function of time t are coded implicitly in a stand-alone fashion. The connection of symbolic linguistic units to continuous-valued vectors is made via a one-to-one mapping, denoted by $\mathbf{t}_{l_t}$ in Eq. 13.41, to the hidden dynamic's asymptotic "targets" denoted by vector $\mathbf{t}$. This type of mapping is common in phonetic-oriented phonology literature, and is called the "interface between phonology and phonetics" in a functional computational model of speech production [15]. Further, the HDM uses the linguistic labels that are represented in a localist manner to index separate sets of time-varying parameters $\mathbf{W}_{l_t}$ and Ω_{l_t}, leading to "switching" dynamics which considerably complicates the decoding computation. This kind of parameter specification isolates the parameter interactions across different linguistic labels, gaining the advantage of explicit interpretation of the model but losing on direct discrimination across linguistic labels.

In contrast, in the state equation of the RNN model shown in Eq. 13.45, the symbolic linguistic units are directly represented as vectors of $\mathbf{y}_t$ (one-hot or otherwise) as a function of time t. No mapping to separate continuous-valued "phonetic" vectors are needed. Even if the one-hot coding of $\mathbf{y}_t$ vectors is localist, the hidden state vector $\mathbf{h}$ provides a distributed representation and thus allows the model to store a lot of information about the past in a highly efficient manner. Importantly, there is no longer a notion of label-specific parameter sets of $\mathbf{W}_{l_t}$ and Ω_{l_t} as in the HDM. The weight parameters in the RNN are shared across all linguistic label classes. This enables direct discriminative learning for the RNN. In addition, the distributed representation used by the hidden layer of the RNN allows efficient and redundant storage of information, and has the capacity to automatically disentangle variation factors embedded in the data. However, as inherent in distributed representations discussed earlier, the RNN also carries with them the difficulty of interpreting the parameters and hidden states, and the difficulties of modeling structure and of exploiting explicit knowledge of articulatory dynamics into the model which we discuss next.

13.6.3 Interpretability: Inferring Latent Layers versus End-to-End Learning

An obvious strength of the localist representation as adopted by the HDM for modeling deep speech structure is that the model parameters and the latent-state variables are explainable and easy to diagnose. In fact, one main motivation of many of such models is that the knowledge of hierarchical structure in speech production in terms of articulatory and vocal tract resonance dynamics can be directly (but approximately with a clear sense of the degree of approximation) incorporated into the design of the models [9, 15, 25, 30, 76, 80, 105]. Practical benefits of using interpretable, localist representation of hidden state vectors include sensible ways of initializing the parameters to be learned—e.g., with extracted formants to initialize hidden variables composed of vocal tract resonances. Another obvious benefit is the ease with which one can diagnose and analyze errors during model implementation via examination of the inferred latent variables. Since localist representations, unlike their distributed counterpart, do not superimpose patterns for signaling the presence of different linguistic labels, the hidden state variables not only are explanatory but also unambiguous. Further, the interpretable nature of the models allows complex causal and structured relationships to be built into them, free from the difficulty of doing so that is associated with distributed representations. In fact, several versions of the HDM has been constructed with many layers in the hierarchical latent space, all with clear physical embodiment in human speech production; e.g., Chap. 2 in [18]. However, the complex structure makes it very difficult to do discriminative parameter learning. As a result, nearly all versions of HDMs have adopted maximum-likelihood learning or data fitting approaches. For example, the use of linear or nonlinear Kalman filtering (E-step of the EM algorithm) for learning the parameters in the generative state-space models has been applied to only maximum-likelihood estimates [93, 97].

In contrast, the learning algorithm of BPTT commonly used for end-to-end training of the RNN with distributed representations for the hidden states performs discriminative training by directly minimizing linguistic label prediction errors. It is straightforward to do so in the formulation of the learning objective because each element in the hidden state vector contributes to all linguistic labels due to the very nature of the distributed representation. It is very unnatural and difficult to do so in the generative HDM based on localist representations of the hidden states, where each state and the associated model parameters typically contribute to only one particular linguistic unit, which is used to index the set of model parameters in most generative models including HDMs and HMMs.

13.6.4 Parameterization: Parsimonious Conditionals versus Massive Weight Matrices

The next aspect of comparisons between the HDM and the RNN concerns very different ways to parameterize the HDM and the RNN—Characterization of conditional distributions using parsimonious sets of parameters in the conditionals or representing the complex mapping using largely nonstructured massive weight matrices. Due to the interpretable latent states in the HDM as well as the parameters associated with them, speech knowledge can be used in the design of the model, leaving the size of free parameters to be relatively small. For example, when vocal tract resonance vectors are used to represent the hidden dynamics, a dimension of eight appears to be sufficient to capture the prominent dynamic properties responsible for generating the observed acoustic feature sequences. Somewhat higher dimensionality, about 13, is needed with the use of the hidden dynamic vectors associated with the articulators' configuration in speech production. The use of such parsimonious parameter sets to characterize conditional distributions in the HDM, a special of the dynamic Bayesian network, which is sometimes called "small is good," is also facilitated by the localist representation of hidden state components and the related parameters that are connected or indexed to a specific linguistic unit. This contrasts with the distributed representation adopted by the RNN where both the hidden state vector elements and the connecting weights are shared across all linguistic units, thereby demanding many folds more model parameters.

The ability to use speech-domain knowledge to construct generative models with a parsimonious set of parameters is both a blessing and a curse. On the one hand, such knowledge has been usefully exploited to constrain the target-directed and smooth (i.e., nonoscillatory) hidden dynamics within each phone segment [63, 64], to represent the analytical relationship between the latent vocal tract resonance vector (both resonance frequencies and bandwidths) and cepstral coefficients [1, 20, 34], and to model both anticipatory and regressive types of coarticulation expressed in the latent space as a result of hidden dynamics [15, 33]. With the right prediction of time-varying trajectories in the hidden space and then causally in the observed acoustic space, powerful constraints can be placed in the model formulation to reduce overgeneration in the model space and to avoid unnecessarily large model capacity. On the other hand, the use of speech knowledge limits the growth of the model size as more data are made available in training. For example, when the dimensionality of the vocal tract resonance vectors goes beyond eight, many advantages of interpretable hidden vectors no longer hold. Since speech knowledge is necessarily incomplete, the constraints imposed on the model space may be outweighed by the opportunity lost with increasingly large amounts of training data and by the incomplete knowledge.

In contrast, the RNN usually do not use any speech knowledge to constrain the model space due to the inherent difficulty of interpreting the ambiguous hidden state represented in a distributed manner. As such, the RNN in principle has the freedom to use increasingly larger parameters in keeping with the growing size of the training data. Lack of constraints may cause the model to overgeneralize. This, together with the known difficulties of the various learning algorithms for the RNN as analyzed in [5, 78], has limited the progress of using RNNs in speech recognition for many years until recently. More recent progresses of applying RNNs to speech recognition have involved various methods of introducing constraints either in the model construction or in the implementation of learning algorithms. For example, in the studies reported in [45, 46, 88, 89], the RNN's hidden layers are designed with memory units based on a very clever LSTM structure. While strongly constraining possible variations of hidden layer activities, the LSTM-RNN nevertheless allows massive weight parameters to be used by simply increasing the total number of the memory units as well as the complexity in the LSTM cells. Separately, the RNN can also be constrained during the learning stage, where the size of the gradient computed by BPTT is limited by a threshold to avoid explosion as reported in [5, 70] or where the range of the permissible RNN parameters are constrained to be within what the "echo-state property" would allow as explored in [10] and we reviewed in Sect. 13.4 in this chapter.

The different ways of parameterizing the HDM and the RNN also lead to distinct styles of computation associated with both training and run-time of the two models. In particular, the computation of the RNN is structured to be highly regular, large matrix multiplications. This is well matched with the capability of GPUs, which is designed for commodity high-performance supercomputing with hardware optimizability. This computational advantage of the RNN and other neural network-based deep learning methods, unfortunately, is lacking in the HDM and in most other deep generative models.

13.6.5 Methods of Model Learning: Variational Inference versus Gradient Descent

The final aspect of contrastive analysis between the HDM and the RNN is on the very different methods in learning model parameters.[1]

The HDM is a deep and directed generative model, also known as deep, dynamic Bayesian network or belief network, with heavily loopy structure and with both discrete-valued and continuous-valued latent variables, and is hence highly intractable in inference and learning. Many empirically motivated approximate solutions have been explored both in model structure and in learning methods; see comprehensive reviews in [18, 31] and Chaps. 10 and 12 of [29]. The more principled approximate approach, called variational inference and learning [41, 57, 79], has also been explored in learning the HDM. In [19, 61], it was found that variational inference works surprisingly well in inferring or estimating the intermediate, continuous-valued latent variables, i.e., vocal tract resonances or formants. But it does not do so well in inferring or decoding the top-level, discrete-valued latent variables, i.e., the phonetic label sequences, both in terms of decoding accuracy and computational cost. More recent developments in variational inference, especially the exploitation of the DNN in implementing efficient exact sampling from the variational posterior [4, 54, 58, 59, 74], hold promise to overcome some earlier weaknesses of HDM inference and learning.

On the other hand, rather than a wide range of inference and learning algorithms available to deep generative models, the RNN, as with most other neural network models, usually exploits a single, unchallenged inference and learning algorithm of backpropagation with at most some not so drastic variants. To bridge the above completely different styles of learning with strengths derived from both, the RNN and HDM need to be re-parameterized so that model parameterization will become similar to each other. The preceding subsection on parameterization touched on this topic from the general computational and modeling viewpoint. Recent work in machine learning also discussed a way of transformation between Bayesian networks and neural networks [58, 59]. The main idea is that although the posterior required in the E-step of the variational learning is often hard to approximate well in many intractable Bayesian network (such as the HDM), one can exploit a powerful DNN with high capacity to drastically improve the degree of the approximation.

13.6.6 Recognition Accuracy Comparisons

Given the analysis on and comparisons presented so far in this section between the generative HDM using localist representations and the discriminative RNN using

---

[1] Most of the contrasts discussed here can be generalized to the differences in learning general deep generative models (i.e., those with latent variables) and in learning deep discriminative models with neural network architectures.

distributed representations, the respective strengths and weaknesses associated with these two types of models are apparent. Here, let us compile and compare the empirical performance of the HDM and the RNN in terms of speech recognition accuracy. For calibration reasons, we use the standard TIMIT phone recognition task for the comparison since no other common tasks have been used to assess both types of models in a consistent manner. It is worthwhile mentioning that both types of the dynamic models are more difficult to implement than other models in use for speech recognition such as the GMM-HMM and DNN-HMM. While the HDM has been evaluated on the large vocabulary tasks involving Switchboard databases [9, 62, 63, 65, 80], the RNN has been mainly evaluated on the TIMIT task [10, 21, 45, 46, 81], and most recently a nonstandard large task [88, 89].

One particular version of the HDM, called the hidden trajectory model, was developed and evaluated after careful design with approximations aimed to overcome various difficulties associated with localist representations as discussed earlier in this section [32, 34]. The main approximation involves using the finite impulse response filter to replace the infinite impulse response filter with recurrent structure as in the original state equation of the state-space formulation of the model. This version gives 75.2 % phone recognition accuracy as reported in [32], somewhat higher than 73.9 % obtained by the basic RNN (with very careful engineering) as reported in Table I (on page 303) of [81] and somewhat lower than 76.1 % obtained by an elaborated version of the RNN with LSTM memory units without stacking as reported in Table I (on page 4) of [46].[2] This comparison shows that the top-down generative HDM based on localist representation of the hidden state performs similarly to the bottom-up discriminative RNN based on distributed representation of the hidden state. This is understandable due to the pros and cons of these different types of models analyzed throughout this section.

13.7 Discussions

Deep hierarchical structure with multiple layers of hidden space in human speech is intrinsically connected to its dynamic character manifested in all levels of speech production and perception. The desire and an attempt to capitalize on a (superficial) understanding of this deep speech structure helped ignite the recent surge of interest in the deep learning approach to speech recognition and related applications [23, 24, 49, 106], and a more thorough understanding of the deep structure of speech dynamics and their representations is expected to further advance the research progress in speech modeling and in ASR. In Chap. 3 (Sect. 3.7 on HMM variants), we surveyed a series of deep as well as shallow generative models incorporating speech dynamics at various levels, including notably the HDM. In this chapter, we study the

---

[2] With less careful engineering, the basic RNN accepting inputs of raw speech features could only achieve 71.8 % accuracy as reported in [21] before using DNN-derived input features.

discriminative counterpart of the HDM, the RNN, both expressed mathematically in the state-space formalism. With detailed examinations of and comparisons between these two types of models, we focus on the top-down versus bottom-up information flows and localist versus distributed representations as their respective hallmarks and defining characteristics.

In the distributed representation adopted by the RNNs, we cannot interpret the meaning of activity on a single unit or neuron in isolation. Rather, the meaning of the activity on any particular unit depends on the activities of other units. Using distributed representations, multiple concepts (i.e., phonological/linguistic symbols) can be represented at the same time on the same set of neuronal units by superimposing their patterns together. The strengths of distributed representations used by the RNN include robustness, representational and mapping efficiency, and the embedding of symbols into continuous-valued vector spaces which enable the use of powerful gradient-based learning methods. The localist representation adopted by the generative HDM has very different properties. It offers very different advantages—easy to interpret, understand, diagnose, and easy to work with.

The interpretability of the generative HDM allows the design of appropriate structure in the dynamic system matrix that governs the articulatory-like dynamic behavior, where the constraint is imposed that no oscillatory dynamics occur within a regime of phone-like units.[3] It is much harder to develop and impose structural constraints in the discriminative RNN, where the hidden layer does not lend itself to physical interpretation. The LSTM structure described in Sect. 13.5 is a rare and interesting exception, motivated by very different considerations from those in structuring the HDM.

Both the HDM and the RNN are characterized by the use of dynamic recursion in the hidden space not directly observed. The temporal unfolding of these dynamic sequence models make the related architectures deep, with the depth being the length of the speech feature sequence to be modeled. In the HDM, the hidden state adopts the localist representation with explicit physical interpretation and the model parameters are indexed with respect to each separate linguistic/phonetic class in a parsimonious manner. In the RNN, the hidden state adopts the distributed representation where each unit in the hidden layer contributes to all linguistic classes via the shared use of regular and massive weight matrices.

The comprehensive comparisons between the RNN and the HDM conducted in Sect. 13.6 and summarized above are aimed to shed insights into the question of how to leverage the strengths of both types of models while overcoming their respective weaknesses. The integration of these two distinct types of generative and discriminative models may be done blindly as in the case of using the generative DBN to pretrain the discriminative DNN. However, much better strategies can be pursued a future research direction, given our sufficient understanding by now of the nature of the pros and cons associated with the two model types. As an example, one weakness

---

[3] For example, in [15], second-order dynamics with critical damping were used to incorporate such constraints.

associated with the discriminative RNN is that distributed representations are not suitable for providing direct input to the network. This difficulty has been circumvented in the preliminary work reported in [21] by first using the DNN to extract input features, which gains the advantages of distributed representations embedded in the hidden layers of the DNN. Then, the DNN-extracted features equipped with distributed representations of the data are fed into the subsequent RNN, producing dramatic improvement of phone recognition accuracy from 71.8 % to as high as 81.2 %. Other ways to overcome the difficulties associated with localist representations in the generative, deep, and dynamic model and those with distributed representations in the discriminative model counterpart are expected to also improve ASR performance. As a further example, given the strength of the localist representation in interpreting the hidden space of the model and thus in integrating domain knowledge, we can exploit the generative model and create new features extracted from the latent variables or even the generated visible variables that may be effectively combined with other features based on distributed representations. We expect more advanced deep learning architectures in the future for more effective ASR to be superior to the current best RNN discussed in this chapter in several ways—Realistic properties of speech dynamics (based on human speech production and perception) in the continuous, latent, articulatory-like space will be beneficially exploited in the integrated generative and discriminative deep architectures; the learning of such integrated architecture will have more than a simple pass of BPTT but rather several iterative steps of top-down followed by bottom-up, and left-to-right followed by right-to-left; and a variety of effective deep generative-discriminative learning algorithms generalizing, expanding, or integrating into the current BPTT, along the line of [14, 59, 74, 95, 104], will be developed for the integrated generative-discriminative model emulating the deep and dynamic process of human speech in a functionally effective and computationally efficient manner.

References

1. Bazzi, I., Acero, A., Deng, L.: An expectation-maximization approach for formant tracking using a parameter-free nonlinear predictor. In: Proceedings of the International Conference on Acoustics, Speech and Signal Processing (ICASSP) (2003)
2. Beck, A., Teboulle, M.: A fast iterative shrinkage-thresholding algorithm for linear inverse problems. SIAM J. Imaging Sci. **2**(1), 183–202 (2009)
3. Bengio, Y.: Practical recommendations for gradient-based training of deep architectures. In: Neural Networks. Tricks of the Trade, pp. 437–478. Springer (2012)
4. Bengio, Y.: Estimating or propagating gradients through stochastic neurons. CoRR (2013)
5. Bengio, Y., Boulanger, N., Pascanu, R.: Advances in optimizing recurrent networks. In: Proceedings of the International Conference on Acoustics, Speech and Signal Processing (ICASSP). Vancouver, Canada (2013)
6. Bengio, Y., Boulanger-Lewandowski, N., Pascanu, R.: Advances in optimizing recurrent networks. In: Proceeding of the International Conference on Acoustics, Speech and Signal Processing (ICASSP). Vancouver, Canada (2013)
7. Boden, M.: A guide to recurrent neural networks and backpropagation. Technical Report T2002:03, SICS (2002)

8. Boyd, S.P., Vandenberghe, L.: Convex Optimization. Cambridge University Press (2004)
9. Bridle, J., Deng, L., Picone, J., Richards, H., Ma, J., Kamm, T., Schuster, M., Pike, S., Reagan, R.: An investigation fo segmental hidden dynamic models of speech coarticulation for automatic speech recognition. Final Report for 1998 Workshop on Langauge Engineering, CLSP, Johns Hopkins (1998)
10. Chen, J., Deng, L.: A primal-dual method for training recurrent neural networks constrained by the echo-state property. In: Proceeding of the ICLR (2014)
11. Cho, K., van Merrienboer, B., Gulcehre, C., Bougares, F., Schwenk, H., Bengio, Y.: Learning phrase representations using rnn encoder-decoder for statistical machine translation. In: Conference on Empirical Methods in Natural Language Processing (EMNLP) (2014)
12. Dahl, G.E., Yu, D., Deng, L., Acero, A.: Large vocabulary continuous speech recognition with context-dependent DBN-HMMs. In: Proceeding of the International Conference on Acoustics, Speech and Signal Processing (ICASSP), pp. 4688–4691 (2011)
13. Dahl, G.E., Yu, D., Deng, L., Acero, A.: Context-dependent pre-trained deep neural networks for large-vocabulary speech recognition. IEEE Trans. Audio, Speech Lang. Process. **20**(1), 30–42 (2012)
14. Danilo Jimenez Rezende Shakir Mohamed, D.W.: Stochastic backpropagation and approximate inference in deep generative models. In: Proceedings of the International Conference on Machine Learning (ICML) (2014)
15. Deng, L.: A dynamic, feature-based approach to the interface between phonology and phonetics for speech modeling and recognition. Speech Commun. **24**(4), 299–323 (1998)
16. Deng, L.: Computational models for speech production. In: Computational Models of Speech Pattern Processing, pp. 199–213. Springer, New York (1999)
17. Deng, L.: Switching dynamic system models for speech articulation and acoustics. In: Mathematical Foundations of Speech and Language Processing, pp. 115–134. Springer, New York (2003)
18. Deng, L.: Dyamic Speech Models—Theory, Algorithm, and Applications. Morgan and Claypool (2006)
19. Deng, L., Attias, H., Lee, L., Acero, A.: Adaptive kalman smoothing for tracking vocal tract resonances using a continuous-valued hidden dynamic model. IEEE Trans. Audio, Speech Lang. Process. **15**, 13–23 (2007)
20. Deng, L., Bazzi, I., Acero, A.: Tracking vocal tract resonances using an analytical nonlinear predictor and a target-guided temporal constraint. In: Proceedings of the Annual Conference of International Speech Communication Association (INTERSPEECH) (2003)
21. Deng, L., Chen, J.: Sequence classification using high-level features extracted from deep neural networks. In: Proceedings of the International Conference on Acoustics, Speech and Signal Processing (ICASSP) (2014)
22. Deng, L., Hassanein, K., Elmasry, M.: Analysis of correlation structure for a neural predictive model with application to speech recognition. Neural Netw. **7**, 331–339 (1994)
23. Deng, L., Hinton, G., Kingsbury, B.: New types of deep neural network learning for speech recognition and related applications: An overview. In: Proceedings of the International Conference on Acoustics, Speech and Signal Processing (ICASSP). Vancouver, Canada (2013)
24. Deng, L., Hinton, G., Yu, D.: Deep learning for speech recognition and related applications. In: NIPS Workshop. Whistler, Canada (2009)
25. Deng, L., Lee, L., Attias, H., Acero, A.: Adaptive kalman filtering and smoothing for tracking vocal tract resonances using a continuous-valued hidden dynamic model. IEEE Trans. Audio, Speech Lang. Process. **15**(1), 13–23 (2007)
26. Deng, L., Li, J., Huang, J.T., Yao, K., Yu, D., Seide, F., Seltzer, M., Zweig, G., He, X., Williams, J., Gong, Y., Acero, A.: Recent advances in deep learning for speech research at microsoft. In: Proceedings of the International Conference on Acoustics, Speech and Signal Processing (ICASSP). Vancouver, Canada (2013)
27. Deng, L., Li, X.: Machine learning paradigms in speech recognition: An overview. IEEE Trans. Audio, Speech Lang. Process. **21**(5), 1060–1089 (2013)

28. Deng, L., Ma, J.: Spontaneous speech recognition using a statistical coarticulatory model for the hidden vocal-tract-resonance dynamics. J. Acoust. Soc. Am. **108**, 3036–3048 (2000)

29. Deng, L., O'Shaughnessy, D.: Speech Processing—A Dynamic and Optimization-Oriented Approach. Marcel Dekker Inc, NY (2003)

30. Deng, L., Ramsay, G., Sun, D.: Production models as a structural basis for automatic speech recognition. Speech Commun. **33**(2–3), 93–111 (1997)

31. Deng, L., Togneri, R.: Deep dynamic models for learning hidden representations of speech features. In: Speech and Audio Processing for Coding, Enhancement and Recognition. Springer (2014)

32. Deng, L., Yu, D.: Use of differential cepstra as acoustic features in hidden trajectory modelling for phonetic recognition. In: Proceedings of the International Conference on Acoustics, Speech and Signal Processing (ICASSP), pp. 445–448 (2007)

33. Deng, L., Yu, D., Acero, A.: A bidirectional target filtering model of speech coarticulation: two-stage implementation for phonetic recognition. IEEE Trans. Speech Audio Process. **14**, 256–265 (2006)

34. Deng, L., Yu, D., Acero, A.: Structured speech modeling. IEEE Trans. Speech Audio Process. **14**, 1492–1504 (2006)

35. Divenyi, P., Greenberg, S., Meyer, G.: Dynamics of Speech Production and Perception. IOS Press (2006)

36. Fan, Y., Qian, Y., Xie, F., Soong, F.K.: TTS synthesis with bidirectional lstm based recurrent neural networks. In: Proceedings of the Annual Conference of International Speech Communication Association (INTERSPEECH) (2014)

37. Fernandez, R., Rendel, A., Ramabhadran, B., Hoory, R.: Prosody contour prediction with long short-term memory, bi-directional, deep recurrent neural networks. In: Proceedings of the Annual Conference of International Speech Communication Association (INTERSPEECH) (2014)

38. Geiger, J., Zhang, Z., Weninger, F., Schuller, B., Rigoll, G.: Robust speech recognition using long short-term memory recurrent neural networks for hybrid acoustic modelling. In: Proceedings of the Annual Conference of International Speech Communication Association (INTERSPEECH) (2014)

39. Gers, F., Schmidhuber, J., Cummins, F.: Learning to forget: continual prediction with lstm. Neural Comput. **12**, 2451–2471 (2000)

40. Gers, F., Schraudolph, N., Schmidhuber, J.: Learning precise timing with lstm recurrent networks. J. Mach. Learn. Res. **3**, 115–143 (2002)

41. Ghahramani, Z., Hinton, G.E.: Variational learning for switching state-space models. Neural Comput. **12**, 831–864 (2000)

42. Gonzalez, J., Lopez-Moreno, I., Sak, H., Gonzalez-Rodriguez, J., Moreno, P.: Automatic language identification using long short-term memory recurrent neural networks. In: Proceedings of the Annual Conference of International Speech Communication Association (INTERSPEECH) (2014)

43. Graves, A.: Sequence transduction with recurrent neural networks. In: ICML Representation Learning Workshop (2012)

44. Graves, A.: Generating sequences with recurrent neural networks. arXvi preprint. arXiv:1308.0850 (2013)

45. Graves, A., Jaitly, N., Mahamed, A.: Hybrid speech recognition with deep bidirectional lstm. In: Proceeding of the International Conference on Acoustics, Speech and Signal Processing (ICASSP). Vancouver, Canada (2013)

46. Graves, A., Mahamed, A., Hinton, G.: Speech recognition with deep recurrent neural networks. In: Proceedings of the International Conference on Acoustics, Speech and Signal Processing (ICASSP). Vancouver, Canada (2013)

47. Heigold, G., Vanhoucke, V., Senior, A., Nguyen, P., Ranzato, M., Devin, M., Dean, J.: Multilingual acoustic models using distributed deep neural networks. In: Proceedings of the International Conference on Acoustics, Speech and Signal Processing (ICASSP) (2013)

48. Hermans, M., Schrauwen, B.: Training and analysing deep recurrent neural networks. In: Proceedings of the Neural Information Processing Systems (NIPS) (2013)
49. Hinton, G., Deng, L., Yu, D., Dahl, G., Mohamed, A., Jaitly, N., Senior, A., Vanhoucke, V., Nguyen, P., Sainath, T., Kingsbury, B.: Deep neural networks for acoustic modeling in speech recognition. IEEE Signal Process. Mag. **29**(6), 82–97 (2012)
50. Hinton, G., Deng, L., Yu, D., Dahl, G.E.: Mohamed, A.r., Jaitly, N., Senior, A., Vanhoucke, V., Nguyen, P., Sainath, T.N., et al.: Deep neural networks for acoustic modeling in speech recognition: The shared views of four research groups. IEEE Signal Process. Mag. **29**(6), 82–97 (2012)
51. Hinton, G., Osindero, S., Teh, Y.: A fast learning algorithm for deep belief nets. Neural Comput. **18**, 1527–1554 (2006)
52. Hinton, G., Salakhutdinov, R.: Reducing the dimensionality of data with neural networks. Science **313**(5786), 504–507 (2006)
53. Hochreiter, S., Schmidhuber, J.: Long short-term memory. Neural Comput. **9**(8), 1735–1780 (1997)
54. Hoffman, M.D., Blei, D.M., Wang, C., Paisley, J.: Stochastic variational inference
55. Jaeger, H.: Short term memory in echo state networks. GMD Report 152,GMD—German National Research Institute for Computer Science (2001)
56. Jaeger, H.: Tutorial on training recurrent neural networks, covering BPPT, RTRL, EKF and the "echo state network" approach. GMD Report 159, GMD—German National Research Institute for Computer Science (2002)
57. Jordan, M., Sudderth, E., Wainwright, M., Wilsky, A.: Major advances and emerging developments of graphical models, special issue. IEEE Signal Process. Mag. **27**(6), 17,138 (2010)
58. Kingma, D., Welling, M.: Auto-encoding variational bayes. In: arXiv:1312.6114v10 (2014)
59. Kingma, D., Welling, M.: Efficient gradient-based inference through transformations between bayes nets and neural nets. In: Proceedings of the International Conference on Machine Learning (ICML) (2014)
60. Kingsbury, B., Sainath, T.N., Soltau, H.: Scalable minimum bayes risk training of deep neural network acoustic models using distributed hessian-free optimization. In: Proceedings of the Annual Conference of International Speech Communication Association (INTERSPEECH) (2012)
61. Lee, L., Attias, H., Deng, L.: Variational inference and learning for segmental switching state space models of hidden speech dynamics. In: Proceedings of the International Conference on Acoustics, Speech and Signal Processing (ICASSP), vol. 1, pp. I-872–I-875 (2003)
62. Ma, J., Deng, L.: A path-stack algorithm for optimizing dynamic regimes in a statistical hidden dynamic model of speech. Comput. Speech Lang. **14**, 101–104 (2000)
63. Ma, J., Deng, L.: Efficient decoding strategies for conversational speech recognition using a constrained nonlinear state-space model. IEEE Trans. Audio Speech Process. **11**(6), 590–602 (2003)
64. Ma, J., Deng, L.: Efficient decoding strategies for conversational speech recognition using a constrained nonlinear state-space model. IEEE Trans. Audio, Speech Lang. Process. **11**(6), 590–602 (2004)
65. Ma, J., Deng, L.: Target-directed mixture dynamic models for spontaneous speech recognition. IEEE Trans. Audio Speech Process. **12**(1), 47–58 (2004)
66. Maas, A.L., Le, Q., O'Neil, T.M., Vinyals, O., Nguyen, P., Ng, A.Y.: Recurrent neural networks for noise reduction in robust asr. In: Proceedings of the Annual Conference of International Speech Communication Association (INTERSPEECH). Portland, OR (2012)
67. Mesnil, G., He, X., Deng, L., Bengio, Y.: Investigation of recurrent-neural-network architectures and learning methods for spoken language understanding. In: Proceedings of the Annual Conference of International Speech Communication Association (INTERSPEECH). Lyon, France (2013)
68. Mikolov, T.: Rnntoolkit http://www.fit.vutbr.cz/imikolov/rnnlm/ (2012). http://www.fit.vutbr.cz/~imikolov/rnnlm/

69. Mikolov, T.: Statistical Language Models Based on Neural Networks. Ph.D. thesis, Brno University of Technology (2012)
70. Mikolov, T., Deoras, A., Povey, D., Burget, L., Cernocky, J.: Strategies for training large scale neural network language models. In: Proceedings of the IEEE Workshop on Automfatic Speech Recognition and Understanding (ASRU), pp. 196–201. IEEE, Honolulu, HI (2011)
71. Mikolov, T., Karafiát, M., Burget, L., Cernockỳ, J., Khudanpur, S.: Recurrent neural network based language model. In: Proceedings of the Annual Conference of International Speech Communication Association (INTERSPEECH), pp. 1045–1048. Makuhari, Japan (2010)
72. Mikolov, T., Kombrink, S., Burget, L., Cernocky, J., Khudanpur, S.: Extensions of recurrent neural network language model. In: Proceedings of the International Conference on Acoustics, Speech and Signal Processing (ICASSP), pp. 5528–5531. Prague, Czech (2011)
73. Mikolov, T., Zweig, G.: Context dependent recurrent neural network language model. In: Proceedings of the IEEE Spoken Language Technology Workshop (SLT), pp. 234–239 (2012)
74. Mnih, A., Gregor, K.: Neural variational inference and learning in belief networks. In: Proceedings of the International Conference on Machine Learning (ICML) (2014)
75. Mohamed, A.r., Dahl, G.E., Hinton, G.: Deep belief networks for phone recognition. In: NIPS Workshop on Deep Learning for Speech Recognition and Related Applications (2009)
76. Ozkan, E., Ozbek, I., Demirekler, M.: Dynamic speech spectrum representation and tracking variable number of vocal tract resonance frequencies with time-varying dirichlet process mixture models. IEEE Trans. Audio, Speech Lang. Process. **17**(8), 1518–1532 (2009)
77. Pascanu, R., Gulcehre, C., Cho, K., Bengio, Y.: How to construct deep recurrent neural networks. In: The 2nd International Conference on Learning Representation (ICLR) (2014)
78. Pascanu, R., Mikolov, T., Bengio, Y.: On the difficulty of training recurrent neural networks. In: Proceedings of the International Conference on Machine Learning (ICML). Atlanta, GA (2013)
79. Pavlovic, V., Frey, B., Huang, T.: Variational learning in mixed-state dynamic graphical models. In: UAI, pp. 522–530. Stockholm (1999)
80. Picone, J., Pike, S., Regan, R., Kamm, T., bridle, J., Deng, L., Ma, Z., Richards, H., Schuster, M.: Initial evaluation of hidden dynamic models on conversational speech. In: Proceedings of the International Conference on Acoustics, Speech and Signal Processing (ICASSP) (1999)
81. Robinson, A.J.: An application of recurrent nets to phone probability estimation. IEEE Trans. Neural Netw. **5**(2), 298–305 (1994)
82. Robinson, A.J., Cook, G., Ellis, D.P., Fosler-Lussier, E., Renals, S., Williams, D.: Connectionist speech recognition of broadcast news. Speech Commun. **37**(1), 27–45 (2002)
83. Rumelhart, D.E., Hintont, G.E., Williams, R.J.: Learning representations by back-propagating errors. Nature **323**(6088), 533–536 (1986)
84. Sainath, T., Kingsbury, B., Soltau, H., Ramabhadran, B.: Optimization techniques to improve training speed of deep neural networks for large speech tasks. IEEE Trans. Audio, Speech, Lang. Process. **21**(11), 2267–2276 (2013)
85. Sainath, T.N., Kingsbury, B., Mohamed, A.r., Dahl, G.E., Saon, G., Soltau, H., Beran, T., Aravkin, A.Y., Ramabhadran, B.: Improvements to deep convolutional neural networks for lvcsr. In: Proceedings of the IEEE Workshop on Automfatic Speech Recognition and Understanding (ASRU), pp. 315–320 (2013)
86. Sainath, T.N., Kingsbury, B., Mohamed, A.r., Ramabhadran, B.: Learning filter banks within a deep neural network framework. In: Proceedings of the IEEE Workshop on Automfatic Speech Recognition and Understanding (ASRU) (2013)
87. Sainath, T.N., Kingsbury, B., Sindhwani, V., Arisoy, E., Ramabhadran, B.: Low-rank matrix factorization for deep neural network training with high-dimensional output targets. In: Proceedings of the International Conference on Acoustics, Speech and Signal Processing (ICASSP), pp. 6655–6659 (2013)
88. Sak, H., Senior, A., Beaufays, F.: Long short-term memory recurrent neural network architectures for large scale acoustic modeling. In: Proceedings of the Annual Conference of International Speech Communication Association (INTERSPEECH) (2014)

89. Sak, H., Vinyals, O., Heigold, G., Senior, A., McDermott, E., Monga, R., Mao, M.: Sequence discriminative distributed training of long short-term memory recurrent neural networks. In: Proceedings of the Annual Conference of International Speech Communication Association (INTERSPEECH) (2014)
90. Schmidhuber, J.: Deep learning in neural networks: an overview. CoRR abs/1404.7828 (2014)
91. Seide, F., Fu, H., Droppo, J., Li, G., Yu, D.: On parallelizability of stochastic gradient descent for speech dnns. In: Proceedings of the International Conference on Acoustics, Speech and Signal Processing (ICASSP) (2014)
92. Seide, F., Li, G., Yu, D.: Conversational speech transcription using context-dependent deep neural networks. In: Proceedings of the Annual Conference of International Speech Communication Association (INTERSPEECH), pp. 437–440 (2011)
93. Shen, X., Deng, L.: Maximum likelihood in statistical estimation of dynamical systems: Decomposition algorithm and simulation results. Signal Process. **57**, 65–79 (1997)
94. Stevens, K.: Acoustic Phonetics. MIT Press (2000)
95. Stoyanov, V., Ropson, A., Eisner, J.: Empirical risk minimization of graphical model parameters given approximate inference, decoding, and model structure. In: Proceedings of the International Conference on Artificial Intelligence and Statistics (AISTATS) (2011)
96. Sutskever, I.: Training Recurrent Neural Networks. Ph.D. thesis, University of Toronto (2013)
97. Togneri, R., Deng, L.: Joint state and parameter estimation for a target-directed nonlinear dynamic system model. IEEE Trans. Signal Process. **51**(12), 3061–3070 (2003)
98. Togneri, R., Deng, L.: A state-space model with neural-network prediction for recovering vocal tract resonances in fluent speech from mel-cepstral coefficients. Speech Commun. **48**(8), 971–988 (2006)
99. Triefenbach, F., Jalalvand, A., Demuynck, K., Martens, J.P.: Acoustic modeling with hierarchical reservoirs. IEEE Trans. Audio, Speech, Lang. Process. **21**(11), 2439–2450 (2013)
100. Vanhoucke, V., Devin, M., Heigold, G.: Multiframe deep neural networks for acoustic modeling
101. Vanhoucke, V., Senior, A., Mao, M.Z.: Improving the speed of neural networks on CPUs. In: Proceedings of the NIPS Workshop on Deep Learning and Unsupervised Feature Learning (2011)
102. Waibel, A., Hanazawa, T., Hinton, G., Shikano, K., Lang, K.J.: Phoneme recognition using time-delay neural networks. IEEE Trans. Speech Audio Process. **37**(3), 328–339 (1989)
103. Weninger, F., Geiger, J., Wollmer, M., Schuller, B., Rigoll, G.: Feature enhancement by deep lstm networks for ASR in reverberant multisource environments. Comput. Speech and Lang. 888–902 (2014)
104. Xing, E., Jordan, M., Russell, S.: A generalized mean field algorithm for variational inference in exponential families. In: Proceedings of the Uncertainty in Artificial Intelligence (2003)
105. Yu, D., Deng, L.: Speaker-adaptive learning of resonance targets in a hidden trajectory model of speech coarticulation. Comput. Speech Lang. **27**, 72–87 (2007)
106. Yu, D., Deng, L.: Deep-structured hidden conditional random fields for phonetic recognition. In: Proceedings of the International Conference on Acoustics, Speech and Signal Processing (ICASSP) (2010)
107. Yu, D., Deng, L., Acero, A.: A lattice search technique for a long-contextual-span hidden trajectory model of speech. Speech Commun. **48**, 1214–1226 (2006)
108. Yu, D., Deng, L., Dahl, G.: Roles of pre-training and fine-tuning in context-dependent DBN-HMMs for real-world speech recognition. In: Proceedings of the Neural Information Processing Systems (NIPS) Workshop on Deep Learning and Unsupervised Feature Learning (2010)

Chapter 14
Computational Network

Abstract In the previous chapters, we have discussed various deep learning models for automatic speech recognition (ASR). In this chapter, we introduce computational network (CN), a unified framework for describing arbitrary learning machines, such as deep neural networks (DNNs), computational neural networks (CNNs), recurrent neural networks (RNNs), long short term memory (LSTM), logistic regression, and matrixum entropy model, that can be illustrated as a series of computational steps. A CN is a directed graph in which each leaf node represents an input value or a parameter and each nonleaf node represents a matrix operation upon its children. We describe algorithms to carry out forward computation and gradient calculation in CN and introduce most popular computation node types used in a typical CN.

14.1 Computational Network

There is a common property in key machine learning models, such as deep neural networks (DNNs) [7, 13, 16, 23, 24, 31], convolutional neural networks (CNNs) [1, 2, 5, 6, 8, 17–19, 21, 22], and recurrent neural networks (RNNs) [14, 20, 25–27]. All these models can be described as a series of computational steps. For example, a one-hidden-layer sigmoid neural network can be described as the computational steps listed in Algorithm 14.1. If we know how to compute each step and in which order the steps are computed, we have an implementation of the neural network. This observation suggests that we can unify all these models under the framework of computational network (CN), part of which has been implemented in toolkits such as Theano [3], CNTK [12] and RASR/NN [29].

A computational network is a directed graph $\{\mathbb{V}, \mathbb{E}\}$, where $\mathbb{V}$ is a set of vertices and $\mathbb{E}$ is a set of directed edges. Each vertex, called a computation node, represents a computation. Vertices with edges toward a computation node are the operands of the associated computation and sometimes called the children of the computation node. Here, the order of operands matters for some operations such as matrix multiplication. Leaf nodes do not have children and are used to represent input values or model

This chapter has been published as part of the CNTK document [30].

Algorithm 14.1 Computational Steps Involved in an One-Hidden-Layer Sigmoid Neural Network.

1: **procedure** ONEHIDDENLAYERNNCOMPUTATION($\mathbf{X}$)

               ▷ Each column of $\mathbf{X}$ is an observation vector

2:    $\mathbf{T}^{(1)} \leftarrow \mathbf{W}^{(1)}\mathbf{X}$

3:    $\mathbf{P}^{(1)} \leftarrow \mathbf{T}^{(1)} + \mathbf{B}^{(1)}$ ▷ Each column of $\mathbf{B}^{(1)}$ is the bias $\mathbf{b}^{(1)}$

4:    $\mathbf{S}^{(1)} \leftarrow \sigma\left(\mathbf{P}^{(1)}\right)$ ▷ σ (.) is the sigmoid function applied element-wise

5:    $\mathbf{T}^{(2)} \leftarrow \mathbf{W}^{(2)}\mathbf{S}^{(1)}$

6:    $\mathbf{P}^{(2)} \leftarrow \mathbf{T}^{(2)} + \mathbf{B}^{(2)}$ ▷ Each column of $\mathbf{B}^{(2)}$ is the bias $\mathbf{b}^{(2)}$

7:    $\mathbf{O} \leftarrow$ softmax $\left(\mathbf{P}^{(2)}\right)$ ▷ Apply softmax column-wise to get output $\mathbf{O}$

8: **end procedure**

---

parameters that are not result of some computation. A CN can be easily represented as a set of computation nodes n and their children $\{n : c_1, \ldots, c_{K_n}\}$, where K_n is the number of children of node n. For leaf nodes $K_n = 0$. Each computation node knows how to compute the value of itself given the input operands (children).

Figure 14.1 is the one-hidden-layer sigmoid neural network of Algorithm 14.1 represented as a CN in which each node n is identified by a {*nodename:operatortype*} pair and takes its ordered children as the operator's inputs. From the figure, we can observe that in CN, there is no concept of layers. Instead, a computation node is the basic element of operations. This makes the description of a simple model such as DNN more cumbersome, but this can be alleviated by grouping computation nodes together with macros. In return, CN provides us with greater flexibility in describing arbitrary networks and allows us to build almost all models we are interested in within the same unified framework. For example, we can easily modify the network illustrated in Fig. 14.1 to use rectified linear unit instead of a sigmoid nonlinearity. We can also build a network that has two input nodes as shown in Fig. 14.2 or a network with shared model parameters as shown in Fig. 14.3.

Fig. 14.1 Represent the one-hidden-layer sigmoid neural network of Algorithm 14.1 with a computational network

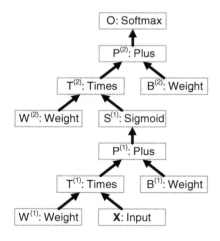

Fig. 14.2 A computational network with two input nodes (*highlighted*)

Fig. 14.3 A computational network with shared model parameters (*highlighted*). Here, we assume the input **X**, hidden layer $\mathbf{S}^{(1)}$, and output layer **O** have the same dimension

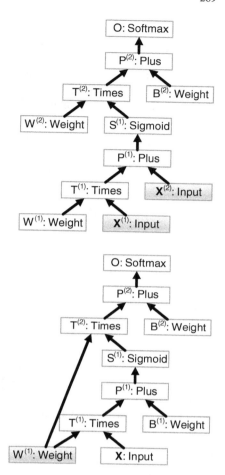

14.2 Forward Computation

When the model parameters (i.e., weight nodes in Fig. 14.1) are known, we can compute the value of any node given the new input values. Unlike in the DNN case, where the computation order can be trivially determined as layer-by-layer computation from bottom up, in CN different network structure comes with a different computation order. When the CN is a directed acyclic graph (DAG), the computation order can be determined with a depth-first traverse over the DAG. Note that in a DAG, there is no directed cycles (i.e., no recurrent loop). However, there might be loops if we do not consider edge directions, for which Fig. 14.3 is an example. This is because the same computation node may be a child of several other nodes. Algorithm 14.2 determines the computation order of a DAG and takes care of this condition. Once the order is decided, it will remain the same for all the subsequent runs, regardless of the computational environment. In other words, this algorithm only needs to be executed per output node and then cache the computation order. Following the order determined by Algorithm 14.2, the forward computation of the CN is carried out synchronously. The computation of the next node starts only after the computation of the previous node has finished. It is suitable for environments where single computing device, such as one GPGPU or one CPU host, is used, or the CN itself is inherently sequential, e.g., when the CN represents a DNN.

Algorithm 14.2 Synchronous forward computation of a CN. The computation order is determined by a depth-first traverse over the DAG.

1: **procedure** DECIDEFORWARDCOMPUTATIONORDER($root$, $visited$, $order$)
                                         ▷ Enumerate nodes in the DAG in the depth-first order.
                 ▷ $visited$ is initialized as an empty set. $order$ is initialized as an empty queue
2:    **if** $root \notin visited$ **then**   ▷ the same node may be a child of several nodes.
3:       $visited \leftarrow visited \cup root$
4:       **for** each $c \in root.children$ **do**   ▷ apply to children recursively
5:          call DECIDEFORWARDCOMPUTATIONORDER(c, $visited$, $order$)
6:          $order \leftarrow order + root$ ▷ Add $root$ to the end of $order$
7:       **end for**
8:    **end if**
9: **end procedure**

The forward computation can also be carried out asynchronously with which the order of the computation is determined dynamically. This can be helpful when the CN has many parallel branches and there are more than one computing device to compute these branches in parallel. Algorithm 14.3 shows an algorithm that carries out the forward computation of a CN asynchronously. In this algorithm, all the nodes whose children have not been computed are in the waiting set and those whose children are computed are in the ready set. At the beginning, all nonleaf descendents of *root* are in the waiting set and all leaf descendents are in the ready set. The scheduler picks a node from the ready set based on some policy, removes it from the ready set, and dispatches it for computation. Popular policies include first-come/first-serve, shortest task first, and least data movement. When the computation of the node finishes, the system calls the SIGNALCOMPLETE method to signal to all its parent nodes. If all the children of a node have been computed, the node is moved from the waiting set to the ready set. The algorithm stops when all nodes are computed. Although not explicitly indicated in the Algorithm 14.3, the SIGNALCOMPLETE procedure is called under concurrent threads and should be guarded for thread safety. This algorithm can be exploited to carry out computation on any DAG instead of just a CN.

In many cases, we may need to compute the value for a node with changing input values. To avoid duplicate computation of shared branches, we can add a time stamp to each node and only recompute the value of the node if at least one of the children has newer value. This can be easily implemented by updating the time stamp whenever a new value is provided or computed, and by excluding nodes whose children is older from the actual computation.

Algorithm 14.3 Asynchronous forward computation of a CN. A node is moved to the ready set when all its children have been computed. A scheduler monitors the ready set and decides where to compute each node in the set.

1: **procedure** SIGNALCOMPLETE(*node*, *waiting*, *ready*)

▷ Called when the computation of the *node* is finished. Needs to be thread safe.

▷ *waiting* is initialized to include all nonleaf descendents of *root*.

▷ *ready* is initialized to include all leaf descendents of *root*.

2:     **for** each $p \in node.parents \land p \in waiting$ **do**

3:         $p.numFinishedChildren + +$

4:         **if** $p.numFinishedChildren == p.numChildren$ **then**

▷ all children have been computed

5:             $waiting \leftarrow waiting - node$

6:             $ready \leftarrow ready \cup node$

7:         **end if**

8:     **end for**

9: **end procedure**

10: **procedure** SCHEDULECOMPUTATION(*ready*)

▷ Called by the job scheduler when a new node is ready or computation resource is available.

11:     pick $node \in ready$ according to some policy

12:     $ready \leftarrow ready - node$

13:     dispatch *node* for computation.

14: **end procedure**

In both Algorithms 14.2 and 14.3 each computation node needs to know how to compute its value when the operands are known. The computation can be as simple as matrix summation or element-wise application of sigmoid function or as complex as whatever it may be. We will describe the evaluation functions for popular computation node types in Sect. 14.4.

14.3 Model Training

To train a CN, we need to define a training criterion J. Popular criteria include cross-entropy (CE) for classification and mean square error (MSE) for regression as have been discussed in Chap. 4. Since the training criterion is also a result of some computation, it can be represented as a computation node and inserted into the CN. Figure 14.4 illustrates a CN that represents a one-hidden-layer sigmoid neural network augmented with a CE training criterion node. If the training criterion contains regularization terms, the regularization terms can also be implemented as computation nodes and the final training criterion node is a weighted sum of the main criterion and the regularization term.

The model parameters in a CN can be optimized over a training set $\mathbb{S} = \{(\mathbf{x}^m, \mathbf{y}^m) | 0 \leq m < M\}$ using the minibatch based backpropagation (BP) algorithm similar to that described in Algorithm 4.2. More specifically, we improve

Fig. 14.4 The
one-hidden-layer sigmoid
neural network augmented
with a cross-entropy training
criterion node J and a label
node L

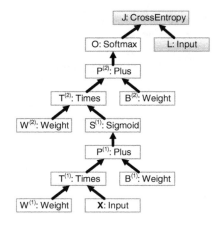

the model parameter $\mathbf{W}$ at each step $t + 1$ as

$$\mathbf{W}_{t+1} \leftarrow \mathbf{W}_{\mathrm{t}} - \varepsilon \triangle \mathbf{W}_{t}, \tag{14.1}$$

where

$$\triangle \mathbf{W}_{t} = \frac{1}{M_{b}} \sum_{m=1}^{M_{b}} \nabla_{\mathbf{W}_{t}} J \left(\mathbf{W}; \mathbf{x}^{m}, \mathbf{y}^{m} \right), \tag{14.2}$$

and M_b is the minibatch size. The key here is the computation of $\nabla_{\mathbf{W}_t} J \left(\mathbf{W}; \mathbf{x}^m, \mathbf{y}^m \right)$ which we will simplify as $\nabla_{\mathbf{W}}^{J}$. Since a CN can have arbitrary structure, we cannot use the exact same BP algorithm described in Algorithm 4.2 to compute $\nabla_{\mathbf{W}}^{J}$.

A naive solution to compute $\nabla_{\mathbf{W}}^{J}$ is illustrated in Fig. 14.5, in which $\mathbf{W}^{(1)}$ and $\mathbf{W}^{(2)}$ are model parameters. In this solution, each edge is associated with a partial derivative, and

$$\nabla_{\mathbf{W}^{(1)}}^{J} = \frac{\partial J}{\partial \mathbf{V}^{(1)}} \frac{\partial \mathbf{V}^{(1)}}{\partial \mathbf{V}^{(2)}} \frac{\partial \mathbf{V}^{(2)}}{\partial \mathbf{W}^{(1)}} + \frac{\partial J}{\partial \mathbf{V}^{(3)}} \frac{\partial \mathbf{V}^{(3)}}{\partial \mathbf{V}^{(4)}} \frac{\partial \mathbf{V}^{(4)}}{\partial \mathbf{W}^{(1)}} \tag{14.3}$$

$$\nabla_{\mathbf{W}^{(2)}}^{J} = \frac{\partial J}{\partial \mathbf{V}^{(1)}} \frac{\partial \mathbf{V}^{(1)}}{\partial \mathbf{V}^{(2)}} \frac{\partial \mathbf{V}^{(2)}}{\partial \mathbf{W}^{(2)}}. \tag{14.4}$$

This solution comes with two major drawbacks. First, each derivative can have very high dimension. If $\mathbf{V} \in \mathbb{R}^{N_1 \times N_2}$ and $\mathbf{W} \in \mathbb{R}^{N_3 \times N_4}$ then $\frac{\partial \mathbf{V}}{\partial \mathbf{W}} \in \mathbb{R}^{(N_1 \times N_2) \times (N_3 \times N_4)}$. This means a large amount of memory is needed to keep the derivatives. Second, there are many duplicated computations. For example, $\frac{\partial J}{\partial \mathbf{V}^{(1)}} \frac{\partial \mathbf{V}^{(1)}}{\partial \mathbf{V}^{(2)}}$ is computed twice in this example, once for $\nabla_{\mathbf{W}^{(1)}}^{J}$ and once for $\nabla_{\mathbf{W}^{(2)}}^{J}$.

Fortunately, there is a much simpler and more efficient approach to compute the gradient as illustrated in Fig. 14.6. In this approach, each node n keeps two values: the evaluation (forward computation) result $\mathbf{V}_n$ and the gradient ∇_n^{J}. Note that the

Fig. 14.5 The naive gradient computation in CN. $\mathbf{W}^{(1)}$ and $\mathbf{W}^{(2)}$ are model parameters and each edge is associated with a partial derivative

Fig. 14.6 An efficient gradient computation algorithm in CN. $\mathbf{W}^{(1)}$ and $\mathbf{W}^{(2)}$ are model parameters. Each node n stores both the value of the node υ_n and the gradient ∇_n^J

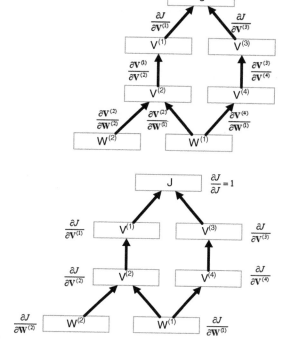

training criterion J is always a scalar, if $\mathbf{V}_n \in \mathbb{R}^{N_1 \times N_2}$ then $\nabla_n^J \in \mathbb{R}^{N_1 \times N_2}$. This requires significantly less memory than that required in the naive solution illustrated in Fig. 14.5. This approach also allows for factorizing out the common prefix terms and making computation linear in the number of nodes in the graph. For example, $\frac{\partial J}{\partial \mathbf{V}^{(2)}}$ is computed only once and used twice when computing $\frac{\partial J}{\partial \mathbf{W}^{(1)}}$ and $\frac{\partial J}{\partial \mathbf{W}^{(2)}}$ with this approach. This is analogous to common subexpression elimination in a conventional expression graph, only here the common subexpressions are the parents of the nodes rather than the children.

Automatic differentiation has been an active research area for decades and many techniques have been proposed [10]. A simple recursive algorithm for deciding the gradient computation order, so that the gradient computation can be efficiently carried out is shown in Algorithm 14.4 to which a similar recursive algorithm for scalar functions has been provided [4, 11]. This algorithm assumes that each node has a ComputePartialGradient(*child*) function which computes the gradient of the training criterion with regard to the node's child *child* and is called in the order that is decided by the algorithm. Before the algorithm is computed, the gradient ∇_n^J at each node n is set to 0, the queue *order* is set to empty, and *parentsLeft* is set to the number of parents of each node. This function is then called on a criterion node that evaluates to a scalar. Similar to the forward computation, an asynchronous algorithm can be derived. Once the gradient is known, the minibatch-based stochastic gradient descent

(SGD) algorithm discussed in Chap. 4 and other training algorithms that depend on gradient only can be used to train the model.

Alternatively, the gradients can be computed by following the reverse order of the forward computation and calling each node's parents' ComputePartialGradient (*child*) function and passing itself as the *child* argument. This approach, however, requires additional bookkeeping, e.g., keeping pointers to a node's parents which can introduce additional overhead when manipulating the network architecture.

Algorithm 14.4 Reverse automatic gradient computation algorithm. At the top level *node* must be a training criterion node that evaluates to a scalar.

1: **procedure** DECIDEGRADIENTCOMPUTATIONORDER(*node, parentsLeft, order*)
   ▷ Decide the order to compute the gradient at all descendents of *node*.
   ▷ *parentsLeft* is initialized to the number of parents of each node.
   ▷ *order* is initialized to an empty queue.
2:   **if** IsNotLeaf(*node*) **then**
3:     *parentsLeft*[*node*] − −
4:     **if** *parentsLeft*[*node*] == 0 ∧ *node* ∉ *order* **then**   ▷ All parents have been computed.
5:       *order* ← *order* + *node*   ▷ Add *node* to the end of *order*
6:       **for** each *c* ∈ *node.children* **do**
7:         call DECIDEGRADIENTCOMPUTATIONORDER(*c*, *parentsLeft, order*)
8:       **end for**
9:     **end if**
10: **end if**
11: **end procedure**

In many cases, not all the gradients need to be computed. For example, the gradient with regard to the input value is never needed. When adapting the model, some of the model parameters do not need to be updated and thus it is unnecessary to compute the gradients with regard to these parameters. We can reduce the gradient computation by keeping a *needGradient* flag for each node. Once the flags of leaf nodes (either input values or model parameters) are known, the flags of the nonleaf nodes can be determined using Algorithm 14.5, which is essentially a depth-first traversal over the DAG. Since both Algorithms 14.2 and 14.5 are essentially depth-first traversal over the DAG and both only need to be executed once they may be combined in one function.

Since every instantiation of a CN is task dependent and different, it is critical to have a way to check and verify the gradients computed automatically. A simple technique to estimate the gradient numerically is:

$$\frac{\partial J}{\partial w_{ij}} \approx \frac{J\left(w_{ij} + \varepsilon\right) - J\left(w_{ij} - \varepsilon\right)}{2\varepsilon}, \tag{14.5}$$

where w_{ij} is the (i, j)th element of a model parameter $\mathbf{W}$, ε is a small constant typically set to 10^{-4}, and $J\left(w_{ij} + \varepsilon\right)$ and $J\left(w_{ij} - \varepsilon\right)$ are the objective function values evaluated with all other parameters fixed and w_{ij} changed to $w_{ij} + \varepsilon$ and

---

Algorithm 14.5 Update the *needGradient* flag recursively.

1: **procedure** UPDATENEEDGRADIENTFLAG(*root*, *visited*)
            ▷ Enumerate nodes in the DAG in the depth-first order.
            ▷ *visited* is initialized as an empty set.
2:   **if** *root* $\notin$ *visited* **then**   ▷ The same node may be a child of several nodes and revisited.
3:     *visited* $\leftarrow$ *visited* $\cup$ *root*
4:     **for** each $c \in root.children$ **do**
5:       call UPDATENEEDGRADIENTFLAG(*c*, *visited*, *order*)
6:       **if** *IsNotLeaf*(*node*) **then**
7:         **if** *node.AnyChildNeedGradient*() **then**
8:           *node.needGradient* $\leftarrow$ *true*
9:         **else**
10:           *node.needGradient* $\leftarrow$ *false*
11:         **end if**
12:       **end if**
13:     **end for**
14:   **end if**
15: **end procedure**

$w_{ij} - \varepsilon$, respectively. In most cases, the numerically estimated gradient and the gradient computed from the automatic gradient computation agree to at least four significant digits if double precision computation is used. Note that this technique works well with a large range of ε values, except extremely small values such as 10^{-20} which would lead to numerical round-off errors.

14.4 Typical Computation Nodes

For the forward computation and gradient calculation algorithms described above to work, we assume that each type of computation node implements a function *Evaluate* to evaluate the value of the node given the values of its child nodes, and the function *ComputePartialGradient*(*child*) to compute the gradient of the training criterion with regard to the child node *child* given the node value $\mathbf{V}_n$ and the gradient ∇_n^J of the node n and values of all its child nodes. For simplicity, we will remove the subscript in the following discussion.

In this section, we introduce the most widely used computation node types and the corresponding *Evaluate* and *ComputePartialGradient*(*child*) functions. In the following discussion, we use symbols listed in Table 14.1 to describe the computations. We treat each minibatch of input values as a matrix in which each column is a sample. In all the derivations of the gradients, we use the identity

$$\frac{\partial J}{\partial x_{ij}} = \sum_{m,n} \frac{\partial J}{\partial v_{mn}} \frac{\partial v_{mn}}{\partial x_{ij}}, \qquad (14.6)$$

Table 14.1 Symbols used in describing the computation nodes

Symbols	Description
λ	A scalar
$\mathbf{d}$	Column vector that represents the diagonal of a square matrix
$\mathbf{X}$	Matrix of the first operand
$\mathbf{Y}$	Matrix of the second operand
V	Value of current node
∇_n^J (or ∇_V^J)	Gradient of the current node
$\nabla_{\mathbf{X}}^J$	Gradient of the child node (operand) $\mathbf{X}$
$\nabla_{\mathbf{Y}}^J$	Gradient of the child node (operand) $\mathbf{Y}$
$\bullet$	Element-wise product
$\oslash$	Element-wise division
$\circ$	Inner product of vectors applied on matrices column-wise
$\odot$	Inner product applied to each row
$\delta(.)$	Kronecker delta
$\mathbf{1}_{m,n}$	An $m \times n$ matrix with all 1's
X^α	Element-wise power
vec $(\mathbf{X})$	Vector formed by concatenating columns of $\mathbf{X}$

where υ_{mn} is the (m, n)th element of the matrix $\mathbf{V}$, and x_{ij} is the (i, j)th element of matrix $\mathbf{X}$.

14.4.1 Computation Node Types with No Operand

The values of a computation node that has no operand are given instead of computed. As a result, both *Evaluate* and *ComputePartialGradient(child)* functions for these computation node types are empty.

- *Parameter*: Used to represent model parameters that need to be saved as part of the model.
- *InputValue*: used to represent features, labels, or control parameters that are provided by users at run time.

14.4.2 Computation Node Types with One Operand

In these computation node types, $Evaluate = V(\mathbf{X})$ and $ComputePartial$ $Gradient(\mathbf{X}) = \nabla_{\mathbf{X}}^J$

- *Negate*: reverse the sign of each element in the operand $\mathbf{X}$.

$$V(\mathbf{X}) \leftarrow -\mathbf{X} \tag{14.7}$$

$$\nabla_{\mathbf{X}}^J \leftarrow \nabla_{\mathbf{X}}^J - \nabla_{\mathbf{n}}^J. \tag{14.8}$$

The gradient can be derived by observing

$$\frac{\partial v_{mn}}{\partial x_{ij}} = \begin{cases} -1 & m = i \wedge n = j \\ 0 & \text{else} \end{cases} \tag{14.9}$$

and

$$\frac{\partial J}{\partial x_{ij}} = \sum_{m,n} \frac{\partial J}{\partial v_{mn}} \frac{\partial v_{mn}}{\partial x_{ij}} = -\frac{\partial J}{\partial v_{ij}}. \tag{14.10}$$

- *Sigmoid*: apply the sigmoid function element-wise to the operand $\mathbf{X}$.

$$V(\mathbf{X}) \leftarrow \frac{1}{1 + e^{-\mathbf{X}}} \tag{14.11}$$

$$\nabla_{\mathbf{X}}^J \leftarrow \nabla_{\mathbf{X}}^J + \nabla_{\mathbf{n}}^J \bullet [V \bullet (1 - V)]. \tag{14.12}$$

The gradient can be derived by observing

$$\frac{\partial v_{mn}}{\partial x_{ij}} = \begin{cases} v_{ij}(1 - v_{ij}) & m = i \wedge n = j \\ 0 & \text{else} \end{cases} \tag{14.13}$$

and

$$\frac{\partial J}{\partial x_{ij}} = \sum_{m,n} \frac{\partial J}{\partial v_{mn}} \frac{\partial v_{mn}}{\partial x_{ij}} = \frac{\partial J}{\partial v_{ij}} v_{ij}(1 - v_{ij}). \tag{14.14}$$

- *Tanh*: apply the tanh function element-wise to the operand $\mathbf{X}$.

$$V(\mathbf{X}) \leftarrow \frac{e^{\mathbf{X}} - e^{-\mathbf{X}}}{e^{\mathbf{X}} + e^{-\mathbf{X}}} \tag{14.15}$$

$$\nabla_{\mathbf{X}}^J \leftarrow \nabla_{\mathbf{X}}^J + \nabla_{\mathbf{n}}^J \bullet (1 - V \bullet V). \tag{14.16}$$

The gradient can be derived by observing

$$\frac{\partial v_{mn}}{\partial x_{ij}} = \begin{cases} 1 - v_{ij}^2 & m = i \wedge n = j \\ 0 & \text{else} \end{cases} \tag{14.17}$$

and

$$\frac{\partial J}{\partial x_{ij}} = \sum_{m,n} \frac{\partial J}{\partial v_{mn}} \frac{\partial v_{mn}}{\partial x_{ij}} = \frac{\partial J}{\partial v_{ij}} \left(1 - v_{ij}^2 \right) \tag{14.18}$$

- *ReLU*: apply the rectified linear operation element-wise to the operand **X**.

$$V (\mathbf{X}) \leftarrow \max (0, \mathbf{X}) \tag{14.19}$$

$$\nabla_{\mathbf{X}}^J \leftarrow \nabla_{\mathbf{X}}^J + \nabla_{\mathbf{n}}^J \bullet \delta (\mathbf{X} > 0) . \tag{14.20}$$

The gradient can be derived by observing

$$\frac{\partial v_{mn}}{\partial x_{ij}} = \begin{cases} \delta \left(x_{ij} > 0 \right) & m = i \wedge n = j \\ 0 & \text{else} \end{cases} \tag{14.21}$$

we have

$$\frac{\partial J}{\partial x_{ij}} = \sum_{m,n} \frac{\partial J}{\partial v_{mn}} \frac{\partial v_{mn}}{\partial x_{ij}} = \frac{\partial J}{\partial v_{ij}} \delta \left(x_{ij} > 0 \right) . \tag{14.22}$$

- *Log*: apply the log function element-wise to the operand **X**.

$$V (\mathbf{X}) \leftarrow \log (\mathbf{X}) \tag{14.23}$$

$$\nabla_{\mathbf{X}}^J \leftarrow \nabla_{\mathbf{X}}^J + \nabla_{\mathbf{n}}^J \bullet \frac{1}{\mathbf{X}} . \tag{14.24}$$

The gradient can be derived by observing

$$\frac{\partial v_{mn}}{\partial x_{ij}} = \begin{cases} \frac{1}{x_{ij}} & m = i \wedge n = j \\ 0 & \text{else} \end{cases} \tag{14.25}$$

and

$$\frac{\partial J}{\partial x_{ij}} = \sum_{m,n} \frac{\partial J}{\partial v_{mn}} \frac{\partial v_{mn}}{\partial x_{ij}} = \frac{\partial J}{\partial v_{ij}} \frac{1}{x_{ij}} . \tag{14.26}$$

- *Exp*: apply the exponent function element-wise to the operand **X**.

$$V(\mathbf{X}) \leftarrow \exp(\mathbf{X}) \tag{14.27}$$

$$\nabla_{\mathbf{X}}^{J} \leftarrow \nabla_{\mathbf{X}}^{J} + \nabla_{\mathbf{n}}^{J} \bullet V. \tag{14.28}$$

The gradient can be derived by observing

$$\frac{\partial v_{mn}}{\partial x_{ij}} = \begin{cases} v_{ij} & m = i \wedge n = j \\ 0 & \text{else} \end{cases} \tag{14.29}$$

we have

$$\frac{\partial J}{\partial x_{ij}} = \sum_{m,n} \frac{\partial J}{\partial v_{mn}} \frac{\partial v_{mn}}{\partial x_{ij}} = \frac{\partial J}{\partial v_{ij}} v_{ij}. \tag{14.30}$$

- *Softmax*: apply the softmax function column-wise to the operand $\mathbf{X}$. Each column is treated as a separate sample.

$$m_j(\mathbf{X}) \leftarrow \max_i x_{ij} \tag{14.31}$$

$$e_{ij}(\mathbf{X}) \leftarrow e^{x_{ij} - m_j(\mathbf{X})} \tag{14.32}$$

$$s_j(\mathbf{X}) \leftarrow \sum_i e_{ij}(\mathbf{X}) \tag{14.33}$$

$$v_{ij}(\mathbf{X}) \leftarrow \frac{e_{ij}(\mathbf{X})}{s_j(\mathbf{X})} \tag{14.34}$$

$$\nabla_{\mathbf{X}}^{J} \leftarrow \nabla_{\mathbf{X}}^{J} + \left[\nabla_{\mathbf{n}}^{J} - \nabla_{\mathbf{n}}^{J} \circ V\right] \bullet V. \tag{14.35}$$

The gradient can be derived by observing

$$\frac{\partial v_{mn}}{\partial x_{ij}} = \begin{cases} v_{ij}(1 - v_{ij}) & m = i \wedge n = j \\ -v_{mj} v_{ij} & n = j \\ 0 & \text{else} \end{cases} \tag{14.36}$$

we have

$$\frac{\partial J}{\partial x_{ij}} = \sum_{m,n} \frac{\partial J}{\partial v_{mn}} \frac{\partial v_{mn}}{\partial x_{ij}} = \left(\frac{\partial J}{\partial v_{ij}} - \sum_m \frac{\partial J}{\partial v_{mj}} v_{mj}\right) v_{ij}. \tag{14.37}$$

- *Sum*Elements: sum over all elements in the operand $\mathbf{X}$.

$$v(\mathbf{X}) \leftarrow \sum_{i,j} x_{ij} \tag{14.38}$$

$$\nabla_{\mathbf{X}}^{J} \leftarrow \nabla_{\mathbf{X}}^{J} + \nabla_{\mathbf{n}}^{J}. \tag{14.39}$$

The gradient can be derived by noting that υ and ∇_n^J are scalars,

$$\frac{\partial \upsilon}{\partial x_{ij}} = 1 \tag{14.40}$$

and

$$\frac{\partial J}{\partial x_{ij}} = \frac{\partial J}{\partial \upsilon} \frac{\partial \upsilon}{\partial x_{ij}} = \frac{\partial J}{\partial \upsilon}. \tag{14.41}$$

- *L1Norm*: take the matrix L_1 norm of the operand X.

$$\upsilon\,(\mathbf{X}) \leftarrow \sum_{i,j} |x_{ij}| \tag{14.42}$$

$$\nabla_{\mathbf{X}}^J \leftarrow \nabla_{\mathbf{X}}^J + \nabla_n^J \mathrm{sgn}\,(\mathbf{X})\,. \tag{14.43}$$

The gradient can be derived by noting that υ and ∇_n^J are scalars,

$$\frac{\partial \upsilon}{\partial x_{ij}} = \mathrm{sgn}\,\left(x_{ij}\right) \tag{14.44}$$

and

$$\frac{\partial J}{\partial x_{ij}} = \frac{\partial J}{\partial \upsilon} \frac{\partial \upsilon}{\partial x_{ij}} = \frac{\partial J}{\partial \upsilon} \mathrm{sgn}\,\left(x_{ij}\right)\,. \tag{14.45}$$

- *L2Norm*: take the matrix L_2 norm (Frobenius norm) of the operand $\mathbf{X}$.

$$\upsilon\,(\mathbf{X}) \leftarrow \sqrt{\sum_{i,j} \left(x_{ij}\right)^2} \tag{14.46}$$

$$\nabla_{\mathbf{X}}^J \leftarrow \nabla_{\mathbf{X}}^J + \frac{1}{\upsilon} \nabla_n^J \mathbf{X}. \tag{14.47}$$

The gradient can be derived by noting that υ and ∇_n^J are scalars,

$$\frac{\partial \upsilon}{\partial x_{ij}} = \frac{x_{ij}}{\upsilon} \tag{14.48}$$

and

$$\frac{\partial J}{\partial x_{ij}} = \frac{\partial J}{\partial \upsilon} \frac{\partial \upsilon}{\partial x_{ij}} = \frac{1}{\upsilon} \frac{\partial J}{\partial \upsilon} x_{ij}\,. \tag{14.49}$$

14.4.3 Computation Node Types with Two Operands

In these computation node types, $Evaluate = V(a, \mathbf{Y})$, where a can be $\mathbf{X}$, λ or $\mathbf{d}$, and $Compute\,Partial\,Gradient(\mathbf{b}) = \nabla_{\mathbf{b}}^J$ where $\mathbf{b}$ can be $\mathbf{X}$, $\mathbf{Y}$ or $\mathbf{d}$.

- *Scale*: scale each element of $\mathbf{Y}$ by λ.

$$V(\lambda, \mathbf{Y}) \leftarrow \lambda \mathbf{Y} \tag{14.50}$$

$$\nabla_\lambda^J \leftarrow \nabla_\lambda^J + \text{vec}\left(\nabla_{\mathbf{n}}^J\right) \circ \text{vec}(\mathbf{Y}) \tag{14.51}$$

$$\nabla_{\mathbf{Y}}^J \leftarrow \nabla_{\mathbf{Y}}^J + \lambda \nabla_{\mathbf{n}}^J. \tag{14.52}$$

The gradient ∇_λ^J can be derived by observing

$$\frac{\partial v_{mn}}{\partial \lambda} = y_{mn} \tag{14.53}$$

and

$$\frac{\partial J}{\partial \lambda} = \sum_{m,n} \frac{\partial J}{\partial v_{mn}} \frac{\partial v_{mn}}{\partial \lambda} = \sum_{m,n} \frac{\partial J}{\partial v_{mn}} y_{mn}. \tag{14.54}$$

Similarly to derive the gradient $\nabla_{\mathbf{y}}^J$, we note that

$$\frac{\partial v_{mn}}{\partial y_{ij}} = \begin{cases} \lambda & m = i \wedge n = j \\ 0 & \text{else} \end{cases} \tag{14.55}$$

and get

$$\frac{\partial J}{\partial y_{ij}} = \sum_{m,n} \frac{\partial J}{\partial v_{mn}} \frac{\partial v_{mn}}{\partial y_{ij}} = \lambda \frac{\partial J}{\partial v_{ij}} \tag{14.56}$$

- *Times*: matrix product of operands $\mathbf{X}$ and $\mathbf{Y}$. Must satisfy $\mathbf{X}.cols = \mathbf{Y}.rows$.

$$V(\mathbf{X}, \mathbf{Y}) \leftarrow \mathbf{X}\mathbf{Y} \tag{14.57}$$

$$\nabla_{\mathbf{X}}^J \leftarrow \nabla_{\mathbf{X}}^J + \nabla_{\mathbf{n}}^J \mathbf{Y}^T \tag{14.58}$$

$$\nabla_{\mathbf{Y}}^J \leftarrow \nabla_{\mathbf{Y}}^J + \mathbf{X}^T \nabla_{\mathbf{n}}^J. \tag{14.59}$$

The gradient $\nabla_{\mathbf{X}}^J$ can be derived by observing

$$\frac{\partial v_{mn}}{\partial x_{ij}} = \begin{cases} y_{jn} & m = i \\ 0 & \text{else} \end{cases} \tag{14.60}$$

and

$$\frac{\partial J}{\partial x_{ij}} = \sum_{m,n} \frac{\partial J}{\partial v_{mn}} \frac{\partial v_{mn}}{\partial x_{ij}} = \sum_{n} \frac{\partial J}{\partial v_{in}} y_{jn}. \qquad (14.61)$$

Similarly to derive the gradient $\nabla_{\mathbf{Y}}^{J}$, we note that

$$\frac{\partial v_{mn}}{\partial y_{ij}} = \begin{cases} x_{mi} & n = j \\ 0 & \text{else} \end{cases} \qquad (14.62)$$

and get

$$\frac{\partial J}{\partial y_{ij}} = \sum_{m,n} \frac{\partial J}{\partial v_{mn}} \frac{\partial v_{mn}}{\partial y_{ij}} = \sum_{m} \frac{\partial J}{\partial v_{mj}} x_{mi} \qquad (14.63)$$

• *ElementTimes*: element-wise product of two matrices. Must satisfy $\mathbf{X}.rows = \mathbf{Y}.rows$ and $\mathbf{X}.cols = \mathbf{Y}.cols$.

$$v_{ij}(\mathbf{X}, \mathbf{Y}) \leftarrow x_{ij} y_{ij} \qquad (14.64)$$
$$\nabla_{\mathbf{X}}^{J} \leftarrow \nabla_{\mathbf{X}}^{J} + \nabla_{\mathbf{n}}^{J} \bullet \mathbf{Y} \qquad (14.65)$$
$$\nabla_{\mathbf{Y}}^{J} \leftarrow \nabla_{\mathbf{Y}}^{J} + \nabla_{\mathbf{n}}^{J} \bullet \mathbf{X}. \qquad (14.66)$$

The gradient $\nabla_{\mathbf{X}}^{J}$ can be derived by observing

$$\frac{\partial v_{mn}}{\partial x_{ij}} = \begin{cases} y_{ij} & m = i \land n = j \\ 0 & \text{else} \end{cases} \qquad (14.67)$$

and

$$\frac{\partial J}{\partial x_{ij}} = \sum_{m,n} \frac{\partial J}{\partial v_{mn}} \frac{\partial v_{mn}}{\partial y_{ij}} = \frac{\partial J}{\partial v_{ij}} y_{ij}. \qquad (14.68)$$

The gradient $\nabla_{\mathbf{Y}}^{J}$ can be derived exactly the same way due to symmetry.

• *Plus*: sum of two matrices $\mathbf{X}$ and $\mathbf{Y}$. Must satisfy $\mathbf{X}.rows = \mathbf{Y}.rows$. If $\mathbf{X}.cols \neq \mathbf{Y}.cols$ but one of them is a multiple of the other, the smaller matrix needs to be expanded by repeating itself.

$$V(\mathbf{X}, \mathbf{Y}) \leftarrow \mathbf{X} + \mathbf{Y} \qquad (14.69)$$

$$\nabla_{\mathbf{X}}^{J} \leftarrow \begin{cases} \nabla_{\mathbf{X}}^{J} + \nabla_{\mathbf{n}}^{J} & \mathbf{X}.rows = \mathbf{V}.rows \land \mathbf{X}.cols = \mathbf{V}.cols \\ \nabla_{\mathbf{X}}^{J} + \mathbf{1}_{1,\mathbf{V}.rows} \nabla_{\mathbf{n}}^{J} & \mathbf{X}.rows = 1 \land \mathbf{X}.cols = \mathbf{V}.cols \\ \nabla_{\mathbf{X}}^{J} + \nabla_{\mathbf{n}}^{J} \mathbf{1}_{\mathbf{V}.cols,1} & \mathbf{X}.rows = \mathbf{V}.rows \land \mathbf{X}.cols = 1 \\ \nabla_{\mathbf{X}}^{J} + \mathbf{1}_{1,\mathbf{V}.rows} \nabla_{\mathbf{n}}^{J} \mathbf{1}_{v.cols,1} & \mathbf{X}.rows = 1 \land \mathbf{X}.cols = 1 \end{cases} \qquad (14.70)$$

$$\nabla_{\mathbf{Y}}^J \leftarrow \begin{cases} \nabla_{\mathbf{Y}}^J + \nabla_{\mathbf{n}}^J & \mathbf{Y}.rows = \mathbf{V}.rows \wedge \mathbf{Y}.cols = \mathbf{V}.cols \\ \nabla_{\mathbf{Y}}^J + \mathbf{1}_{1,\mathbf{V}.rows}\nabla_{\mathbf{n}}^J & \mathbf{Y}.rows = 1 \wedge \mathbf{Y}.cols = \mathbf{V}.cols \\ \nabla_{\mathbf{Y}}^J + \nabla_{\mathbf{n}}^J \mathbf{1}_{\mathbf{V}.cols,1} & \mathbf{Y}.rows = \mathbf{V}.rows \wedge \mathbf{Y}.cols = 1 \\ \nabla_{\mathbf{Y}}^J + \mathbf{1}_{1,\mathbf{V}.rows}\nabla_{\mathbf{n}}^J \mathbf{1}_{\mathbf{V}.cols,1} & \mathbf{Y}.rows = 1 \wedge \mathbf{Y}.cols = 1 \end{cases} \quad (14.71)$$

The gradient $\nabla_{\mathbf{X}}^J$ can be derived by observing that when $\mathbf{X}$ has the same dimension as $\mathbf{V}$, we have

$$\frac{\partial v_{mn}}{\partial x_{ij}} = \begin{cases} 1 & m = i \wedge n = j \\ 0 & \text{else} \end{cases} \quad (14.72)$$

and

$$\frac{\partial J}{\partial x_{ij}} = \sum_{m,n} \frac{\partial J}{\partial v_{mn}} \frac{\partial v_{mn}}{\partial x_{ij}} = \frac{\partial J}{\partial v_{ij}} \quad (14.73)$$

If $\mathbf{X}.rows = 1 \wedge \mathbf{X}.cols = \mathbf{V}.cols$ we have

$$\frac{\partial v_{mn}}{\partial x_{ij}} = \begin{cases} 1 & n = j \\ 0 & \text{else} \end{cases} \quad (14.74)$$

and

$$\frac{\partial J}{\partial x_{ij}} = \sum_{m,n} \frac{\partial J}{\partial v_{mn}} \frac{\partial v_{mn}}{\partial x_{ij}} = \sum_{m} \frac{\partial J}{\partial v_{mj}} \quad (14.75)$$

We can derive $\nabla_{\mathbf{X}}^J$ and $\nabla_{\mathbf{Y}}^J$ under other conditions similarly.

- *Minus*: difference of two matrices $\mathbf{X}$ and $\mathbf{Y}$. Must satisfy $\mathbf{X}.rows = \mathbf{Y}.rows$. If $\mathbf{X}.cols \neq \mathbf{Y}.cols$ but one of them is a multiple of the other, the smaller matrix needs to be expanded by repeating itself.

$$V(\mathbf{X}, \mathbf{Y}) \leftarrow \mathbf{X} - \mathbf{Y} \quad (14.76)$$

$$\nabla_{\mathbf{X}}^J \leftarrow \begin{cases} \nabla_{\mathbf{X}}^J + \nabla_{\mathbf{n}}^J & \mathbf{X}.rows = \mathbf{V}.rows \wedge \mathbf{X}.cols = \mathbf{V}.cols \\ \nabla_{\mathbf{X}}^J + \mathbf{1}_{1,\mathbf{V}.rows}\nabla_{\mathbf{n}}^J & \mathbf{X}.rows = 1 \wedge \mathbf{X}.cols = \mathbf{V}.cols \\ \nabla_{\mathbf{X}}^J + \nabla_{\mathbf{n}}^J \mathbf{1}_{\mathbf{V}.cols,1} & \mathbf{X}.rows = \mathbf{V}.rows \wedge \mathbf{X}.cols = 1 \\ \nabla_{\mathbf{X}}^J + \mathbf{1}_{1,\mathbf{V}.rows}\nabla_{\mathbf{n}}^J \mathbf{1}_{v.cols,1} & \mathbf{X}.rows = 1 \wedge \mathbf{X}.cols = 1 \end{cases} \quad (14.77)$$

$$\nabla_{\mathbf{Y}}^J \leftarrow \begin{cases} \nabla_{\mathbf{Y}}^J - \nabla_{\mathbf{n}}^J & \mathbf{Y}.rows = \mathbf{V}.rows \wedge \mathbf{Y}.cols = \mathbf{V}.cols \\ \nabla_{\mathbf{Y}}^J - \mathbf{1}_{1,\mathbf{V}.rows}\nabla_{\mathbf{n}}^J & \mathbf{Y}.rows = 1 \wedge \mathbf{Y}.cols = \mathbf{V}.cols \\ \nabla_{\mathbf{Y}}^J - \nabla_{\mathbf{n}}^J \mathbf{1}_{\mathbf{V}.cols,1} & \mathbf{Y}.rows = \mathbf{V}.rows \wedge \mathbf{Y}.cols = 1 \\ \nabla_{\mathbf{Y}}^J - \mathbf{1}_{1,\mathbf{V}.rows}\nabla_{\mathbf{n}}^J \mathbf{1}_{v.cols,1} & \mathbf{Y}.rows = 1 \wedge \mathbf{Y}.cols = 1 \end{cases} \quad (14.78)$$

The derivation of the gradients is similar to that for the *Plus* node.

- *DiagTimes*: the product of a diagonal matrix (whose diagonal equals to **d**) and an arbitrary matrix **Y**. Must satisfy **d**.*rows* = **Y**.*rows*.

$$v_{ij}(\mathbf{d}, \mathbf{Y}) \leftarrow d_i y_{ij} \tag{14.79}$$

$$\nabla_{\mathbf{d}}^J \leftarrow \nabla_{\mathbf{d}}^J + \nabla_{\mathbf{n}}^J \odot \mathbf{Y} \tag{14.80}$$

$$\nabla_{\mathbf{Y}}^J \leftarrow \nabla_{\mathbf{Y}}^J + \text{DiagTimes}\left(\mathbf{d}, \nabla_{\mathbf{n}}^J\right). \tag{14.81}$$

The gradient $\nabla_{\mathbf{d}}^J$ can be derived by observing

$$\frac{\partial v_{mn}}{\partial d_i} = \begin{cases} y_{in} & m = i \\ 0 & \text{else} \end{cases} \tag{14.82}$$

and

$$\frac{\partial J}{\partial d_i} = \sum_{m,n} \frac{\partial J}{\partial v_{mn}} \frac{\partial v_{mn}}{\partial d_i} = \sum_n \frac{\partial J}{\partial v_{in}} y_{in} \tag{14.83}$$

Similarly to derive the gradient $\nabla_{\mathbf{Y}}^J$ we note that

$$\frac{\partial v_{mn}}{\partial y_{ij}} = \begin{cases} d_i & m = i \wedge n = j \\ 0 & \text{else} \end{cases} \tag{14.84}$$

and get

$$\frac{\partial J}{\partial y_{ij}} = \sum_{m,n} \frac{\partial J}{\partial v_{mn}} \frac{\partial v_{mn}}{\partial y_{ij}} = \frac{\partial J}{\partial v_{ij}} d_i \tag{14.85}$$

- *Dropout*: randomly set λ percentage of values of **Y** to be zero and scale the rest, so that the expectation of the sum is not changed:

$$m_{ij}(\lambda) \leftarrow \begin{cases} 0 & \text{rand}(0, 1) \leq \lambda \\ \frac{1}{1-\lambda} & \text{else} \end{cases} \tag{14.86}$$

$$v_{ij}(\lambda, \mathbf{Y}) \leftarrow m_{ij} y_{ij} \tag{14.87}$$

$$\nabla_{\mathbf{Y}}^J \leftarrow \nabla_{\mathbf{Y}}^J + \begin{cases} \nabla_{\mathbf{n}}^J & \lambda = 0 \\ \nabla_{\mathbf{n}}^J \bullet \mathbf{M} & \text{else} \end{cases} \tag{14.88}$$

Note that λ is a given value instead of part of the model. We only need to get the gradient with regard to **Y**. If $\lambda = 0$ then **V** = **X** which is a trivial case. Otherwise it is equivalent to the *ElementTimes* node with a randomly set mask **M**.

- *KhatriRaoProduct*: column-wise cross product of two matrices **X** and **Y**. Must satisfy **X**.*cols* = **Y**.*cols*. Useful for constructing tensor networks.

$$\boldsymbol{v}_{.j}\left(\mathbf{X},\mathbf{Y}\right) \leftarrow \mathbf{x_j} \otimes \mathbf{y}_{.j} \tag{14.89}$$

$$\left[\nabla_{\mathbf{X}}^{J}\right]_{.j} \leftarrow \left[\nabla_{\mathbf{X}}^{J}\right]_{.j} + \left[\left[\nabla_{\mathbf{n}}^{J}\right]_{.j}\right]_{\mathbf{X}.rows,\mathbf{Y}.rows} \mathbf{Y} \tag{14.90}$$

$$\left[\nabla_{\mathbf{Y}}^{J}\right]_{.j} \leftarrow \left[\nabla_{\mathbf{Y}}^{J}\right]_{.j} + \left[\left[\nabla_{\mathbf{n}}^{J}\right]_{.j}\right]_{\mathbf{X}.rows,\mathbf{Y}.rows}^{T} \mathbf{X}, \tag{14.91}$$

where $[\mathbf{X}]_{m,n}$ reshapes $\mathbf{X}$ to become an $m \times n$ matrix. The gradient $\nabla_{\mathbf{X}}^{J}$ can be derived by observing

$$\frac{\partial v_{mn}}{\partial x_{ij}} = \begin{cases} y_{kj} & n = j \wedge i = m/\mathbf{Y}.rows \wedge k = \text{modulus}(m, \mathbf{Y}.rows) \\ 0 & \text{else} \end{cases} \tag{14.92}$$

and

$$\frac{\partial J}{\partial x_{ij}} = \sum_{m,n} \frac{\partial J}{\partial v_{mn}} \frac{\partial v_{mn}}{\partial x_{ij}} = \sum_{i,k} \frac{\partial J}{\partial v_{i\times y.rows+k,j}} y_{kj}. \tag{14.93}$$

The gradient $\nabla_{\mathbf{y}}^{J}$ can be derived similarly.

- *Cos*: column-wise cosine distance of two matrices $\mathbf{X}$ and $\mathbf{Y}$. Must satisfy $\mathbf{X}.cols = \mathbf{Y}.cols$. The result is a row vector. Frequently used in natural language processing tasks.

$$\boldsymbol{v}_{.j}\left(\mathbf{X},\mathbf{Y}\right) \leftarrow \frac{\mathbf{x}_{.j}^{T}\mathbf{y}_{.j}}{\|\mathbf{x}_j\|\,\|\mathbf{y}_j\|} \tag{14.94}$$

$$\left[\nabla_{\mathbf{X}}^{J}\right]_{.j} \leftarrow \left[\nabla_{\mathbf{X}}^{J}\right]_{.j} + \left[\nabla_{\mathbf{n}}^{J}\right]_{.j} \bullet \left[\frac{y_{ij}}{\|\mathbf{x}_j\|\,\|\mathbf{y}_j\|} - \frac{x_{ij}v_{.j}}{\|\mathbf{x}_j\|^2}\right] \tag{14.95}$$

$$\left[\nabla_{\mathbf{Y}}^{J}\right]_{.j} \leftarrow \left[\nabla_{\mathbf{Y}}^{J}\right]_{.j} + \left[\nabla_{\mathbf{n}}^{J}\right]_{.j} \bullet \left[\frac{x_{ij}}{\|\mathbf{x}_j\|\,\|\mathbf{y}_j\|} - \frac{y_{ij}v_{.j}}{\|\mathbf{y}_j\|^2}\right]. \tag{14.96}$$

The gradient $\nabla_{\mathbf{X}}^{J}$ can be derived by observing

$$\frac{\partial v_{.n}}{\partial x_{ij}} = \begin{cases} \dfrac{y_{ij}}{\|\mathbf{x}_j\|\,\|\mathbf{y}_j\|} - \dfrac{x_{ij}\left(\mathbf{x}_{.j}^{T}y_{.j}\right)}{\|\mathbf{x}_j\|^3\,\|\mathbf{y}_j\|} & n = j \\ 0 & \text{else} \end{cases} \tag{14.97}$$

and

$$\frac{\partial J}{\partial x_{ij}} = \sum_n \frac{\partial J}{\partial \upsilon_{.n}} \frac{\partial \upsilon_{.n}}{\partial x_{ij}} = \frac{\partial J}{\partial \upsilon_{.,j}} \left[\frac{y_{ij}}{\|\mathbf{x}_{.j}\| \|\mathbf{y}_{.j}\|} - \frac{x_{ij} \left(\mathbf{x}_{.j}^T \mathbf{y}_{.j} \right)}{\|\mathbf{x}_{.j}\|^3 \|\mathbf{y}_{.j}\|} \right]. \quad (14.98)$$

$$= \frac{\partial J}{\partial \upsilon_{.,j}} \left[\frac{y_{ij}}{\|\mathbf{x}_{.j}\| \|\mathbf{y}_{.j}\|} - \frac{x_{ij} \upsilon_{.,j}}{\|\mathbf{x}_{.j}\|^2} \right]. \quad (14.99)$$

The gradient $\nabla_{\mathbf{y}}^J$ can be derived similarly.

- *ClassificationError*: compute the total number of columns in which the indexes of the maximum values disagree. Each column is considered as a sample and δ is the Kronecker delta. Must satisfy $\mathbf{X}.cols = \mathbf{Y}.cols$.

$$a_j (\mathbf{X}) \leftarrow \arg\max_i x_{ij} \quad (14.100)$$

$$b_j (\mathbf{Y}) \leftarrow \arg\max_i y_{ij} \quad (14.101)$$

$$v (\mathbf{X}, \mathbf{Y}) \leftarrow \sum_j \delta \left(a_j (\mathbf{X}) \neq b_j (\mathbf{Y}) \right) \quad (14.102)$$

This node type is only used to compute classification errors during the decoding time and is not involved in the model training. For this reason, calling *ComputePartialGradient* (**b**) should just raise an error.

- *SquareError*: compute the square of Frobenius norm of the difference $\mathbf{X} - \mathbf{Y}$. Must satisfy $\mathbf{X}.rows = \mathbf{Y}.rows$ and $\mathbf{X}.cols = \mathbf{Y}.cols$.

$$v (\mathbf{X}, \mathbf{Y}) \leftarrow \frac{1}{2} \mathrm{Tr} \left((\mathbf{X} - \mathbf{Y}) (\mathbf{X} - \mathbf{Y})^T \right) \quad (14.103)$$

$$\nabla_{\mathbf{X}}^J \leftarrow \nabla_{\mathbf{X}}^J + \nabla_{\mathbf{n}}^J (\mathbf{X} - \mathbf{Y}) \quad (14.104)$$

$$\nabla_{\mathbf{Y}}^J \leftarrow \nabla_{\mathbf{Y}}^J - \nabla_{\mathbf{n}}^J (\mathbf{X} - \mathbf{Y}). \quad (14.105)$$

Note that v is a scalar. The derivation of the gradients is trivial given

$$\frac{\partial v}{\partial \mathbf{X}} = \mathbf{X} - \mathbf{Y} \quad (14.106)$$

$$\frac{\partial v}{\partial \mathbf{Y}} = - (\mathbf{X} - \mathbf{Y}). \quad (14.107)$$

- *CrossEntropy*: compute the sum of cross-entropy computed column-wise (over samples) where each column of $\mathbf{X}$ and $\mathbf{Y}$ is a probability distribution. Must satisfy $\mathbf{X}.rows = \mathbf{Y}.rows$ and $\mathbf{X}.cols = \mathbf{Y}.cols$.

$$\mathbf{R} (\mathbf{Y}) \leftarrow \log (\mathbf{Y}) \quad (14.108)$$

$$v (\mathbf{X}, \mathbf{Y}) \leftarrow -\mathrm{vec} (\mathbf{X}) \circ \mathrm{vec} (\mathbf{R} (\mathbf{Y})) \quad (14.109)$$

$$\nabla_{\mathbf{X}}^J \leftarrow \nabla_{\mathbf{X}}^J - \nabla_{\mathbf{n}}^J \mathbf{R} (\mathbf{Y}) \quad (14.110)$$

$$\nabla_{\mathbf{Y}}^J \leftarrow \nabla_{\mathbf{Y}}^J - \nabla_{\mathbf{n}}^J (\mathbf{X} \oslash \mathbf{Y}). \quad (14.111)$$

Note that v is a scalar. The gradient $\nabla_{\mathbf{X}}^J$ can be derived by observing

$$\frac{\partial v}{\partial x_{ij}} = -\log\left(y_{ij}\right) = -r_{ij}\left(\mathbf{Y}\right) \tag{14.112}$$

and

$$\frac{\partial J}{\partial x_{ij}} = \frac{\partial J}{\partial v}\frac{\partial v}{\partial x_{ij}} = -\frac{\partial J}{\partial v}r_{ij}\left(\mathbf{Y}\right) \tag{14.113}$$

Similarly to derive the gradient $\nabla_{\mathbf{Y}}^J$ we note that

$$\frac{\partial v}{\partial y_{ij}} = -\frac{x_{ij}}{y_{ij}} \tag{14.114}$$

and get

$$\frac{\partial J}{\partial y_{ij}} = \frac{\partial J}{\partial v}\frac{\partial v}{\partial y_{ij}} = -\frac{\partial J}{\partial v}\frac{x_{ij}}{y_{ij}}. \tag{14.115}$$

- *CrossEntropyWithSoftmax*: same as *CrossEntropy* except that $\mathbf{Y}$ contains values before the softmax operation (i.e., unnormalized).

$$\mathbf{P}\left(\mathbf{Y}\right) \leftarrow \text{Softmax}\left(\mathbf{Y}\right) \tag{14.116}$$
$$\mathbf{R}\left(\mathbf{Y}\right) \leftarrow \log\left(\mathbf{P}\left(\mathbf{Y}\right)\right) \tag{14.117}$$
$$v\left(\mathbf{X}, \mathbf{Y}\right) \leftarrow \text{vec}\left(\mathbf{X}\right) \circ \text{vec}\left(\mathbf{R}\left(\mathbf{Y}\right)\right) \tag{14.118}$$
$$\nabla_{\mathbf{X}}^J \leftarrow \nabla_{\mathbf{X}}^J - \nabla_{\mathbf{n}}^J \mathbf{R}\left(\mathbf{Y}\right) \tag{14.119}$$
$$\nabla_{\mathbf{Y}}^J \leftarrow \nabla_{\mathbf{Y}}^J + \nabla_{\mathbf{n}}^J \left(\mathbf{P}\left(\mathbf{Y}\right) - \mathbf{X}\right) \tag{14.120}$$

The gradient $\nabla_{\mathbf{X}}^J$ is the same as in the *CrossEntropy* node. To derive the gradient $\nabla_{\mathbf{Y}}^J$ we note that

$$\frac{\partial v}{\partial y_{ij}} = \mathbf{p}_{ij}\left(\mathbf{Y}\right) - x_{ij} \tag{14.121}$$

and get

$$\frac{\partial J}{\partial y_{ij}} = \frac{\partial J}{\partial v}\frac{\partial v}{\partial y_{ij}} = \frac{\partial J}{\partial v}\left(\mathbf{p}_{ij}\left(\mathbf{Y}\right) - x_{ij}\right). \tag{14.122}$$

14.4.4 Computation Node Types for Computing Statistics

Sometimes, we only want to get some statistics of the input values (either input features or labels). For example, to normalize the input features, we need to compute the mean and standard deviation of the input feature. In speech recognition, we need to compute the frequencies (mean) of the state labels to convert state posterior probability to the scaled likelihood as explained in Chap. 6. Unlike other computation node types we just described, computation node types for computing statistics do not require a gradient computation function (i.e., this function should not be called for these types of nodes) because they are not learned and often need to be precomputed before model training starts. Here, we list the most popular computation node types in this category.

- *Mean*: compute the mean of the operand $\mathbf{X}$ across the whole training set. When the computation is finished, it needs to be marked so to avoid recomputation. When a minibatch of input $\mathbf{X}$ is fed in

$$k \leftarrow k + \mathbf{X}.cols \tag{14.123}$$

$$v\left(\mathbf{X}\right) \leftarrow \frac{1}{k}\mathbf{X}\mathbf{1}_{\mathbf{X}.cols,1} + \frac{k - \mathbf{X}.cols}{k}v\left(\mathbf{X}\right) \tag{14.124}$$

Note here $\mathbf{X}.cols$ is the number of samples in the minibatch.

- *InvStdDev*: compute the invert standard deviation of the operand $\mathbf{X}$ element-wise across the whole training set. When the computation is finished, it needs to be marked so to avoid recomputation. In the accumulation step

$$k \leftarrow k + \mathbf{X}.cols \tag{14.125}$$

$$v\left(\mathbf{X}\right) \leftarrow \frac{1}{k}\mathbf{X}\mathbf{1}_{\mathbf{X}.cols,1} + \frac{k - \mathbf{X}.cols}{k}v\left(\mathbf{X}\right) \tag{14.126}$$

$$\omega\left(\mathbf{X}\right) \leftarrow \frac{1}{k}\left(\mathbf{X} \bullet \mathbf{X}\right)\mathbf{1} + \frac{k - \mathbf{X}.cols}{k}\omega\left(\mathbf{X}\right) \tag{14.127}$$

When the end of the training set is reached,

$$v \leftarrow \left(\omega - \left(v \bullet v\right)\right)^{1/2} \tag{14.128}$$

$$v \leftarrow 1 \oslash v. \tag{14.129}$$

- *PerDimMeanVarNorm*: compute the normalized operand $\mathbf{X}$ using mean $\mathbf{m}$ and invert standard deviation $\mathbf{s}$ for each sample. Here, $\mathbf{X}$ is matrix whose number of columns equals to the number of samples in the minibatch and $\mathbf{m}$ and $\mathbf{s}$ are vectors that needs to be expanded before element-wise product is applied.

$$V\left(\mathbf{X}\right) \leftarrow \left(\mathbf{X}-\mathbf{m}\right) \bullet \mathbf{s}. \tag{14.130}$$

14.5 Convolutional Neural Network

A Convolutional neural network (CNN) [1, 2, 5, 6, 8, 17–19, 21, 22] provides shift invariance over time and space and is critical to achieve state-of-the-art performance on image recognition. It has also been shown to improve speech recognition accuracy over pure DNNs on some tasks [1, 2, 8, 21, 22]. To support CNN, we need to implement several new computation nodes.

- *Convolution*: convolve element-wise products of a kernel to an image. An example of a convolution operation is shown in Fig. 14.7, where the input to the convolution node has three channels (represented by three 3×3 matrices) and the output has two channels (represented by two 2×2 matrices at top). A channel is a view of the same image. For example, an RGB image can be represented with three channels: R, G, B. Each channel is of the same size.

There is a kernel for each output and input channel pair. The total number of kernels equals to the product of the number of input channels C_x and the number of output channels C_v. In Fig. 14.7 $C_x = 3$, $C_v = 2$ and the total number of kernels is 6. Each kernel $\mathbf{K}_{k\ell}$ of input channel k and output channel ℓ is a matrix. The kernel moves along (and thus shared across) the input with strides (or subsampling rate) S_r and S_c at the vertical (row) and horizontal (column) direction, respectively. For each output channel ℓ and input slice (i, j) (the ith step along the vertical direction and jth step along the horizontal direction)

$$v_{\ell i j} (\mathbf{K}, \mathbf{Y}) = \sum_k \text{vec} (\mathbf{K}_{k\ell}) \circ \text{vec} (\mathbf{Y}_{kij}), \qquad (14.131)$$

where $\mathbf{Y}_{kij}$ has the same size as $\mathbf{K}_{k\ell}$.

This evaluation function involves many small matrix operations and can be slow. Chellapilla et al. [5] proposed a technique to convert all these small matrix operations to a large matrix product as shown at the bottom of Fig. 14.7. With this trick, all the kernel parameters are combined into a big kernel matrix $\mathbf{W}$ as shown at left bottom of Fig. 14.7. Note that to allow for the output of the convolution node to be used by another convolution node, in Fig. 14.7 we have organized the conversion slightly differently from what proposed by Chellapilla et al. [5] by transposing both the kernel matrix and the input features matrix as well as the order of the matrices in the product. By doing so, each sample in the output can be represented by $O_r \times O_c$ columns of $C_v \times 1$ vectors which can be reshaped to become a single column, where

$$O_r = \begin{cases} \frac{I_r - K_r}{S_r} + 1 & \text{no padding} \\ \frac{(I_r - \text{mod}(K_r, 2))}{S_r} + 1 & \text{zero padding} \end{cases} \qquad (14.132)$$

is the number of rows in the output image, and

$$O_c = \begin{cases} \frac{I_c - K_c}{S_c} + 1 & \text{no padding} \\ \frac{(I_c - \text{mod}(K_c, 2))}{S_c} + 1 & \text{zero padding} \end{cases} \qquad (14.133)$$

is the number of rows in the output image, where I_r and I_c are, respectively, the number of rows and columns of the input image, K_r and K_c are, respectively, the number of rows and columns in each kernel. The combined kernel matrix

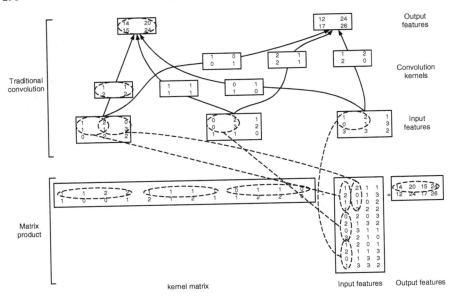

Fig. 14.7 Example convolution operations. *Top* The original convolution operations. *Bottom* The same operations represented as a large matrix product. Our matrix organization is different from what proposed by Chellapilla et al. [5] to allow for stacked convolution operations (Modified from the figure in Chellapilla et al. [5], permitted to use by Simard)

is of size $C_v \times (O_r \times O_c \times C_x)$, and the packed input feature matrix is of size $(O_r \times O_c \times C_x) \times (K_r \times K_c)$.

With this conversion, the related computations of the convolution node with operands **W**, **Y** become

$$\mathbf{X}(\mathbf{Y}) \leftarrow \text{Pack}(\mathbf{Y}) \tag{14.134}$$

$$V(\mathbf{W}, \mathbf{Y}) \leftarrow \mathbf{WX} \tag{14.135}$$

$$\nabla_{\mathbf{W}}^J \leftarrow \nabla_{\mathbf{W}}^J + \nabla_{\mathbf{n}}^J \mathbf{X}^T \tag{14.136}$$

$$\nabla_{\mathbf{X}}^J \leftarrow \mathbf{W}^T \nabla_{\mathbf{n}}^J \tag{14.137}$$

$$\nabla_{\mathbf{Y}}^J \leftarrow \nabla_{\mathbf{Y}}^J + \text{Unpack}\left(\nabla_{\mathbf{X}}^J\right). \tag{14.138}$$

Note that this technique enables better parallelization with large matrix operations but introduces additional cost to pack and unpack the matrices. In most conditions, the gain outweighs the cost. By composing convolution node with plus nodes and element-wise nonlinear functions, we can add bias and nonlinearity to the convolution operations.

- *MaxPooling*: apply the maximum pooling operation to input **X** inside a window with size $K_r \times K_c$ for each channel. The operation window moves along the input with strides (or subsampling rate) S_r and S_c at the vertical (row) and horizontal (column) direction, respectively. The pooling operation does not change the number of channels and so $C_v = C_x$. For each output channel ℓ and the (i, j)th input slice $\mathbf{X}_{\ell ij}$ of size $K_r \times K_c$ we have

$$v_{\ell ij}(\mathbf{X}) \leftarrow \max\left(\mathbf{X}_{\ell ij}\right) \tag{14.139}$$

$$\left[\nabla_{\mathbf{X}}^J\right]_{\ell, i_m, j_n} \leftarrow \begin{cases} \left[\nabla_{\mathbf{X}}^J\right]_{\ell, i_m, j_n} + \left[\nabla_{\mathbf{n}}^J\right]_{\ell, i_m, j_n} & (m, n) = arg\max_{m,n} x_{\ell, i_m, j_n} \\ \left[\nabla_{\mathbf{X}}^J\right]_{\ell, i_m, j_n} & \text{else} \end{cases} \tag{14.140}$$

where $i_m = i \times S_r + m$, and $j_n = j \times S_c + n$.

- *AveragePooling*: same as *MaxPooling* except average instead of maximum is applied to input $\mathbf{X}$ inside a window with size $K_r \times K_c$ for each channel. The operation window moves along the input with strides (or subsampling rate) S_r and S_c at the vertical (row) and horizontal (column) direction, respectively. For each output channel ℓ and the (i, j)th input slice $\mathbf{X}_{\ell ij}$ of size $K_r \times K_c$ we have

$$v_{\ell ij}(\mathbf{X}) \leftarrow \frac{1}{K_r \times K_c} \sum_{m,n} x_{\ell, i_m, j_n} \tag{14.141}$$

$$\left[\nabla_{\mathbf{X}}^J\right]_{\ell ij} \leftarrow \left[\nabla_{\mathbf{X}}^J\right]_{\ell ij} + \frac{1}{K_r \times K_c}\left[\nabla_{\mathbf{n}}^J\right]_{\ell ij}, \tag{14.142}$$

where $i_m = i \times S_r + m$, and $j_n = j \times S_c + n$.

14.6 Recurrent Connections

In the above sections, we assumed that the CN is a DAG. However, when there are recurrent connections in the CN, this assumption is no longer true. The recurrent connection can be implemented using a *Delay* node that retrieves the value λ samples to the past where each column of $\mathbf{Y}$ is a separate sample stored in the ascending order of time.

$$v_{.j}(\lambda, \mathbf{Y}) \leftarrow \mathbf{Y}_{.(j-\lambda)} \tag{14.143}$$

$$\left[\nabla_{\mathbf{Y}}^J\right]_{.j} \leftarrow \left[\nabla_{\mathbf{Y}}^J\right]_{.j} + \left[\nabla_{\mathbf{n}}^J\right]_{.j+\lambda}. \tag{14.144}$$

When $j - \lambda < 0$ some default values need to be set for $\mathbf{Y}_{.(j-\lambda)}$. The gradient can be derived by observing

$$\frac{\partial v_{mn}}{\partial y_{ij}} = \begin{cases} 1 & m = i \wedge n = j + \lambda \\ 0 & \text{else} \end{cases} \tag{14.145}$$

and

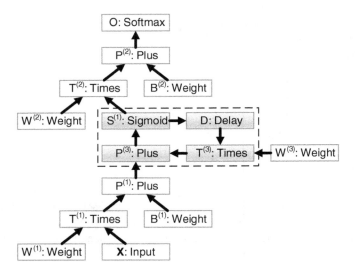

Fig. 14.8 An example CN with a delay node. The *shaded nodes* form a recurrent loop which can be treated as a composite node

$$\frac{\partial J}{\partial y_{ij}} = \sum_{m,n} \frac{\partial J}{\partial v_{mn}} \frac{\partial v_{mn}}{\partial y_{ij}} = \frac{\partial J}{\partial v_{ij+\lambda}}. \tag{14.146}$$

An example CN that contains a delay node is shown in Fig. 14.8. Different from the CN without a directed loop, a CN with a loop cannot be computed for a sequence of samples as a batch since the next sample's value depends on the previous samples. A simple way to do forward computation and backpropagation in a recurrent network is to unroll all samples in the sequence over time. Once unrolled, the graph is expanded into a DAG and the forward computation and gradient calculation algorithms we just discussed can be directly used. This means, however, all computation nodes in the CN need to be computed sample by sample and this significantly reduces the potential of parallelization.

There are two approaches to speed up the computation of a CN with directed loops. In the next two subsections, we will discuss them.

14.6.1 Sample by Sample Processing Only Within Loops

The first approach identifies the loops in the CN and only applies the sample-by-sample computation for nodes inside the loops. For the rest of the computation nodes, all samples in the sequence can be computed in parallel as a single matrix operation. For example, in Fig. 14.8 all the nodes included in the loop of $\mathbf{T}^{(3)} \rightarrow \mathbf{P}^{(3)} \rightarrow \mathbf{S}^{(1)} \rightarrow \mathbf{D} \rightarrow \mathbf{T}^{(3)}$ need to be computed sample by sample. All the rest of the nodes can be computed in batches. A popular technique is to identify the strongly connected components (SCC) in the graph, in which there is a path between each pair of vertices and adding any edges or vertices would violate this property, using algorithms such as Tarjan's strongly connected components algorithm [28].[1] Once the loops are identified, they can be treated as a composite node in the CN and the CN is reduced to a DAG. All the nodes inside each loop (or composite node) can be unrolled over time and also reduced to a DAG. For all these DAGs, the forward computation and backpropagation algorithms we discussed in the previous sections can be applied. The detailed procedure in determining the forward computation order in the CN with arbitrary recurrent connections is described in Algorithm 14.6. Since the input to the delay nodes are computed in the past, they can be considered as the leaf if we only consider one time slice. This makes the order decision inside loops much easier.

14.6.2 Processing Multiple Utterances Simultaneously

The second approach to speed up the processing in the recurrent CN is to process multiple sequences at a time. To implement this, we need to organize sequences in a way that the frames with the same frame id from different sequences are grouped together as shown in Fig. 14.9. By organizing sequences in this way, we can compute frames from different sequences in batch when inside a loop and compute all samples in all utterances in one batch outside loops. For example, in Fig. 14.9 we can compute four frames together for each time step. If sequences have different lengths, we can truncate them to the same length and save the final state of the sequences that are not finished yet. The remaining frames can be grouped with other sequences for further processing.

14.6.3 Building Arbitrary Recurrent Neural Networks

With the inclusion of delay nodes, we can easily construct complicated recurrent networks and dynamic systems. For example, the long short term memory (LSTM) [9, 14] neural network that is widely used to recognize and generate handwritten characters involves the following operations:

---

[1] Tarjan's algorithm is favored over others such as Kosaraju's algorithm [15] since it only requires one depth-first traverse, has a complexity of $O(|\mathbb{V}| + |\mathbb{E}|)$, and does not require reversing arcs in the graph.

Algorithm 14.6 Forward computation of an arbitrary CN.

1: **procedure** DECIDEFORWARDCOMPUTATIONORDERWITHRECCURENTLOOP($G = (V, E)$)
2:      StronglyConnectedComponentsDetection(G, G')   ▷ G' is a DAG of strongly connected components (SCC)
3:      Call DecideForwardComputationOrder on $G' \rightarrow order$ for DAG
4:      **for** $v \in G, v \in V$ **do**
5:          Set the *order* of v equal the max order of the SCC V ▷ This guarantee the forward order of SCC is correct
6:      **end for**
7:      **for** each SCC V in G **do**
8:          Call GetLoopForwardOrder($root$ of V) $\rightarrow order$ for each SCC
9:      **end for**
10:     **return** $order$ for DAG and $order$ for each SCC (loop)
11: **end procedure**
12: **procedure** GETLOOPFORWARDORDER($root$)
13:     Treat all the $delayNode$ as leaf and Call DecideForwardComponentionOrder
14: **end procedure**
15: **procedure** STRONGLYCONNECTEDCOMPONENTSDETECTION($G = (V, E), DAG$)
16:     $index = 0, S = empty$
17:     **for** $v \in V$ **do**
18:         **if** $v.index$ is undefined **then** StrongConnectComponents(v, DAG)
19:         **end if**
20:     **end for**
21: **end procedure**
22: **procedure** STRONGCONNECTCOMPONENT(v, DAG)
23:     $v.index = index, v.lowlink = index, index = index + 1$
24:     $S.push(v)$
25:     **for** $(v, w) \in E$ **do**
26:         **if** $w.index$ is undefined **then**
27:             StrongConnectComponent(w)
28:             $v.lowlink = \min(v.lowlink, w.lowlink)$
29:         **else if** $w \in S$ **then**
30:             $v.lowlink = \min(v.lowlink, w.index)$
31:         **end if**
32:     **end for**
33:     **if** $v.lowlink = v.index$ **then**   ▷ If v is a root node, pop the stack and generate an SCC
34:         start a new strongly connected component
35:         **repeat**
36:             $w = S.pop()$
37:             add w to current strongly connected component
38:         **until** $w == v$
39:         Save current strongly connected component to DAG
40:     **end if**
41: **end procedure**

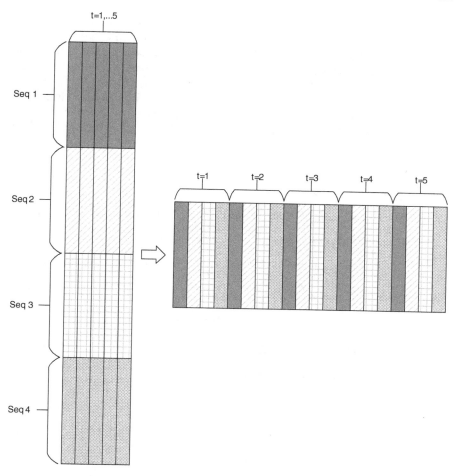

Fig. 14.9 Process multiple sequences in a batch. Shown in the figure is an example with four sequences. Each *color* represents one sequence. Frames with the same frame id from different sequences are grouped together and computed in batch. The right matrix is a reshape of the left matrix

$$\mathbf{i}_t = \sigma \left(\mathbf{W}^{(xi)}\mathbf{x}_t + \mathbf{W}^{(hi)}\mathbf{h}_{t-1} + \mathbf{W}^{(ci)}\mathbf{c}_{t-1} + \mathbf{b}^{(i)} \right) \qquad (14.147)$$

$$\mathbf{f}_t = \sigma \left(\mathbf{W}^{(xf)}\mathbf{x}_t + \mathbf{W}^{(hf)}\mathbf{h}_{t-1} + \mathbf{W}^{(cf)}\mathbf{c}_{t-1} + \mathbf{b}^{(f)} \right) \qquad (14.148)$$

$$\mathbf{c}_t = \mathbf{f}_t \bullet \mathbf{c}_{t-1} + \mathbf{i}_t \bullet \tanh \left(\mathbf{W}^{(xc)}\mathbf{x}_t + \mathbf{W}^{(hc)}\mathbf{h}_{t-1} + \mathbf{b}^{(c)} \right) \qquad (14.149)$$

$$\mathbf{o}_t = \sigma \left(\mathbf{W}^{(xo)}\mathbf{x}_t + \mathbf{W}^{(ho)}\mathbf{h}_{t-1} + \mathbf{W}^{(co)}\mathbf{c}_t + \mathbf{b}^{(o)} \right) \qquad (14.150)$$

$$\mathbf{h}_t = \mathbf{o}_t \bullet \tanh \left(\mathbf{c}_t \right), \qquad (14.151)$$

where σ (.) is the logistic sigmoid function, $\mathbf{i}_t$, $\mathbf{f}_t$, $\mathbf{o}_t$, $\mathbf{c}_t$ and $\mathbf{h}_t$ are vectors with same size (since there is one and only one of each in every cell) to represent values at time t of the input gate, forget gate, output gate, cell, cell input activation, and hidden layer, respectively, $\mathbf{W}$'s are the weight matrices connecting different gates, and $\mathbf{b}$'s are the corresponding bias vectors. All the weight matrices are full except the weight matrix $\mathbf{W}^{(ci)}$ from the cell to gate vectors which is diagonal. It is obvious that the whole LSTM can be described as a CN with following node types: *Times, Plus, Sigmoid, Tanh, DiagTimes, ElementTimes* and *Delay*. More specifically, using these computational nodes, the LSTM can be described as:

$$\mathbf{H}^{(d)} = Delay\ (\mathbf{H}) \tag{14.152}$$

$$\mathbf{C}^{(d)} = Delay\ (\mathbf{C}) \tag{14.153}$$

$$\mathbf{T}^{(1)} = Macro2W1b(\mathbf{W}^{(xi)}, \mathbf{X}, \mathbf{W}^{(hi)}, \mathbf{H}^{(d)}, \mathbf{b}^{(i)}) \tag{14.154}$$

$$\mathbf{I} = Sigmoid\left(Plus\left(\mathbf{T}^{(1)}, DiagTimes\left(\mathbf{d}^{(ci)}, \mathbf{C}^{(d)}\right)\right)\right) \tag{14.155}$$

$$\mathbf{T}^{(2)} = Macro3W1b\left(\mathbf{W}^{(xf)}, \mathbf{X}, \mathbf{W}^{(hf)}, \mathbf{H}^{(d)}, \mathbf{W}^{(cf)}\mathbf{C}^{(d)}, \mathbf{b}^{(f)}\right) \tag{14.156}$$

$$\mathbf{F} = Sigmoid\left(\mathbf{T}^{(2)}\right) \tag{14.157}$$

$$\mathbf{T}^{(3)} = Macro2W1b\left(\mathbf{W}^{(xc)}, \mathbf{X}, \mathbf{W}^{(hc)}, \mathbf{H}^{(d)}, \mathbf{b}^{(c)}\right) \tag{14.158}$$

$$\mathbf{T}^{(4)} = ElementTime\left(\mathbf{F}, \mathbf{C}^{(d)}\right) \tag{14.159}$$

$$\mathbf{T}^{(5)} = ElementTimes\left(\mathbf{I}, Tanh\left(\mathbf{T}^{(3)}\right)\right) \tag{14.160}$$

$$\mathbf{C} = Plus\left(\mathbf{T}^{(4)}, \mathbf{T}^{(5)}\right) \tag{14.161}$$

$$\mathbf{T}^{(6)} = Macro3W1b\left(\mathbf{W}^{(xo)}, \mathbf{X}, \mathbf{W}^{(ho)}, \mathbf{H}^{(d)}, \mathbf{W}^{(co)}\mathbf{C}, \mathbf{b}^{(o)}\right) \tag{14.162}$$

$$\mathbf{O} = Sigmoid\left(\mathbf{T}^{(6)}\right) \tag{14.163}$$

$$\mathbf{H} = ElementTime\ (\mathbf{O}, Tanh\ (\mathbf{C})), \tag{14.164}$$

where we have defined the macro $Macro2W1b\left(\mathbf{W}^{(1)}, \mathbf{I}^{(1)}, \mathbf{W}^{(2)}, \mathbf{I}^{(2)}, \mathbf{b}\right)$ as

$$\mathbf{S}^{(1)} = Plus\left(Times\left(\mathbf{W}^{(1)}, \mathbf{I}^{(1)}\right), Times\left(\mathbf{W}^{(2)}, \mathbf{I}^{(2)}\right)\right) \tag{14.165}$$

$$Macro2W1b = Plus\left(\mathbf{S}^{(1)}, \mathbf{b}\right) \tag{14.166}$$

and macro $Macro3W1b\left(\mathbf{W}^{(1)}, \mathbf{I}^{(1)}, \mathbf{W}^{(2)}, \mathbf{I}^{(2)}, \mathbf{W}^{(3)}, \mathbf{I}^{(3)}, \mathbf{b}\right)$ as

$$\mathbf{S}^{(1)} = Macro2W1b\left(\mathbf{W}^{(1)}, \mathbf{I}^{(1)}, \mathbf{W}^{(2)}, \mathbf{I}^{(2)}, \mathbf{b}\right) \tag{14.167}$$

$$Macro3W1b = Plus\left(\mathbf{S}^{(1)}, Times\left(\mathbf{W}^{(3)}, \mathbf{I}^{(3)}\right)\right) \tag{14.168}$$

Using CN, we can easily build arbitrary network architectures to solve different problems. For example, it will be easy to add a masking layer to reduce noises in the input features in an otherwise conventional DNN. CN provides the needed tool to explore new speech recognition architectures and we believe additional advancement in the field will attribute at least partially to CN.

References

1. Abdel-Hamid, O., Deng, L., Yu, D.: Exploring convolutional neural network structures and optimization techniques for speech recognition pp. 3366–3370 (2013)
2. Abdel-Hamid, O., Mohamed, A.r., Jiang, H., Penn, G.: Applying convolutional neural networks concepts to hybrid nn-hmm model for speech recognition. In: Proceedings of the International Conference on Acoustics, Speech and Signal Processing (ICASSP), pp. 4277–4280. IEEE (2012)
3. Bergstra, J., Breuleux, O., Bastien, F., Lamblin, P., Pascanu, R., Desjardins, G., Turian, J., Warde-Farley, D., Bengio, Y.: Theano: a CPU and GPU math expression compiler. In: Proceedings of the Python for Scientific Computing Conference (SciPy), vol. 4 (2010)
4. Bischof, C., Roh, L., Mauer-Oats, A.: ADIC: an extensible automatic differentiation tool for ANSI-C. Urbana 51, 61,802 (1997)
5. Chellapilla, K., Puri, S., Simard, P.: High performance convolutional neural networks for document processing. In: Tenth International Workshop on Frontiers in Handwriting Recognition (2006)
6. Ciresan, D.C., Meier, U., Schmidhuber, J.: Transfer learning for Latin and Chinese characters with deep neural networks. In: Proceedings of the International Conference on Neural Networks (IJCNN), pp. 1–6 (2012)
7. Dahl, G.E., Yu, D., Deng, L., Acero, A.: Context-dependent pre-trained deep neural networks for large-vocabulary speech recognition. IEEE Trans. Audio, Speech Lang. Process. 20(1), 30–42 (2012)
8. Deng, L., Abdel-Hamid, O., Yu, D.: A deep convolutional neural network using heterogeneous pooling for trading acoustic invariance with phonetic confusion. In: Proceedings of the International Conference on Acoustics, Speech and Signal Processing (ICASSP), pp. 6669–6673 (2013)
9. Graves, A.: Generating sequences with recurrent neural networks. arXiv preprint arXiv:1308.0850 (2013)
10. Griewank, A., Walther, A.: Evaluating derivatives: principles and techniques of algorithmic differentiation. Siam (2008)
11. Guenter, B.: Efficient symbolic differentiation for graphics applications. In: ACM Transactions on Graphics (TOG), vol. 26, p. 108 (2007)
12. Guenter, B., Yu, D., Eversole, A., Kuchaiev, O., Seltzer, M.L., "Stochastic Gradient Descent Algorithm in the Computational Network Toolkit", OPT2013: NIPS 2013 Workshop on Optimization for Machine Learning (2013)
13. Hinton, G., Deng, L., Yu, D., Dahl, G.E., Mohamed, Ar, Jaitly, N., Senior, A., Vanhoucke, V., Nguyen, P., Sainath, T.N., et al.: Deep neural networks for acoustic modeling in speech recognition: the shared views of four research groups. IEEE Signal Process. Mag. 29(6), 82–97 (2012)
14. Hochreiter, S., Schmidhuber, J.: Long short-term memory. Neural Comput. 9(8), 1735–1780 (1997)
15. Hopcroft, J.E.: Data Structures and Algorithms. Pearson Education, Boston (1983)
16. Jaitly, N., Nguyen, P., Senior, A.W., Vanhoucke, V.: Application of pretrained deep neural networks to large vocabulary speech recognition. In: Proceedings of the Annual Conference of International Speech Communication Association (INTERSPEECH) (2012)

17. Kavukcuoglu, K., Sermanet, P., Boureau, Y.L., Gregor, K., Mathieu, M., LeCun, Y.: Learning convolutional feature hierarchies for visual recognition. In: NIPS, vol. 1, p. 5 (2010)

18. Krizhevsky, A., Sutskever, I., Hinton, G.E.: Imagenet classification with deep convolutional neural networks. In: NIPS, vol. 1, p. 4 (2012)

19. LeCun, Y., Bengio, Y.: Convolutional networks for images, speech, and time series. The handbook of brain theory and neural networks 3361 (1995)

20. Mikolov, T., Zweig, G.: Context dependent recurrent neural network language model. In: Proceedings of the IEEE Spoken Language Technology Workshop (SLT), pp. 234–239 (2012)

21. Sainath, T.N., Kingsbury, B., Mohamed, A.r., Dahl, G.E., Saon, G., Soltau, H., Beran, T., Aravkin, A.Y., Ramabhadran, B.: Improvements to deep convolutional neural networks for lvcsr. In: Proceedings of the IEEE Workshop on Automfatic Speech Recognition and Understanding (ASRU), pp. 315–320 (2013)

22. Sainath, T.N., Mohamed, A.r., Kingsbury, B., Ramabhadran, B.: Deep convolutional neural networks for LVCSR. In: Proceedings of the International Conference on Acoustics, Speech and Signal Processing (ICASSP), pp. 8614–8618 (2013)

23. Seide, F., Li, G., Chen, X., Yu, D.: Feature engineering in context-dependent deep neural networks for conversational speech transcription. In: Proceedings of the IEEE Workshop on Automfatic Speech Recognition and Understanding (ASRU), pp. 24–29 (2011)

24. Seide, F., Li, G., Yu, D.: Conversational speech transcription using context-dependent deep neural networks. In: Proceedings of the Annual Conference of International Speech Communication Association (INTERSPEECH), pp. 437–440 (2011)

25. Shi, Y., Wiggers, P., Jonker, C.M.: Towards recurrent neural networks language models with linguistic and contextual features. In: Proceedings of the Annual Conference of International Speech Communication Association (INTERSPEECH) (2012)

26. Socher, R., Lin, C.C., Ng, A., Manning, C.: Parsing natural scenes and natural language with recursive neural networks. In: Proceedings of the International Conference on Machine Learning (ICML), pp. 129–136 (2011)

27. Sutskever, I., Martens, J., Hinton, G.E.: Generating text with recurrent neural networks. In: Proceedings of the International Conference on Machine Learning (ICML), pp. 1017–1024 (2011)

28. Tarjan, R.: Depth-first search and linear graph algorithms. SIAM J. Comput. 1(2), 146–160 (1972)

29. Wiesler, S., Richard, A., Golik, P., Schluter, R., Ney, H.: RASR/NN: the RWTH neural network toolkit for speech recognition. In: Proceedings of the International Conference on Acoustics, Speech and Signal Processing (ICASSP), pp. 3305–3309 (2014)

30. Yu, D., Eversole, A., Seltzer, M., Yao, K., Huang, Z., Guenter, B., Kuchaiev, O., Zhang, Y., Seide, F., Wang, H., Droppo, J., Zweig, G., Rossbach, C., Currey, J., Gao, J., May, A., Stolcke, A., Slaney, M.: An introduction to computational networks and the computational network toolkit. Microsoft Technical Report MSR-TR-2014-112 (2014)

31. Yu, D., Seltzer, M.L.: Improved bottleneck features using pretrained deep neural networks. In: Proceedings of the Annual Conference of International Speech Communication Association (INTERSPEECH), pp. 237–240 (2011)

Chapter 15
Summary and Future Directions

Abstract In this chapter, we summarize the book by first listing and analyzing what we view as major milestone studies in the recent history of developing the deep learning-based ASR techniques and systems. We describe the motivations of these studies, the innovations they have engendered, the improvements they have provided, and the impacts they have generated. In this road map, we will first cover the historical context in which the DNN technology made inroad into ASR around 2009 resulting from academic and industry collaborations. Then we select seven main themes in which innovations flourished across-the-board in ASR industry and academic research after the early debut of DNNs. Finally, our belief is provided on the current state-of-the-art of speech recognition systems, and we also discuss our thoughts and analysis on the future research directions.

15.1 Road Map

There are many exciting advancements in the field of automatic speech recognition (ASR) in the past 5 years. However, in this book, we are able to cover only a representative subset of these achievements. Constrained by our limited knowledge, we have selected the topics we believe are useful for the readers to understand these progresses that were described with much more technical detail in many preceding chapters of this book. We feel one way to reasonably summarize these advancements is to provide an outline of major milestone studies achieved historically.

15.1.1 Debut of DNNs for ASR

The application of neural networks on ASR can be dated back to late 1980s. Notable work includes Waibel et al.'s time delay neural network (TDNN) [49, 93] and Morgan and Bourlard's artificial neural network (ANN)/hidden Markov model (HMM) hybrid system [63, 64].

The resurgence of interest in neural network-based ASR started in The 2009 NIPS Workshop on Deep Learning for Speech Recognition and Related Applications [18], where Mohamed et al. from University of Toronto presented a primitive version of

© Springer-Verlag London 2015
D. Yu and L. Deng, *Automatic Speech Recognition*,
Signals and Communication Technology, DOI 10.1007/978-1-4471-5779-3_15

a deep neural network (DNN)[1]—HMM hybrid system for phone recognition [60]. Detailed error analysis and comparisons of the DNN with other speech recognizers were carefully conducted at Microsoft Research (MSR) jointly by MSR and University of Toronto researchers and relative strengths and weaknesses were identified prior to and after the workshop. See [17] for discussions and reflections on this part of early studies that were carried out with "a lot of intuitive guesses without much evidence to support the individual decisions." In [60], the same type of ANN/HMM hybrid architecture was adopted as those developed in early 1990s [63, 64] but it used a DNN to replace the shallow multilayer perceptron (MLP) often used in the early ANN/HMM systems. More specifically, the DNN was constructed to model monophone states and was trained using the frame-level cross-entropy criterion on the conventional MFCC features. They showed that by just using a deeper model they managed to achieve a 23.0% phone error rate (PER) on the TIMIT core test set. This result is significantly better than the 27.7 and 25.6% PER [73] achieved by a monophone and triphone Gaussian mixture model (GMM)-HMM, respectively, trained with the maximum likelihood estimation (MLE) criterion, and is also better than 24.8% PER achieved by a deep, monophone version of generative models of speech [26, 30] developed at MSR but with distinct recognition error patterns (not published). Although their model performs worse than the triphone GMM-HMM system trained using the sequence-discriminative training (SDT) criterion, which achieved 21.7% PER[2] on the same task, and was evaluated only on the phone recognition task, we at MSR noticed its potential because in the past the ANN/HMM hybrid system was hard to beat the context-dependent (CD)-GMM-HMM system trained with the MLE criterion and also because the DNN and the deep generative models were observed to produce very different types of recognition errors with explainable causes based on aspects of human speech production and perception mechanisms. In the mean time, collaborations between MSR and University of Toronto researchers which started in 2009 also looked carefully into the use of raw speech features, one of the fundamental premises of deep learning advocating not to use human-engineered features such as MFCCs. Deep autoencoders were first explored on speech historically, during 2009–2010 at MSR for binary feature encoding and bottleneck feature extraction, where deep architectures were found superior to shallow ones and spectrogram features found superior to MFCCs [21]. All the above kind of insightful and exciting results and progress on speech feature extraction, phone recognition, and error analysis, etc. had never been seen in the speech research history before and have pointed to high promise and practical value of deep learning. This early progress excited MSR researchers to devote more resources to pursue ASR research using deep learning approaches, the DNN approach in particular. A series of studies along this line can be found in [19, 25].

---

[1] Note that while the model was called deep belief network (DBN) at that time, it is in fact a deep neural network (DNN) initialized using the DBN pretraining algorithm. See Chaps. 4 and 5 as well as [25] for discussions on the precise differences between DNNs and DBNs.

[2] The best GMM-HMM system can achieve 20.0% PER on the TIMIT core test set [73].

Our interest at MSR was to improve large vocabulary speech recognition (LVSR) for real-world applications. In early 2010, we started to collaborate with the two student authors of the work [60] to investigate DNN-based ASR techniques. We used the voice search (VS) dataset described in Sect. 6.2.1 to evaluate our new models. We first applied the same architecture used by Mohamed et al. [60], which we refer to as the context-independent (CI)-DNN-HMM, to the LVSR. Similar to the results on the TIMIT phoneme recognition task, this CI-DNN-HMM, trained with 24h of data, achieved 37.3 % sentence error rate (SER) on the VS test set. This result sits in between the 39.6 and 36.2 % SER achieved with the CD-GMM-HMM trained using the MLE and SDT criteria, respectively. The performance breakthrough happened after we adopted the CD-DNN-HMM architecture described in Chap. 6 in which the DNN directly models the tied triphone states (also called senones). The CD-DNN-HMM based on senones achieved 30.1 % SER. It was shown experimentally to cut errors by 17 % over the 36.2 % SER obtained with the CD-GMM-HMM trained using the SDT criterion, and to cut errors by 20 % over the 37.3 % SER obtained using the CI-DNN-HMM in a few papers published by Yu et al. [98] and Dahl et al. [13] . This is the first time the DNN-HMM system was successfully applied to LVSR tasks. In retrospect, it may be possible that other researchers have also thought about similar ideas and even have tried variants of it in the past. However, due to the limited computing power and training data in early days, no one was able to train the models with the large size that we use today.

As successful as it was in retrospect, the above early work on CD-DNN-HMM had not drawn as much attention from speech researchers and practitioners at the time of publications of the studies in 2010 and 2011. This was understandable as the ANN/HMM hybrid system did not win over the GMM-HMM system in mid-1990s and was not considered the right way to go. To turn over this belief, researchers were looking for stronger evidence than that on the Microsoft internal voice search dataset which contains up to 48h of training data in those early experiments.

The CD-DNN-HMM work started to show greater impact after Seide et al. of MSR published their results in September 2011 [79] on applying the same CD-DNN-HMMs as that reported by Yu et al. [98] and Dahl et al. [13] to the Switchboard benchmark dataset [34] described in Sect. 6.2.1. This work scaled CD-DNN-HMMs to 309h of training data and thousands of senones. It demonstrated, quite surprisingly to many people, that the CD-DNN-HMM trained using the frame cross-entropy criterion can achieve as low as 16.1 % word error rate (WER) on the HUB5'00 evaluation set—a 1/3 cut of error over the 23.6 % WER obtained with the CD-GMM-HMM trained using the SDT criterion. This work also confirmed and clarified the findings in [13, 14, 98]: the three key ingredients to make CD-DNN-HMM perform well are: (1) using deep models, (2) modeling senones, and (3) using a contextual window of features as input. It further demonstrated that realignment in training DNN-DMMs helps improve recognition accuracy and that pretraining of DNNs sometimes helps but is not critical. Since then, many ASR research groups shifted their research focus to CD-DNN-HMM and made significant progresses.

15.1.2 Speedup of DNN Training and Decoding

Right after the work [79] was published, many companies started to adopt it in their commercial systems. The first barrier they need to conquer is the decoding speed. With a naive implementation, it takes 3.89 real time on a single CPU core just to compute the DNN score. In late 2011, just several months after [79] was published, Vanhoucke et al. from Google published their work on DNN speedup using engineering optimization techniques [90].[3] They showed that the DNN evaluation time can be reduced to 0.21 real time on a single CPU core by using quantization, SIMD instructions, batching and lazy evaluation—a 20× speedup over the naive implementation. Their work is a great step ahead since it demonstrated that the CD-DNN-HMM can be used in real-time commercial systems without penalty on the decoding speed or throughput.

The second barrier they need to overcome is training speed. Although it has been shown that the CD-DNN-HMM system trained with 309 h of data outperforms the CD-GMM-HMM system trained with 2,000 h of data [79], additional accuracy improvement can be obtained if the DNN system is trained using the same amount of data as that used to train CD-GMM-HMMs. To achieve this goal some sort of parallel training algorithm needs to be developed. At Microsoft, a pipelined GPU training strategy was proposed and evaluated in 2012. The work [10] done by Chen et al. demonstrated that a speedup of 3.3 times can be achieved on 4 GPUs with this approach. Google, on the other hand, adopted the asynchronous stochastic gradient descent (ASGD) algorithm [50, 65] on the CPU clusters.

A different but notable approach to speeding up DNN training and evaluation is the low-rank approximation described in Sect. 7.2.3. In 2013, Sainath et al. from IBM and Xue et al. from Microsoft independently proposed to reduce the model size and training and decoding time by approximating the large weight matrices with the product of smaller ones [71, 97]. This technique can reduce 2/3 of the decoding time. Due to its simplicity and effectiveness it has been widely used in commercial ASR systems.

15.1.3 Sequence Discriminative Training

The exciting results reported in [79] were achieved using the frame-level cross-entropy training criterion. Many research groups noticed that an obvious and low-risk way to further improve ASR accuracy is to use the sequence discriminative training criterion widely adopted for training the state-of-the-art GMM systems.

In fact, back in 2009, before the debut of DNN systems, Brian Kingsbury from IBM Research already proposed a unified framework to train ANN/HMM hybrid systems with SDT [46]. Although the ANN/HMM system he tested in his work

---

[3] Microsoft optimized the DNN evaluation in the internal tool using similar techniques slightly earlier but never published the results.

performs worse than the CD-GMM-HMM system, he did show that the ANN/HMM system trained with the SDT criterion (achieved a 27.7 % WER) performs significantly better than that trained with the frame-level cross-entropy criterion (achieved a 34.0 % WER on the same task). Hence, this work did not attract strong attention at the time since even with the SDT the ANN/HMM system still cannot beat the GMM system.

In 2010, in parallel with the LVSR work, we at MSR clearly realized the importance of sequence training based on the GMM-HMM experience [37, 99, 103] and started the work on sequence discriminative training for CI-DNN-HMM for phone recognition [66]. Unfortunately, we did not find the right approach to control the overfitting problem by then and thus only observed very small improvement by using SDT (22.2 % PER) over the frame cross-entropy training (22.8 % PER)

The breakthrough happened in 2012 when Kingsbury et al. from IBM Research successfully applied the technique described in Kingsbury's 2009 work [46] to the CD-DNN-HMM [45]. Since SDT takes longer time to train than the frame-level cross-entropy training they exploited the Hessian-free training algorithm [55] on a CPU cluster to speed up the training. With SDT the CD-DNN-HMM trained on the SWB 309 h training set obtained a WER of 13.3 % on the Hub5'00 evaluation set. It cuts error by relative 17 % over the already low 16.1 % WER achieved using the frame-level cross-entropy criterion. Their work demonstrated that SDT can be effectively applied to the CD-DNN-HMM and result in great accuracy improvement. More importantly, the 13.3 % WER achieved using the single pass speaker-independent CD-DNN-HMM is also much better than the WER of 14.5 % achieved using the best multipass speaker-adaptive GMM system. Thus, there is obviously no reason not to replace the GMM systems with the DNN systems in the commercial systems given this result.

However, SDT is tricky and not easy to be implemented correctly. In 2013, the work done at MSR by Su et al. [86] and the joint work done by Veselý et al. from Brno University, University of Edinburgh, and Johns Hopkins University [91] proposed a series of practical techniques for making the SDT effective and robust. These techniques, including lattice compensation, frame dropping, and F-smoothing, are now widely used.

15.1.4 Feature Processing

The feature processing pipeline in the conventional GMM systems involves many steps because the GMMs themselves cannot transform features. In 2011, Seide et al. conducted research at MSR on the effect of feature engineering techniques in the CD-DNN-HMM systems [78, 101]. They found that many feature processing steps, such as HLDA [48] and fMLLR [32], that are important in the GMM systems and shallow ANN/HMM hybrid systems are less important in the DNN systems. They explained their results by considering all the hidden layers in the DNN as a powerful nonlinear feature transformation and the softmax layer as a log-linear classifier. The feature transformation and the classifier are jointly optimized. Since DNNs can take correlated inputs many features that cannot be directly used in the GMM systems can now be used in the DNN systems. Because DNNs can approximate complicated feature transformation through many layers of nonlinear operations many of the feature processing steps used in the GMM systems may be removed without sacrificing the accuracy.

In 2012, Mohamed et al. from University of Toronto showed that by using the log Mel-scale filter bank feature instead of MFCC they can reduce PER from 23.7 to 22.6 % on the TIMIT phone recognition task using a two-layer network [61]. Around the same time Li et al. at Microsoft demonstrated that using log Mel-scale filter bank features improves the accuracy on LVSR [52]. They also showed that by using log Mel-scale filter bank features tasks such as mixed-bandwidth speech recognition can be easily implemented in the CD-DNN-HMM systems. Log Mel-scale filter bank features are now the standard in most CD-DNN-HMM systems. Deng et al. reported a series of studies on the theme of backing to the deep learning premise of using spectrogram-like speech features [19].

The efforts to reduce the feature processing pipeline never ceased. For example, the work done at IBM Research by Sainath et al. in 2013 [70] showed that the CD-DNN-HMM systems can directly take the FFT spectrum as the input and learn the Mel-scale filters automatically. Most recently, the use of raw time waveform signals of speech by DNNs (i.e., zero-feature extraction prior to DNN training) was reported in [89]. The study demonstrates the same advantage of learning truly nonstationary patterns of the speech signal localized in time across frame boundaries by the DNN as the earlier waveform-based and HMM-based generative models of speech [31, 82], but very different kinds of challenges remain to be overcome.

15.1.5 Adaptation

When the CD-DNN-HMM system just showed its effectiveness on the Switchboard task in 2011, one of the concerns back then was the lack of effective adaptation techniques, esp. since DNN systems have much more parameters than that in the conventional ANN/HMM hybrid systems. To address this concern, in 2011 in the work done at Microsoft Research by Seide et al. the feature discriminative linear regression (fDLR) adaptation technique was proposed and evaluated on the Switchboard dataset with small improvement in the accuracy [78].

In 2013, Yu et al. conducted a study at Microsoft [102], showing that by using Kullback–Leibler divergence (KLD) regularization they can effectively adapt the CD-DNN-HMM on the short message dictation tasks with 3–20 % relative error reduction over the speaker-independent systems when different number of adaptation utterances are used. Their work indicates that adaptation on CD-DNN-HMM systems can be important and effective.

Later in the same year and in 2014, a series of adaptation techniques based on a similar architecture were developed. In the noise aware training (NaT) [81] developed at Microsoft by Seltzer et al., a noise code is estimated and used as part of the input. In this work, they showed that with NaT they can reduce the WER on the Aurora4 dataset from 13.4 to 12.4%, beating even the most complicated GMM system on the same task. In speaker aware training (SaT) [76] developed at IBM by Saon et al., a speaker code is estimated as the i-vector of the speaker and used as part of the input. They reported the results on the Switchboard dataset and reduced WER from 14.1 to 12.4% on the Hub5'00 evaluation set, a 12% error cut. In the speaker code approach [2, 96] developed at York University by Abdel-Hamid et al., the speaker code is trained for each speaker jointly with the DNN and used as part of the input.

15.1.6 Multitask and Transfer Learning

As pointed out in [9, 16, 78, 101] and discussed in Chap. 12, each hidden layer in the DNN can be considered a new representation of the input feature. This interpretation motivated many studies in sharing the same representation across languages and modalities. In 2012–2013,[4] many groups including Microsoft, IBM, Johns Hopkins University, University of Edinburgh, and Google reported results on using the shared hidden-layer architectures for multilingual and cross-lingual ASR [33, 38, 42, 88], multimodal ASR [41], and multiobjective training of DNNs for speech recognition [11, 54, 80]. These studies indicated that by training the shared hidden layers with data from multiple languages and modalities or with multiple objectives, we can build DNNs that work better for each language or modality than those trained specifically for the language or modality. This approach often helps ASR tasks the most for the languages with very limited training data.

15.1.7 Convolution Neural Networks

Using log Mel-scale filter bank features as the input also opens a door to apply techniques such as convolution neural networks (CNNs) to exploit the structure in the features. In 2012, Abdel-Hamid et al. showed for the first time that by using a CNN along the frequency axis they can normalize speaker differences and further reduce the PER from 20.7 to 20.0% on the TIMIT phone recognition task [4].

These results were later extended to LVSR in 2013 with improved CNN architectures, pretraining techniques, and pooling strategies in the studies by Abdel-Hamid et al. [1, 3] and Deng et al. [15] at Microsoft Research and the study by Sainath et al. [69, 72] at IBM Research. Further studies showed that the CNN helps mostly for the tasks in which the training set size or data variability is relatively small.

---

[4] Some earlier work such as [77] exploited similar ideas but not on the DNN.

For most other tasks, the relative WER reduction is often in the small range of 2–3 %. We believe as the training set size continues to increase the gap between the systems with and without CNN will diminish.

15.1.8 Recurrent Neural Networks and LSTM

Since the inroad of the DNN into ASR starting in 2009, perhaps the most notable new deep architecture is the recurrent neural network (RNN), esp. its long-short-term-memory (LSTM) version. While the RNN as well as the related nonlinear neural predictive models saw its early success in small ASR tasks [28, 68], it was not easy to duplicate due to the intricacy in training, let alone to scale them up for larger ASR tasks. Learning algorithms for the RNN have been dramatically improved since these early days, however, and much stronger and practical results have been obtained recently using the RNN, especially when the bidirectional LSTM architecture is exploited [35, 36] or when the high-level DNN features are used as inputs to the RNN [9, 16].

The LSTM was reported to give the lowest PER on the benchmark TIMIT phone recognition task in 2013 by Grave et al. at University of Toronto [35, 36]. In 2014, Google researchers published the results using the LSTM on large-scale tasks with applications to Google Now, voice search, and mobile dictation with excellent accuracy results [74, 75]. To reduce the model size, the otherwise very large output vectors of LSTM units are linearly projected to smaller-dimensional vectors. Asynchronous stochastic gradient descent (ASGD) algorithm with truncated back-propagation through time (BPTT) is performed across hundreds of machines in CPU clusters. The best accuracy is obtained by optimizing the frame-level cross-entropy objective function followed by sequence discriminative training. With one LSTM stacking on top of another, this deep and recurrent LSTM model produced 9.7 % WER on a large voice search task trained with 3 million utterances. This result is better than 10.7 % WER achieved with frame-level cross entropy training criterion alone. It is also significantly better than the 10.4 % WER obtained with the best DNN-HMM system using rectified linear units. Furthermore, this better accuracy is achieved while the total number of parameters is drastically reduced from 85 millions in the DNN system to 13 millions in the LSTM system. Some recent publications also showed that deep LSTMs are effective in ASR in reverberant mutltisource acoustic environments, as indicated by the strong results achieved by LSTMs in a recent ChiME Challenge task involving ASR in such difficult environments [95].

15.1.9 Other Deep Models

A number of other deep learning architectures have been developed for ASR. These include the deep tensor neural networks [100, 104], deep stacking networks and their kernel version [22–24], tensor deep stacking networks [43, 44], recursive perceptual models [92], sequential deep belief networks [5], and ensemble deep learning architectures [20]. However, although these models have superior theoretical and computational properties over most of the basic deep models discussed above, they have not been explored with as much depth and scope, and are not mainstream methods in ASR so far.

15.2 State of the Art and Future Directions

15.2.1 State of the Art—A Brief Analysis

By combining CNN, DNN, and i-vector based adaptation techniques IBM researchers showed in 2014 that they can reduce the WER on the Switchboard Hub5'00 evaluation set to 10.4 %. Compared to the best possible WER of 14.5 % on the same evaluation set achieved using the GMM systems, DNN systems cut the error by 30 %. This improvement is achieved solely through the acoustic model (AM) improvement. Recent advancements in neural network-based language model (LM) and large-scale n-gram LM can further cut the error by 10–15%. Together the WER on the Switchboard task can be reduced to below 10 %. Also, the LSTM-RNN model developed by Google researchers also demonstrated in 2014 dramatic error reduction in voice search tasks compared with other methods including those based on the feedforward DNN.

In fact, in many commercial systems the word (or character) error rates for the tasks such as mobile short message dictation and voice search are way below 10 %. Some companies are even aiming at reducing the *sentence* error rate to below 10 %. From the practical usage point of view, we can reasonably regard that deep learning has largely solved the close-talk single-speaker ASR problem.

As we relax the constraints we impose on the tasks we are working on, however, we can quickly realize that the ASR systems still perform poorly under the following conditions even given the recent technology advancements:

- ASR with far field microphones; e.g., when the microphone is backgrounded in a living room, meeting room, or field video recordings;
- ASR under very noisy conditions; e.g., when loud music is playing and captured by the microphone;
- ASR with accented speech;
- ASR with multitalker speech or side talks; e.g., in a meeting or in multiparty chatting;
- ASR with spontaneous speech in which the speech is not fluent, with variable speed or with emotions;

For these tasks, the WER of the current best systems is often in the range of 20 %. New technological advances or clever engineering are needed to bring the errors down much further in order to make ASR useful under these difficult yet practical and realistic conditions.

15.2.2 Future Directions

We believe the ASR accuracy under some (not all) of the above conditions may be increased even without substantially new technologies in acoustic modeling for ASR. For example, by using more advanced microphone array techniques we can significantly reduce noise and side-talks and thus improve the recognition accuracy under these conditions. We may also generate or collect more training data for far field microphones and thus improve the performance when similar microphones are used.

However, to ultimately solve the ASR problem so that the ASR system's performance can match or even exceed that of human's under all conditions,[5] new techniques and paradigms in acoustic modeling are needed. We perceive that the next-generation ASR systems can be solely described as a dynamic system that involves many connected components and recurrent feedbacks and can constantly make predictions, corrections, and adaptation. For example, the future ASR system will be able to automatically identify multiple talkers in the mixed speech or resolve speech and noise in noisy speech. The system will then be able to focus on and trace a specific speaker by ignoring other speakers and noises. This is a cognitive function of attention that humans effortlessly equipped with yet conspicuously lacking in today's ASR systems. The future ASR system will also need to be able to learn the key speech characteristics from the training set and generalize well to unseen speakers, accents, noisy conditions.

In order to move toward building such new ASR systems, it is highly desirable to first build powerful tools such as computational network (CN) and computational network toolkit (CNTK) we described in Chap. 14. Such tools would allow large-scale and systematic experimentation on many more advanced deep architectures and algorithms, some of which are outlined in the preceding section, than the basic DNN and RNN. Further, as we discussed in the RNN chapter, new learning algorithms will need to be developed that can integrate the strengths of discriminative dynamic models (e.g., the RNN) with bottom-up information flow and of generative dynamic models with top-down information flow while overcoming their respective weaknesses. Recent progresses in stochastic and neural variational inference shown to be

---

[5] Under some constrained conditions, ASR systems can already perform better than humans. For example, in 2008 ASR systems can already beat the human performance on clean digit recognition with a 0.2 % error rate [103]. In 2006, IBM researchers Kristjansson et al. reported their results on single-channel multitalker speech recognition [47]. Their improved system [12] in 2010 can achieve 21.6 % WER on the challenge task with very constrained language model and closed speaker set. This result is better than 22.3 % WER achieved by humans. In 2014, the DNN-based system developed at Microsoft Research by Weng et al. achieved a WER of 18.8 % [94] on the same task and generated far less errors than humans did.

effective for learning deep generative models [7, 39, 59, 67] have moved us one step closer to the desired bottom-up and top-down learning algorithms with multipasses.

We speculate that the next-generation ASR systems may also seamlessly integrate semantic understanding, for example, to constrain the search space and correct semantically inconsistent hypotheses and thus benefit from research work in semantic understanding. In this direction, one needs to develop better semantic representations for word sequences, which are the output of ASR systems. Recent advances in continuous vector-space distributed representations of words and phrases [40, 58, 83–85], also known as word embedding or phrase embedding, have moved us one step closer to this goal.

Most recently, the concept of word embedding (i.e., with distributed representations of words) is introduced as an alternative to the traditional, phonetic-state-based pronunciation model in ASR, giving improvement in ASR accuracy [6]. This exemplifies an interesting new approach, based on distributed representations in continuous vector space, to modeling the linguistic symbols as the ASR output. It appears to be more powerful than several earlier approaches to distributed representations of word sequences in symbolic vector space—articulatory or phonetic-attribute-based phonological models [8, 27, 29, 51, 53, 87]. Further research along this direction may exploit multiple modalities—speech acoustics and the associated image, gesture, and text—all embedded into the same "semantic" space of phonological nature—to support weakly supervised or even unsupervised learning for ASR.

For a longer term, we believe ASR research can benefit from human brain research projects and research in the areas of representation encoding and learning, recurrent networks with long-range dependency and conditional state switching, multitask and unsupervised learning, and prediction-based methods for temporal/sequential information processing. As examples, effective computational models of attention and of phonetic feature encoding in cortical areas of the human auditory system [56, 57] are expected to help bridge the performance gap between human and computer speech recognition. Modeling perceptual control and interactions between speaker and listener has also been proposed to help improve ASR and spoken language processing performance and its practical use [62]. These capabilities are so far not reachable by current deep learning technology, and require "looking outside" into other fields such as cognitive science, computational linguistics, knowledge representation and management, artificial intelligence, neuroscience, and bio-inspired machine learning.

References

1. Abdel-Hamid, O., Deng, L., Yu, D.: In: Exploring Convolutional Neural Network Structures and Optimization Techniques for Speech Recognition, pp. 3366–3370 (2013)
2. Abdel-Hamid, O., Jiang, H.: Fast speaker adaptation of hybrid NN/HMM model for speech recognition based on discriminative learning of speaker code. In: Proc. International Conference on Acoustics, Speech and Signal Processing (ICASSP), pp. 7942–7946 (2013)
3. Abdel-Hamid, O., Mohamed, A.r., Jiang, H., Deng, L.,Penn, G., Yu, D.: Convolutional neural networks for speech recognition. IEEE Trans. Audio, Speech Lang. Process. (2014)

4. Abdel-Hamid, O., Mohamed, A.r., Jiang, H., Penn, G.: Applying convolutional neural networks concepts to hybrid nn-hmm model for speech recognition. In: Proceedings of the International Conference on Acoustics, Speech and Signal Processing (ICASSP), IEEE, pp. 4277–4280 (2012)
5. Andrew, G., Bilmes, J.: Backpropagation in sequential deep belief networks. In: Neural Information Processing Systems (NIPS) (2013)
6. Bengio, S., Heigold, G.: Word embeddings for speech recognition. In: IINTERSPEECH (2014)
7. Bengio, Y.: Estimating or propagating gradients through stochastic neurons. In: CoRR (2013)
8. Bromberg, I., Qian, Q., Hou, J., Li, J., Ma, C., Matthews, B., Moreno-Daniel, A., Morris, J., Siniscalchi, S.M., Tsao, Y., Wang, Y.: Detection-based ASR in the automatic speech attribute transcription project. In: Proceedings of the Interspeech, pp. 1829–1832 (2007)
9. Chen, J., Deng, L.: A primal-dual method for training recurrent neural networks constrained by the echo-state property. In: Proceedings of the ICLR (2014)
10. Chen, X., Eversole, A., Li, G., Yu, D., Seide, F.: Pipelined back-propagation for context-dependent deep neural networks. In: Proceedings of the Annual Conference of International Speech Communication Association (INTERSPEECH) (2012)
11. Chen, D., Mak, B., Leung, C.C., Sivadas, S.: Joint acoustic modeling of triphones and tri-graphemes by multi-task learning deep neural networks for low-resource speech recognition. In: Proceedings of the International Conference on Acoustics, Speech and Signal Processing (ICASSP) (2014)
12. Cooke, M., Hershey, J.R., Rennie, S.J.: Monaural speech separation and recognition challenge. Comput. Speech Lang. **24**(1), 1–15 (2010)
13. Dahl, G.E., Yu, D., Deng, L., Acero, A.: Large vocabulary continuous speech recognition with context-dependent DBN-HMMs. In: Proceedings of the International Conference on Acoustics, Speech and Signal Processing (ICASSP), pp. 4688–4691 (2011)
14. Dahl, G.E., Yu, D., Deng, L., Acero, A.: Context-dependent pre-trained deep neural networks for large-vocabulary speech recognition. IEEE Trans. Audio, Speech Lang. Process. **20**(1), 30–42 (2012)
15. Deng, L., Abdel-Hamid, O., Yu, D.: A deep convolutional neural network using heterogeneous pooling for trading acoustic invariance with phonetic confusion. In: Proceedings of the International Conference on Acoustics, Speech and Signal Processing (ICASSP), pp. 6669–6673 (2013)
16. Deng, L., Chen, J.: Sequence classification using high-level features extracted from deep neural networks. In: Proceedings of the International Conference on Acoustics, Speech and Signal Processing (ICASSP) (2014)
17. Deng, L., Hinton, G., Kingsbury, B.: New types of deep neural network learning for speech recognition and related applications: an overview. In: Proceedings of the International Conference on Acoustics, Speech and Signal Processing (ICASSP). Vancouver, Canada (2013)
18. Deng, L., Hinton, G., Yu, D.: Deep learning for speech recognition and related applications. In: NIPS Workshop. Whistler, Canada (2009)
19. Deng, L., Li, J., Huang, J.T., Yao, K., Yu, D., Seide, F., Seltzer, M., Zweig, G., He, X., Williams, J., Gong, Y., Acero, A.: Recent advances in deep learning for speech research at microsoft. In: Proceedings of the International Conference on Acoustics, Speech and Signal Processing (ICASSP). Vancouver, Canada (2013)
20. Deng, L., Platt, J.: Ensemble deep learning for speech recognition. In: Proceedings of the Annual Conference of International Speech Communication Association (INTERSPEECH) (2014)
21. Deng, L., Seltzer, M., Yu, D., Acero, A., Mohamed, A., Hinton, G.: Binary coding of speech spectrograms using a deep auto-encoder. In: Proceedings of the Annual Conference of International Speech Communication Association (INTERSPEECH) (2010)
22. Deng, L., Tur, G., He, X., Hakkani-Tur, D.: Use of kernel deep convex networks and end-to-end learning for spoken language understanding. In: Proceedings of the IEEE Spoken Language Technology Workshop (SLT), pp. 210–215 (2012)

23. Deng, L., Yu, D., Platt, J.: Scalable stacking and learning for building deep architectures. In: Proceedings of the International Conference on Acoustics, Speech and Signal Processing (ICASSP) (2012)

24. Deng, L., Yu, D.: Deep convex network: a scalable architecture for speech pattern classification. In: Proceedings of the Annual Conference of International Speech Communication Association (INTERSPEECH) (2011)

25. Deng, L., Yu, D.: Deep Learning: Methods and Applications. NOW Publishers (2014)

26. Deng, L., Yu, D.: Use of differential cepstra as acoustic features in hidden trajectory modelling for phonetic recognition. In: Proceedings of the International Conference on Acoustics, Speech and Signal Processing (ICASSP), pp. 445–448 (2007)

27. Deng, L.: Articulatory features and associated production models in statistical speech recognition. In: Computational Models of Speech Pattern Processing, pp. 214–224. Springer, New York (1999)

28. Deng, L., Hassanein, K., Elmasry, M.: Analysis of the correlation structure for a neural predictive model with application to speech recognition. Neural Netw $7(2)$, 331–339 (1994)

29. Deng, L., Sun, D.: A statistical approach to automatic speech recognition using the atomic speech units constructed from overlapping articulatory features. J. Acoust. Soc. Am. 85, 2702–2719 (1994)

30. Deng, L., Yu, D., Acero, A.: Structured speech modeling. IEEE Trans. Speech Audio Process. 14, 1492–1504 (2006)

31. Ephraim, Y., Roberts, W.J.J.: Revisiting autoregressive hidden markov modeling of speech signals. IEEE Signal Process. Lett. 12, 166–169 (2005)

32. Gales, M.J.: Maximum likelihood linear transformations for HMM-based speech recognition. Comput. Speech Lang. $12(2)$, 75–98 (1998)

33. Ghoshal, A., Swietojanski, P., Renals, S.: Multilingual training of deep-neural networks. In: Proceedings of the International Conference on Acoustics, Speech and Signal Processing (ICASSP) (2013)

34. Godfrey, J.J., Holliman, E.C., McDaniel, J.: Switchboard: Telephone speech corpus for research and development. In: Proceedings of the International Conference on Acoustics, Speech and Signal Processing (ICASSP), vol. 1, pp. 517–520 (1992)

35. Graves, A., Jaitly, N., Mahamed, A.: Hybrid speech recognition with deep bidirectional lstm. In: Proceedings of the International Conference on Acoustics, Speech and Signal Processing (ICASSP). Vancouver, Canada (2013)

36. Graves, A., Mahamed, A., Hinton, G.: Speech recognition with deep recurrent neural networks. In: Proceedings of the International Conference on Acoustics, Speech and Signal Processing (ICASSP). Vancouver, Canada (2013)

37. He, X., Deng, L., Chou, W.: Discriminative learning in sequential pattern recognition—a unifying review for optimization-oriented speech recognition. IEEE Signal Process. Mag. $25(5)$, 14–36 (2008)

38. Heigold, G., Vanhoucke, V., Senior, A., Nguyen, P., Ranzato, M., Devin, M., Dean, J.: Multilingual acoustic models using distributed deep neural networks. In: Proceedings of the International Conference on Acoustics, Speech and Signal Processing (ICASSP) (2013)

39. Hoffman, M.D., Blei, D.M., Wang, C., Paisley, J.: Stochastic variational inference. JMLR $14(1)$, 1303–1347 (2013)

40. Huang, P.S., He, X., Gao, J., Deng, L., Acero, A., Heck, L.: Learning deep structured semantic models for web search using clickthrough data. In: ACM International Conference on Information and Knowledge Management (2013)

41. Huang, J., Kingsbury, B.: Audio-visual deep learning for noise robust speech recognition. In: Proceedings of the International Conference on Acoustics, Speech and Signal Processing (ICASSP), pp. 7596–7599 (2013)

42. Huang, J.T., Li, J., Yu, D., Deng, L., Gong, Y.: Cross-language knowledge transfer using multilingual deep neural network with shared hidden layers. In: Proceedings of the International Conference on Acoustics, Speech and Signal Processing (ICASSP) (2013)

43. Hutchinson, B., Deng, L., Yu, D.: A deep architecture with bilinear modeling of hidden representations: applications to phonetic recognition. In: Proceedings of the International Conference on Acoustics, Speech and Signal Processing (ICASSP) (2012)

44. Hutchinson, B., Deng, L., Yu, D.: Tensor deep stacking networks. IEEE Trans. Pattern Anal. Mach. Intell. (PAMI) (2013)

45. Kingsbury, B., Sainath, T.N., Soltau, H.: Scalable minimum bayes risk training of deep neural network acoustic models using distributed hessian-free optimization. In: Proceedings of the Annual Conference of International Speech Communication Association (INTERSPEECH) (2012)

46. Kingsbury, B.: Lattice-based optimization of sequence classification criteria for neural-network acoustic modeling. In: Proceedings of the International Conference on Acoustics, Speech and Signal Processing (ICASSP), pp. 3761–3764 (2009)

47. Kristjansson, T.T., Hershey, J.R., Olsen, P.A., Rennie, S.J., Gopinath, R.A.: Super-human multi-talker speech recognition: the ibm 2006 speech separation challenge system. In: Proceedings of the Annual Conference of International Speech Communication Association (INTERSPEECH) (2006)

48. Kumar, N., Andreou, A.G.: Heteroscedastic discriminant analysis and reduced rank HMMs for improved speech recognition. Speech Comm. 26(4), 283–297 (1998)

49. Lang, K.J., Waibel, A.H., Hinton, G.E.: A time-delay neural network architecture for isolated word recognition. Neural Netw. 3(1), 23–43 (1990)

50. Le, Q.V., Ranzato, M., Monga, R., Devin, M., Chen, K., Corrado, G.S., Dean, J., Ng, A.Y.: Building high-level features using large scale unsupervised learning. arXiv preprint arXiv:1112.6209 (2011)

51. Lee, C.H.: From knowledge-ignorant to knowledge-rich modeling: A new speech research paradigm for next-generation automatic speech recognition. In: Proceedings of the International Conference on Spoken Language Processing (ICSLP), pp. 109–111 (2004)

52. Li, J., Yu, D., Huang, J.T., Gong, Y.: Improving wideband speech recognition using mixed-bandwidth training data in CD-DNN-HMM. In: Proceedings of the IEEE Spoken Language Technology Workshop (SLT), pp. 131–136 (2012)

53. Lin, H., Deng, L., Yu, D., Gong, Y.f., Acero, A., Lee, C.H.: A study on multilingual acoustic modeling for large vocabulary ASR. In: Proceedings of the International Conference on Acoustics, Speech and Signal Processing (ICASSP), pp. 4333–4336 (2009)

54. Lu, Y., Lu, F., Sehgal, S., Gupta, S., Du, J., Tham, C.H., Green, P., Wan, V.: Multitask learning in connectionist speech recognition. In: Proceedings of the Australian International Conference on Speech Science and Technology (2004)

55. Martens, J.: Deep learning via Hessian-free optimization. In: Proceedings of the International Conference on Machine Learning (ICML), pp. 735–742 (2010)

56. Mesgarani, N., Chang, E.F.: Selective cortical representation of attended speaker in multi-talker speech perception. Nature 485, 233–236 (2012)

57. Mesgarani, N., Cheung, C., Johnson, K., Chang, E.: Phonetic feature encoding in human superior temporal gyrus. Science 343, 1006–1010 (2014)

58. Mikolov, T.: Statistical Language Models Based on Neural Networks. Ph.D. thesis, Brno University of Technology (2012)

59. Mnih, A., Gregor, K.: Neural variational inference and learning in belief networks. In: Proceedings of the International Conference on Machine Learning (ICML) (2014)

60. Mohamed, A.r., Dahl, G.E., Hinton, G.: Deep belief networks for phone recognition. In: NIPS Workshop on Deep Learning for Speech Recognition and Related Applications (2009)

61. Mohamed, A.r., Hinton, G., Penn, G.: Understanding how deep belief networks perform acoustic modelling. In: Proceedings of the International Conference on Acoustics, Speech and Signal Processing (ICASSP), pp. 4273–4276 (2012)

62. Moore, R.: Spoken language processing: time to look outside? In: Second International Conference on Statistical Language and Speech Processing (2014)

63. Morgan, N., Bourlard, H.: Continuous speech recognition using multilayer perceptrons with hidden Markov models. In: Proceedings of the International Conference on Acoustics, Speech and Signal Processing (ICASSP), pp. 413–416 (1990)

64. Morgan, N., Bourlard, H.A.: Neural networks for statistical recognition of continuous speech. Proc. IEEE **83**(5), 742–772 (1995)
65. Niu, F., Recht, B., Ré, C., Wright, S.J.: Hogwild!: a lock-free approach to parallelizing stochastic gradient descent. arXiv preprint arXiv:1106.5730 (2011)
66. rahman Mohamed, A., Yu, D., Deng, L.: Investigation of full-sequence training of deep belief networks for speech recognition. In: Proceedings of the Annual Conference of International Speech Communication Association (INTERSPEECH), pp. 2846–2849 (2010)
67. Rezende, D.J., Mohamed, S., Wierstra, D.: Stochastic backpropagation and approximate inference in deep generative models. In: Proceedings of the International Conference on Machine Learning (ICML) (2014)
68. Robinson, A.J.: An application of recurrent nets to phone probability estimation. IEEE Trans. Neural Netw. **5**(2), 298–305 (1994)
69. Sainath, T.N., Kingsbury, B., Mohamed, A.r., Dahl, G.E., Saon, G., Soltau, H., Beran, T., Aravkin, A.Y., Ramabhadran, B.: Improvements to deep convolutional neural networks for lvcsr. In: Proceedings of the IEEE Workshop on Automfatic Speech Recognition and Understanding (ASRU), pp. 315–320 (2013)
70. Sainath, T.N., Kingsbury, B., Mohamed, A.r., Ramabhadran, B.: Learning filter banks within a deep neural network framework. In: Proceedings of the IEEE Workshop on Automfatic Speech Recognition and Understanding (ASRU) (2013)
71. Sainath, T.N., Kingsbury, B., Sindhwani, V., Arisoy, E., Ramabhadran, B.: Low-rank matrix factorization for deep neural network training with high-dimensional output targets. In: Proceedings of the International Conference on Acoustics, Speech and Signal Processing (ICASSP), pp. 6655–6659 (2013)
72. Sainath, T.N., Mohamed, A.r., Kingsbury, B., Ramabhadran, B.: Deep convolutional neural networks for LVCSR. In: Proceedings of the International Conference on Acoustics, Speech and Signal Processing (ICASSP), pp. 8614–8618 (2013)
73. Sainath, T.N., Ramabhadran, B., Picheny, M.: An exploration of large vocabulary tools for small vocabulary phonetic recognition. In: Proceedings of the IEEE Workshop on Automfatic Speech Recognition and Understanding (ASRU), pp. 359–364 (2009)
74. Sak, H., Senior, A., Beaufays, F.: Long short-term memory recurrent neural network architectures for large scale acoustic modeling. In: Proceedings of the Annual Conference of International Speech Communication Association (INTERSPEECH) (2014)
75. Sak, H., Vinyals, O., Heigold, G., Senior, A., McDermott, E., Monga, R., Mao, M.: Sequence discriminative distributed training of long short-term memory recurrent neural networks. In: Proceedings of the Annual Conference of International Speech Communication Association (INTERSPEECH) (2014)
76. Saon, G., Soltau, H., Nahamoo, D., Picheny, M.: Speaker adaptation of neural network acoustic models using i-vectors. In: Proceedings of the IEEE Workshop on Automfatic Speech Recognition and Understanding (ASRU), pp. 55–59 (2013)
77. Schultz, T., Waibel, A.: Multilingual and crosslingual speech recognition. In: Proceedings of the DARPA Workshop on Broadcast News Transcription and Understanding, pp. 259–262 (1998)
78. Seide, F., Li, G., Chen, X., Yu, D.: Feature engineering in context-dependent deep neural networks for conversational speech transcription. In: Proceedings of the IEEE Workshop on Automfatic Speech Recognition and Understanding (ASRU), pp. 24–29 (2011)
79. Seide, F., Li, G., Yu, D.: Conversational speech transcription using context-dependent deep neural networks. In: Proceedings of the Annual Conference of International Speech Communication Association (INTERSPEECH), pp. 437–440 (2011)
80. Seltzer, M.L., Droppo, J.: Multi-task learning in deep neural networks for improved phoneme recognition. In: Proceedings of the International Conference on Acoustics, Speech and Signal Processing (ICASSP), pp. 6965–6969 (2013)
81. Seltzer, M., Yu, D., Wang, Y.: An investigation of deep neural networks for noise robust speech recognition. In: Proceedings of the International Conference on Acoustics, Speech and Signal Processing (ICASSP) (2013)

82. Sheikhzadeh, H., Deng, L.: Waveform-based speech recognition using hidden filter models: parameter selection and sensitivity to power normalization. IEEE Trans. Speech Audio Process. **2**, 80–91 (1994)
83. Shen, Y., Gao, J., He, X., Deng, L., Mesnil, G.: A latent semantic model with convolutional-pooling structure for information retrieval. In: ACM International Conference on Information and Knowledge Management (2014)
84. Socher, R., Huval, B., Manning, C., Ng, A.: Semantic compositionality through recursive matrix-vector spaces. In: Proceedings of the Joint Conference on Empirical Methods in Natural Language Processing and Computational Natural Language Learning (2012)
85. Socher, R., Lin, C.C., Ng, A., Manning, C.: Parsing natural scenes and natural language with recursive neural networks. In: Proceedings of the International Conference on Machine Learning (ICML), pp. 129–136 (2011)
86. Su, H., Li, G., Yu, D., Seide, F.: Error back propagation for sequence training of context-dependent deep networks for conversational speech transcription. In: Proceedings of the International Conference on Acoustics, Speech and Signal Processing (ICASSP) (2013)
87. Sun, J., Deng, L.: An overlapping-feature based phonological model incorporating linguistic constraints: applications to speech recognition. J. Acoust. Soc. Am. **111**, 1086–1101 (2002)
88. Thomas, S., Ganapathy, S., Hermansky, H.: Multilingual MLP features for low-resource LVCSR systems. In: Proceedings of the International Conference on Acoustics, Speech and Signal Processing (ICASSP), pp. 4269–4272 (2012)
89. Tuske, Z., Golik, P., Schluter, R., Ney, H.: Acoustic modeling with deep neural networks using raw time signal for LVCSR. In: Proceedings of the Annual Conference of International Speech Communication Association (INTERSPEECH) (2014)
90. Vanhoucke, V., Senior, A., Mao, M.Z.: Improving the speed of neural networks on CPUs. In: Proceedings of the NIPS Workshop on Deep Learning and Unsupervised Feature Learning (2011)
91. Veselý, K., Ghoshal, A., Burget, L., Povey, D.: Sequence-discriminative training of deep neural networks. In: Proceedings of the Annual Conference of International Speech Communication Association (INTERSPEECH) (2013)
92. Vinyals, O., Jia, Y., Deng, L., Darrell, T.: Learning with recursive perceptual representations. In: Proceedings of the Neural Information Processing Systems (NIPS), vol. 15 (2012)
93. Waibel, A., Hanazawa, T., Hinton, G., Shikano, K., Lang, K.J.: Phoneme recognition using time-delay neural networks. IEEE Trans. Speech Audio Process. **37**(3), 328–339 (1989)
94. Weng, C., Yu, D., Seltzer, M., Droppo, J.: Single-channel mixed speech recognition using deep neural networks. In: Proceedings of the International Conference on Acoustics, Speech and Signal Processing (ICASSP), pp. 5669–5673 (2014)
95. Weninger, F., Geiger, J., Wollmer, M., Schuller, B., Rigoll, G.: Feature enhancement by deep lstm networks for ASR in reverberant multisource environments. In: Computer Speech and Language, pp. 888–902 (2014)
96. Xue, S., Abdel-Hamid, O., Jiang, H., Dai, L.: Direct adaptation of hybrid DNN/HMM model for fast speaker adaptation in LVCSR based on speaker code. In: Proceedings of the International Conference on Acoustics, Speech and Signal Processing (ICASSP), pp. 6389–6393 (2014)
97. Xue, J., Li, J., Gong, Y.: Restructuring of deep neural network acoustic models with singular value decomposition. In: Proceedings of the Annual Conference of International Speech Communication Association (INTERSPEECH) (2013)
98. Yu, D., Deng, L., Dahl, G.: Roles of pre-training and fine-tuning in context-dependent DBN-HMMs for real-world speech recognition. In: Proceedings of the Neural Information Processing Systems (NIPS) Workshop on Deep Learning and Unsupervised Feature Learning (2010)
99. Yu, D., Deng, L., He, X., Acero, A.: Large-margin minimum classification error training for large-scale speech recognition tasks. In: Proceedings of the International Conference on Acoustics, Speech and Signal Processing (ICASSP), vol. 4, pp. IV-1137 (2007)
100. Yu, D., Deng, L., Seide, F.: Large vocabulary speech recognition using deep tensor neural networks. In: Proceedings of the Annual Conference of International Speech Communication Association (INTERSPEECH) (2012)

101. Yu, D., Seltzer, M.L., Li, J., Huang, J.T., Seide, F.: Feature learning in deep neural networks—studies on speech recognition tasks (2013)
102. Yu, D., Yao, K., Su, H., Li, G., Seide, F.: Kl-divergence regularized deep neural network adaptation for improved large vocabulary speech recognition. In: Proceedings of the International Conference on Acoustics, Speech and Signal Processing (ICASSP), pp. 7893–7897 (2013)
103. Yu, D., Deng, L., He, X., Acero, A.: Large-margin minimum classification error training: a theoretical risk minimization perspective. Compu. Speech Lang. $22(4)$, 415–429 (2008)
104. Yu, D., Deng, L., Seide, F.: The deep tensor neural network with applications to large vocabulary speech recognition. IEEE Trans. Audio Speech Lang. Process. $21(3)$, 388–396 (2013)

Index

A

Accelerated gradient algorithm, 73
Acoustic model (AM), 102
Acoustic modeling, 251
Acoustic scaling factor, 138
Acronyms, list of, xi
Activation, 57
Activation function, 57
Adaptation, 304
AE-BN, 182
Alternating directions method of multipliers (ADMM), 125
ANN-HMM hybrid models, 100
ANN/HMM hybrid system, 299
Articulatory-like dynamics, 253
Artificial neural network (ANN), 99
Artificial neural network/hidden Markov model hybrid system, 299
Asynchronous lock-free update, 122
Asynchronous stochastic gradient descent (ASGD), 121, 302
Atomic summation, 147
Audio-visual speech recognition, 232, 233
Augmented Lagrangian method (ALM), 124
Augmented Lagrangian multiplier, 125
Aurora 4 dataset, 165
Autoencoder bottleneck network, 182
AveragePooling, computation node, 291

B

Backpropagation algorithm, 65
Backpropagation Through Time (BPTT), 238, 251, 256
Backward probability, 30, 32
Batch training, 70
Baum-Welch algorithm, 33, 35

Bayesian network, 258
Bayes rule, 32, 46
Belief network, 258
Bellman optimality principle, 39
Bernoulli-Bernoulli RBM, 79
Boosted MMI (BMMI), 139
Bottleneck, 177
Bottom-up, 253
BP, 61
Broadcast news dataset, 182, 184
Burn-in, Gibbs sampling, 85

C

CD-DNN-HMM, 101, 301
Cell activation, 250
CI-DNN-HMM, 301
ClassificationError, computation node, 286
Coarticulation, 47
Computational network (CN), 267, 308
Computation node, 267
Conditional distributions, 256
Conditional independence, 32
Connectionist, 254
Conservative training (CT), 199
Constrained optimization problem, 246
Contrastive divergence (CD), 84
Convex optimization, 247
Convolutional neural network (CNN), 288, 305
Convolution, computation node, 289
Covariance matrix, 15
Cross-entropy (CE), 60
CrossEntropy, computation node, 286
CrossEntropyWithSoftmax, computation node, 287
Crosslingual acoustic modeling, 222

© Springer-Verlag London 2015
D. Yu and L. Deng, *Automatic Speech Recognition*,
Signals and Communication Technology, DOI 10.1007/978-1-4471-5779-3

CPSIA information can be obtained
at www.ICGtesting.com
Printed in the USA
LVHW081915160419
614231LV00009BA/248/P